The Mandrill

A Case of Extreme Sexual Selection

Living in the remote forests of western central Africa, the mandrill (*Mandrillus sphinx*) is notoriously elusive and has evaded scientific scrutiny for decades. Yet, it is the largest and most sexually dimorphic of all the Old World monkeys, and perhaps the most colourful of all the mammals.

Synthesizing the results of more than 25 years of research, this is the first extensive treatment of the mandrill's reproductive and behavioural biology. Dixson explores in detail the role that sexual selection has played in shaping the mandrill's evolution, covering mechanisms of mate choice, intra-sexual competition, sperm competition and cryptic female choice. Bringing to life, through detailed descriptions and rich illustrations, the mandrill's communicatory biology and the functions of its brightly coloured adornments, this book sheds new light on the evolutionary biology of this fascinating primate.

Alan F. Dixson is a Professor in the School of Biological Sciences at Victoria University of Wellington, New Zealand. He is a world authority on the reproductive biology and evolution of sexuality in primates. During a distinguished career, he has held posts at the Zoological Society of London, Medical Research Council UK, International Medical Research Centre in Gabon, Sub-Department of Animal Behaviour, University of Cambridge and the Zoological Society of San Diego.

The Mandrill

A Case of Extreme Sexual Selection

ALAN F. DIXSON

School of Biological Sciences
Victoria University of Wellington
New Zealand

CAMBRIDGE
UNIVERSITY PRESS

CAMBRIDGE
UNIVERSITY PRESS

University Printing House, Cambridge CB2 8BS, United Kingdom

One Liberty Plaza, 20th Floor, New York, NY 10006, USA

477 Williamstown Road, Port Melbourne, VIC 3207, Australia

314-321, 3rd Floor, Plot 3, Splendor Forum, Jasola District Centre, New Delhi - 110025, India

79 Anson Road, #06-04/06, Singapore 079906

Cambridge University Press is part of the University of Cambridge.

It furthers the University's mission by disseminating knowledge in the pursuit of education, learning and research at the highest international levels of excellence.

www.cambridge.org
Information on this title: www.cambridge.org/9781107535121

First published 2015
First paperback edition 2018

A catalogue record for this publication is available from the British Library

Library of Congress Cataloging in Publication data
Dixson, Alan F.
The mandrill : a case of extreme sexual selection / Alan F. Dixson, School of Biological Sciences, Victoria University of Wellington, New Zealand.
pages cm
Includes bibliographical references and index.
ISBN 978-1-107-11461-6
1. Mandrill. 2. Mandrill – Reproduction.
3. Sexual selection in animals. I. Title.
QL737.P93D59 2015
591.56′2–dc23

2015020637

ISBN 978-1-107-11461-6 Hardback
ISBN 978-1-107-53512-1 Paperback

To Amanda

And at home by the fire,

whenever you look up there I shall be –

and whenever I look up, there will be you.

Thomas Hardy *(Far from the Madding Crowd)*

Contents

The colour plate section appears between pages 112 and 113.

Preface

> I am a kind of farthing dip, unfriendly to the nose and eyes.
> A blue-behinded ape I skip, upon the trees of paradise.
>
> Robert Louis Stevenson

Despite its great size, and colourful secondary sexual adornments, the behavioural biology of the mandrill (*Mandrillus sphinx*) remained shrouded in mystery until comparatively recently. It has given me much pleasure to study the behaviour and reproductive biology of mandrills in Gabon, and to supervise a number of research students who have helped to advance those studies. Indeed, I am indebted to a number of people who made the writing of this book possible. The late Dr Georges Roelants invited me to work at the International Medical Research Centre (CIRMF) in Gabon, and to direct research at its Primate Centre between the years 1989 and 1992. Dr Jean Wickings carried out all the genetic studies and hormone assay work on the mandrills during those years; these contributions were of vital importance. Two research students from the University of Zurich, Thomas Bossi (co-supervised with Professor Bob Martin) and Edi Frei (co-supervised with the late Professor Hans Kummer), helped to study the sexual behaviour and social organization of the mandrill. In later years, after I had returned to the UK to work at the Sub-Department of Animal Behaviour (University of Cambridge), Joanna Setchell joined our group as a PhD student, in order to research the socio-sexual development of male mandrills. After successfully completing her PhD, Jo Setchell continued to work on the mandrill groups at the CIRMF; her research publications, and those of her colleagues, have been of immense value to me.

The drill (*Mandrillus leucophaeus*) is even less well studied than the mandrill, but it is impossible to understand the evolutionary history of the genus *Mandrillus* without considering both these species. Thus, during six years spent in the USA, as Director of Conservation and Science for the Zoological Society of San Diego, I took the opportunity to initiate fieldwork on drills in Cameroon. In this regard, I owe a special debt of gratitude to Dr Bethan Morgan for her tireless efforts to gather data on free-ranging drills, and to Chris Wild who managed field operations in Cameroon from 2000 to 2005.

I should like to thank Martin Griffiths (formerly of Cambridge University Press), Katrina Halliday (Publisher: *Life Sciences*) and Megan Waddington (Editor: *Life Sciences*) for their encouragement and advice during the planning and production of this book. Dr Aimée Komugabe made many of the figures and maps, and she has done an outstanding job. Sincere thanks also go to Dr Jeanette Mitchell, who copy- edited the entire manuscript, and to Susan Leech for the great care she has taken in creating the index.

I am indebted to senior colleagues at Victoria University of Wellington, and especially to Professor Charles Daugherty and Professor David Bibby, who encouraged me to join the School of Biology in 2006. Since that time, I have enjoyed a positive intellectual environment in which to write research papers and books, as well as teaching undergraduates about the mysteries of human evolution and reproduction.

Writing a book is, by its very nature, a protracted and largely solitary process. However, I never feel alone, and this is entirely due to the strong support I receive from my family. My heartfelt thanks go especially to my wife Amanda, to our sons Alexander, Barny and Henry, and our daughter Charis.

Acknowledgements

I am most grateful to the organizations and individuals listed as follows, for permission to reproduce figures, tables and photographs, either in their original form or as redrawn and modified versions.

Cambridge University Press: *Figures 2.2, 2.5, 8.16A, 10.2, 10.8, 10.13 and 10.16;* **Edinburgh University Press:** *Figures 3.12 and 5.2;* **Elsevier Ltd:** *Figures 7.3, 8.24 and 8.26;* **S. Karger AG, Basel:** *Figures 2.3, 10.10, 10.11 and 10.12A;* **Tim Laman and Professor Cheryl Knott:** *Figure 10.14;* **Dr Pierre Moisson:** *Figure 11.8;* **Proceedings of the National Academy of Sciences, USA and Professor John Fleagle:** *Figure 3.7* (copyright 1999. National Academy of Sciences, USA); **Orion Publishing Co:** *Figure 3.8;* **Oxford University Press:** *Figures 3.4, 3.11, 7.6, 8.3, 8.6, 8.10, 8.15, 8.25A, 10.1, 10.3, 10.8, 10.12B and Plate 6;* **Springer:** *Figures 4.4, 4.5, 4.6B, 5.1, 8.13, 8.14, 8.18, 8.21, 11.3, 11.4, 11.5 and Tables 4.1 and 5.1;* **Taylor and Francis Group and the Society for Scientific Study of Sexuality:** *Figure 10.5;* **John Wiley and Sons:** *Figures 2.4, 3.5, 4.2, 4.6A, 7.2, 7.4, 8.17A, 8.23, 8.27, 10.9 and Tables 3.1 and 8.4;* **Dr Kathy Wood and the Tengwood Organization:** *Figures 2.1B, 7.7 and Plates 2 and 8.*

Professor Alan Dixson
Sandwalk
Paraparaumu Beach
New Zealand

Prologue

Among the mammals, sexual selection has sometimes resulted in the evolution of extreme sex differences in body size, weaponry and secondary sexual adornments. Nowhere is this observation more apposite than in the case of the mandrill (*Mandrillus sphinx*), as it offers numerous examples of the effects of sexual selection, especially in adult males. At over 30 kilogrammes in weight, the mature male mandrill is the largest of all the Old World monkeys, and it is more than three times the size of the adult female. The male's enormous jaws are equipped with long, dagger-like canine teeth, notably in the upper jaw. Most extraordinary, however, is the mandrill's colouration. Adult males of this species display large areas of bright blue and red skin (the so-called 'sexual skin') on the face, rump and genitalia. As young males transition to sexual maturity, boney paranasal swellings enlarge on each side of the snout, and cobalt blue sexual skin overlies these swellings in a series of ridges, flanking the scarlet mid-nasal strip and fleshy tip of the nose. Add to these extraordinary secondary sexual traits the possession of a yellow beard, a crest of hair on the scalp and nape of the neck, a mane, a large sternal cutaneous gland and marked enlargement of the colourful rump owing to deposition of fat, and the male mandrill ranks as the most visually striking of all primate species. Although adult female mandrills are certainly much less brightly coloured than the males, they are not lacking in secondary sexual adornments. Thus, there is a female sexual skin covering the perineal and genital areas, and this undergoes marked changes in swelling and colouration during the menstrual cycle.

Biologists have long speculated as to why the mandrill should exhibit such an extreme expression of so many sexually dimorphic traits. When Charles Darwin (1871, 1876) was formulating his ideas concerning evolution by sexual selection, he observed that 'no other member in the whole class of mammals is coloured in so extraordinary a manner as the adult male mandrill'. Darwin regarded the male mandrill as the mammalian equivalent of the peacock, suggesting that its bright colouration had also evolved to 'serve as a sexual ornament and attraction' to females. In Darwin's opinion, the mandrill's colours were thus the products of inter-sexual selection, via female choice, the most attractive males being likely to gain a reproductive advantage by achieving more matings and siring more offspring than less colourful individuals. However, he also noted that 'with mammals the male appears to win the female much more through the law of battle than through the display of his charms'. Thus, the male mandrill's great size and impressive canine teeth might represent the outcome of intra-sexual selection owing to aggressive competition between males for access to mates.

These tentative explanations left unresolved the question of why inter-sexual and intra-sexual selection should have operated in such an extreme fashion in males of this particular primate species. Why should female mandrills also display attractive cues in the form of their sexual skin swellings? A resolution of these problems, and many other questions concerning the mandrill's evolution, requires a detailed knowledge of its ecology and behaviour, social organization, mating system and reproductive biology. How is the sexual behaviour of this species organized, and what governs sexual attractiveness and mate choice? What is the nature of dominance relationships between males and between females? What environmental factors determine the mandrill's social behaviour and the changing composition of its unusual supergroups and subgroups? Exploration of the mechanisms governing sexual selection and reproductive success in the mandrill is of intrinsic interest, as it helps us to understand the biology of this spectacular animal. Moreover, studies of the mandrill will certainly contribute to our wider understanding of how sexual selection operates in other mammals.

Unfortunately, in Darwin's time, almost nothing was known about these matters. Indeed, the natural life of mandrills, within the remote forests of western central Africa, has not been studied until comparatively recently. Super-groups of these magnificent monkeys, numbering as many as 800 individuals, traverse vast areas within the rain-forested interior of Gabon. They travel rapidly on the ground, but frequently take to the trees during daily foraging and in order to seek refuge at night. Hordes also split into smaller subgroups, and for a long time it was assumed that these might represent polygynous one-male units, such as those that occur in geladas and hamadryas baboons. As we shall see in later chapters of this book, such assumptions arose in the absence of detailed information about the mandrill's behaviour and reproductive biology. So difficult are these monkeys to locate and observe in the wild that it is not surprising that the true nature of their behaviour and ecology has evaded scientific scrutiny for so long.

During the 1960s and 1970s, primate field studies were undergoing something of a renaissance. Detailed field investigations of the African and Asian apes, as well as of various monkey species, Malagasy lemurs and other prosimians, were initiated during those two decades. It was an exciting time, and books and journals began to appear devoted to the field of 'Primatology'. Yet, the mandrill (*Mandrillus sphinx*) and its close relative the drill (*M. leucophaeus*) remained unstudied and relatively unknown. Both species were thought of as 'forest baboons', and as such they were assumed to be similar in many ways to the well-studied savannah baboons (*Papio* spp.). John and Prue Napier (1967) had published their *Handbook of Living Primates*, which provided scholars with a valuable guide to current knowledge of the prosimians, monkeys and apes. However, readers of the *Handbook* were informed that 'information on drills and mandrills in the wild is wholly anecdotal'.

This declaration of ignorance of mandrill field biology did not discourage me as a young undergraduate. I was full of naïve enthusiasm but, alas, totally lacking in any practical experience of primatology. I had read George Schaller's (1963) splendid monograph *The Mountain Gorilla: Ecology and Behavior*, and Irven DeVore's (1964) edited volume on *Primate Behavior: Field Studies of Monkeys and Apes* (including work on baboons and macaques, as well as Jane Goodall's ground-breaking observations of chimpanzees at Gombe). If these pioneers could achieve so much, surely it should also be possible to study mandrills in the wild?

At that time, Dr John Hurrel Crook was the leader of a research group in the Department of Psychology at Bristol University and he supervised a number of PhD students who were carrying out fieldwork in Africa. I visited him in 1969, in an attempt to persuade him to take me on, in order to conduct fieldwork on the mandrill. He was very kind and patient with me, as I recall, gently pointing out the near impossibility of finding mandrills in the wild, let alone tracking them in the depths of the rainforest. For an inexperienced student to attempt such a project, working unsupported in remote areas, would be dangerous and also professionally very risky. For how could anyone guarantee that enough data might be generated to justify a PhD?

John Hurrel Crook was, of course, absolutely right and I was obliged to abandon the idea, but not the ambition, to study mandrills in Africa.

Twenty years were to elapse before I had the opportunity to conduct work on mandrill behaviour and reproductive biology at the International Medical Research Centre (CIRMF) in Gabon. Subsequently, the first studies of mandrill supergroups were carried out by fieldworkers based at the CIRMF Field Station in the Lopé National Park. Enough is now known about the mandrill to justify a synthesis of knowledge concerning its behaviour, reproductive biology and evolution. We are presently in a position to examine its natural history and reproductive biology, and to evaluate more critically the role that sexual selection may have played during its remarkable evolutionary history.

Part I

Natural history

1 *Historiae Animalium*

> There is no evidence that any member of the present genus was known to the ancients. Such a spectacular animal as the mandrill would have impressed observers forcibly and called for comment.
>
> Hill, 1970

Sometimes, a picture truly can convey more than a thousand words. This is certainly the case where the discovery of the mandrill is concerned. For, although vague descriptions of monkeys that might possibly have been mandrills can be found in the ancient literature, there are no accurate accounts of the animal. The earliest unequivocal record of the mandrill's existence is a drawing, which appears in the published works of the sixteenth century Swiss naturalist and prolific encyclopaedist Conrad Gesner. Gesner died of the plague in 1565, when he was just 49 years of age. Yet, during his comparatively short life, he had produced intellectually diverse and monumental works. These included the *Bibliotheca Universalis*, which listed in Latin, Greek and Hebrew 1800 authors together with critiques of their various publications. Gesner's *Mithridates De Differentiis Linguarum* gave an account of all the 130 languages that were then known to scholars. However, it is his *Historiae Animalium* (1551–1558) for which he is best remembered. Four volumes of Gesner's great work on natural history were published in Zurich during his lifetime, and a later German edition, *Das Thierbuch* (1606), contained the drawing of a mandrill that is included here in Figure 1.1.

Gesner thought that this rather bristly and dog-like creature might be some type of hyaena! Yet, for all its limitations, the drawing is clearly an adult male mandrill, as evidenced by its stocky build, stumpy tail, and large rump. The hands and feet are plainly those of a monkey. However, the head is poorly rendered, for although the snout is quite prominent, it is foreshortened and lacks the longitudinal paranasal swellings that are so characteristic of mature male mandrills. Nor is there any indication of the bright bare areas of sexual skin, the beard, crest and thick pelage of this species. For comparative purposes, these features are shown in the modern drawing of a male mandrill included in Figure 1.1.

In fairness to Conrad Gesner, it is most unlikely that he ever had the opportunity to examine a living mandrill; it was only much later that accurate descriptions of it appeared, based upon observations of animals that had been brought back to Europe. The Danish anatomist and physician Thomas Bartholin dissected a male specimen and published an illustrated account of his findings (1671–1672). Pennant included illustrations of mandrills in his *Synopsis of Quadrupeds* (1771) and *History of Quadrupeds*

Figure 1.1. Conrad Gesner (1516–1565) (**A**) provided the first published illustration of the mandrill (**B**), as part of his *Historiae Animalium* (1551–1558). A modern drawing of an adult male mandrill (**C**) is included here for comparison with Gesner's illustration, which is taken from the German edition of his work (*Das Thierbuch*, 1606).

(1781), referring to them as 'tufted apes', 'great baboons' and (my particular favourite) 'ribbed-nose baboons'. Buffon (1766) took the opportunity to observe living mandrills in the Paris Menagerie, and so was able to make much more realistic illustrations of both sexes. He was also the first scholar to call this monkey 'the mandrill', perhaps because traders had heard this name applied to the specimens they had acquired in western Africa.

The genus *Mandrillus* (Ritgen, 1824) contains only two species, the mandrill (*M. sphinx*), and the drill (*M. leucophaeus*). Hill (1970) has pointed out that the drill remained unknown to science for much longer than the mandrill. Cuvier (1807, 1833) was the first to describe living specimens of the drill, again based upon animals held in the Paris Menagerie. Prior to Cuvier's account, there had been only conflicting reports that a second species of forest baboon might exist in west Africa. The drill resembles the mandrill in its overall proportions, although adult male drills are less massive than male mandrills. Females are much smaller than males in both species. Female drills and mandrills are also quite similar in appearance; a fact that has led and still leads to incorrect identifications and claims that the two species are sympatric in certain areas. The adult male drill is markedly different to the adult male mandrill, however, as its facial skin is predominantly black, in contrast to the red and blue colouration of the mandrill. Drills also have a lighter brown, or more olive-greenish hue to their pelage.

Ever since their discovery, mandrills and drills have been exhibited in menageries and zoos, and their skins and skeletons have accumulated in museum collections around the world. Yet, as we shall see, only in the last 30 years or so has their ecology, behaviour and reproductive biology been subjected to detailed scientific scrutiny. In what follows, I shall deal firstly with the natural history of the mandrill, including its classification and distribution, as well as basic information concerning its anatomy, behaviour and ecology. Then, Part II focuses on reproductive biology, including the results of long-term studies conducted on semi-free ranging mandrills, as well as fieldwork on supergroups in Gabon. Finally, in Part III, the mandrill's evolutionary biology is reviewed and the role played by sexual selection is discussed in detail. The final chapter considers the conservation status of the mandrill and drill, as both these species now face an increasingly uncertain future.

2 The genus *Mandrillus*: classification and distribution

Mandrills are not baboons

Like all the Old World monkeys, mandrills and drills belong to the Superfamily Cercopithecoidea, which is divisible into two Families; the Colobinae (comprising the African colobus monkeys, Asiatic langurs, leaf monkeys and proboscis monkeys), and the Cercopithecinae (including the guenons, patas monkey, talapoins, macaques, baboons and mangabeys, as well as *Mandrillus* and several other genera). One tribe of the cercopithecine monkeys, the Papionini, comprises the baboons (*Papio*), macaques (*Macaca*), arboreal mangabeys (*Lophocebus*), semi-terrestrial mangabeys (*Cercocebus*), the 'kipunji' (*Rungwecebus*) and the gelada (*Theropithecus*), as well as the genus *Mandrillus* (Table 2.1 and Figure 2.1).

The adaptive radiation of papionin monkeys in Africa and Asia has resulted in the convergent evolution of a number of primarily terrestrial, large and highly sexually dimorphic monkeys, with impressive canine teeth. Because both species of *Mandrillus* are large and superficially baboon-like monkeys, they have traditionally been considered as forest baboons and, as such, they were included in the genus *Papio* (e.g. Stammbach, 1987; Szalay and Delson, 1979). It is only quite recently that these ideas have been challenged and overturned. Similarities between mandrills and baboons are outweighed by many anatomical and genetic differences between the two genera. Comparative studies of mitochondrial DNA (Disotell, 2000; Disotell *et al.*, 1992), as well as of skeletal and other traits (Fleagle and McGraw, 1999, 2002; Groves, 2000) have shown that the genus *Mandrillus* is more closely related to the semi-terrestrial mangabeys (*Cercocebus*) than it is to the true baboons.

The arboreal mangabeys (genus *Lophocebus*) are more closely aligned with members of the genus *Papio* (e.g. see Guevara and Steiper, 2014) rather than with *Cercocebus* or *Mandrillus*. These relationships are made clearer by referring to Figure 2.2.

The phylogenetic position of the 'kipunji' (*Rungwecebus kipunji*) is not shown in Figure 2.2, but this rare and little known species was originally assigned to the arboreal mangabeys (genus *Lophocebus*), based upon studies of a single specimen (Jones *et al.*, 2005). However, subsequent molecular phylogenetic analyses indicate that the kipunji is probably intermediate between *Lophocebus* and *Papio* (Davenport *et al.*, 2006; see also Roberts *et al.*, 2009). As such, it is currently placed in its own genus, *Rungwecebus*, which was named in honour of Mt. Rungwe in Southern Tanzania, where the kipunji was discovered.

Table 2.1. Genera belonging to the Tribe Papionini: numbers of extant species, and sex differences in body weight for selected examples

		Adult body weight (kg)		Body weight ratio
Genus	No. Species	Male	Female	Male/Female
Macaca	21			
M. mulatta		11.0	8.8	1.25
M. nigra		9.89	5.47	1.8
Cercocebus	7			
C. torquatus		8.0	5.5	1.45
C. atys	10.2		5.5	1.85
Mandrillus	2			
M. sphinx		32.21	9.34	3.44
Lophocebus	3			
L. albigena		8.25	6.02	1.37
L. aterrimus		7.84	5.76	1.36
Theropithecus	1			
T. gelada		19.0	11.7	1.6
Papio	5			
P. ursinus		29.8	14.8	2.01
P. hamadryas		16.9	9.9	1.7

The genus *Rungwecebus* is not included in this Table. (Data are from: Dixson, 2012; Smith and Jungers, 1997, and sources cited therein.)

In recognition of the close relationship between *Mandrillus* and the semi-terrestrial mangabeys, Goodman *et al.* (1998) sought to place the mandrill and drill in the same genus as the mangabeys (*Cercocebus*). They did so on the basis that molecular evidence suggests that forms ancestral to *Mandrillus* split from a common ancestor with *Cercocebus* about four million years ago. Goodman *et al.* regarded a time depth of at least seven to eleven million years as being necessary for the recognition of two separate genera, and the *Mandrillus–Cercocebus* split thus fell well outside this time criterion. Yet, it is difficult to understand why it should be necessary for taxa to share a last common ancestor seven to eleven million years ago, rather than four or five million years ago, before a generic separation is justified. Colin Groves (2000) expressed similar concerns, and he also doubted 'whether the world is quite ready for *Cercocebus sphinx* and *Cercocebus leucophaeus*'. In this book, I adopt the position that it is more constructive to avoid causing unnecessary confusion by combining these two genera and radically changing the Latin names of the mandrill and drill.

Recognition of the phylogenetic relationship between *Cercocebus* and *Mandrillus* is important because it helps us to better appreciate the evolutionary affinities of the mandrill and drill, as the descendants of a smaller-bodied and more arboreal ancestor. They share this ancestor with extant mangabeys, and especially those West African species belonging to the *torquatus* species group (*Cercocebus torquatus, C. atys* and *C. lunulatus*). Comparative studies of craniodental morphology indicate that

Figure 2.1. **A:** Male and female mandrills (*Mandrillus sphinx*). **B:** Male and female drills (*M.leucophaeus*), as compared to two other large-bodied terrestrial representatives of the Papionini. **C:** Male and female hamadryas baboons (*Papio hamadryas*). **D:** A male stump-tail macaque (*Macaca arctoides*). (Photographs: A and D: Author's collection; B: Dr Kathy Wood; C: F. Bond.)

C. torquatus is likely to represent the basal member of the *Cercocebus* 'clade' (Devreese and Gilbert, 2013).

These advances in our understanding of the classification and origins of the genus *Mandrillus* lead to a discussion of the current distribution of the mandrill and drill, and to a consideration of how changes in climate, and associated contractions and expansions of the African rainforest, may have affected their evolution during the last four to five million years.

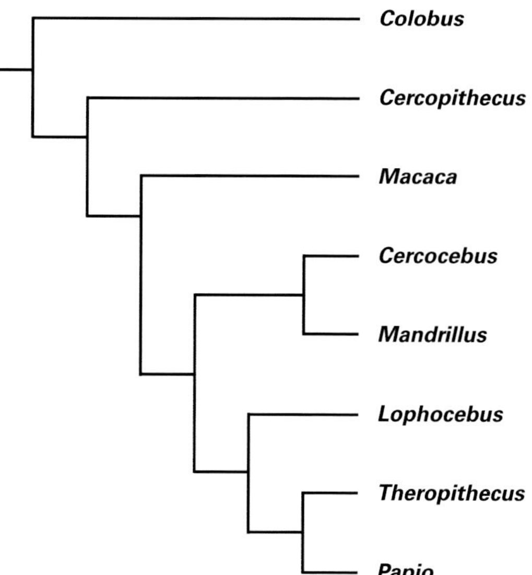

Figure 2.2. Phylogenetic relationships among the Papionini and some other Old World monkey genera. A maximum parsimony tree of mitochondrial cytochrome oxidase subunit II sequences. (Redrawn from Disotell, 2000.)

Historic distribution range and speciation

The distribution ranges of the mandrill and drill in Western Central Africa are shown in Figure 2.3. The term 'historic distribution range' is used advisedly here because rainforest destruction, hunting and other human activities have greatly reduced the ranges of both species, especially so in the case of the drill. This topic will be addressed in the final section of this book, in relation to the conservation status of the mandrill and drill. The distribution map shown in Figure 2.3 is based primarily upon the classic work of Peter Grubb (1973), who assembled compelling evidence that the two *Mandrillus* species are allopatric; the drill occurs in forests to the north of the Sanaga River in Cameroon, whereas the mandrill occurs to the south of this river. In addition to Grubb's data, I have added information derived from more recent reports (as listed in the caption to Figure 2.3); each point on the map refers to well-documented evidence of mandrill or drill presence at a given site. This is, I believe, preferable to showing these species' distributions as shaded areas on the map, as if the animals were uniformly present across the landscape.

The drill is found in S. E. Nigeria (in Cross River State), in Cameroon in forests from the north bank of the Sanaga to the Cross River, and on Bioko Island (Equatorial Guinea). The mandrill's range extends from southern Cameroon (south of the Sanaga River), the mainland of Equatorial Guinea (formerly Rio Muni), throughout Gabon (to the west of the upper reaches of the Ivindo and Ogooué Rivers, which constitute barriers to the eastward dispersal of the species) and in some areas of Congo Brazzaville, in forests to the north of the Congo river.

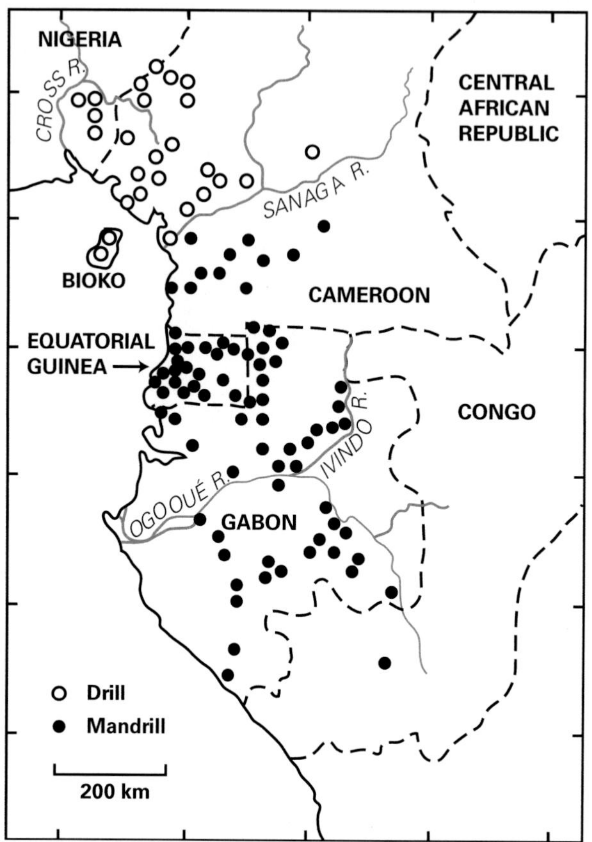

Figure 2.3. The historical distribution range of the mandrill and drill, based upon Grubb's (1973) distribution map, and incorporating additional records from Bossi, 1991; Harrison, 1988; Jouventin, 1975a; Morgan *et al.*, 2013; Telfer, 2006, and Wild *et al.*, 2005.

Prior to the publication of Grubb's (1973) paper in *Folia Primatologica*, mandrills and drills were considered to be partially sympatric in their distribution, as both species were thought to inhabit forests on each side of the Sanaga River. For example, in Volume 8 of his monograph *Primates: Comparative Anatomy and Taxonomy*, Hill (1970) followed Dobroruka (1966) in recognizing three subspecies of *M. sphinx* and three subspecies of *M. leucophaeus*, as follows:

Mandrillus sphinx sphinx* and *M. leucophaeus mundamensis: north of the Sanaga River.

M. sphinx madarogaster* and *M. leucophaeus leucophaeus: south of the Sanaga River.

M. sphinx insularis* and *M. leucophaeus poensis: on Bioko Island.

It is now clear that mandrills are definitely not found on Bioko Island (which was formerly known as Fernando Po). Specimens collected on the African mainland, but shipped via Bioko, probably gave rise to the mistaken impression that the Island was their point of origin. Only a small population of drills currently survives on Bioko, a remnant of the founder population, which must have become isolated there when the island became separated from mainland Africa by rising sea levels at the close of the last ice age, approximately 10,000–15,000 years ago. These drills are still classified as members of a distinct subspecies (*M. l. poensis*), although they differ very little from the mainland populations, either anatomically (Hill, 1970) or on the basis of comparisons of their mitochondrial DNA (Telfer, 2006; Ting *et al.*, 2012).

In 1973, Peter Grubb also analysed the available evidence concerning records of drills to the south of the Sanaga River, in Cameroon, or even further to the south, in Gabon. He judged most of these records to be dubious at best. The same conclusion applied to evidence concerning the existence of mandrills to the north of the Sanaga River. He did not entirely dismiss the possibility that rare incursions might occur across the river, by either species, but no certainty could be established regarding such events. The position has remained unchanged until the present day. During surveys of the status of drill populations in Cameroon (Morgan *et al.*, 2013), we found that forests to the north of the Sanaga River in the Douala–Edea region are now highly degraded, making it unlikely that drills could still survive there. Such deforestation thus makes it very difficult to determine whether the two *Mandrillus* species might, at some time in the past, have been sympatric in forests bordering the Sanaga River, or whether a hybrid zone between them existed. Hybridization between mandrills and drills is certainly possible, as it has been recorded a number of times in captive animals. I know this to be true from my own experience, as many years ago I had the opportunity to study the behaviour of a drill–mandrill group at the Yerkes Primate Research Center in the USA. The group included five (male drill X female mandrill) hybrids; the female hybrids were fertile. An adolescent male hybrid lacked the red and blue facial pigmentation typical of male mandrills; instead, its face was black and the paranasal swellings were less pronounced than those of male mandrills. I was unable to determine whether this hybrid male was fertile.

Putative sightings of groups of drills in the rainforests of Gabon are even more doubtful (Malbrant and Maclatchy, 1949). It should be kept in mind that females and immature individuals of *M. leucophaeus* and *M. sphinx* are very similar morphologically. Some female mandrills exhibit only minimal blue and red facial pigmentation; instead, they have dark faces quite similar to those of female drills (see Plate 10 for an example of this phenomenon). Encounters with wild mandrills or drills are typically fleeting, and it would be unwise to make a firm judgement of drill presence in Gabon on the basis of such contacts. Tutin and Fernandez (1987) reported, however, that several primatologists have sighted drill groups (including adult males) in what is now the Lopé National Park. These sightings have not been confirmed by more recent and more detailed field studies of mandrills in the Lopé National Park.

Currently, only two subspecies of the drill are recognized (*M. leucophaeus leucophaeus*, and *M.l. poensis*), while all mandrills are placed in a single subspecies (*M. sphinx sphinx*). However, Telfer *et al.* (2003) have provided good evidence for a significant and ancient divergence between those mandrill populations situated to the north and the south of the Ogooué River, in Gabon. Studies of *cytochrome-b* sequences from 53 mandrills of known provenance, as well as from additional specimens, led

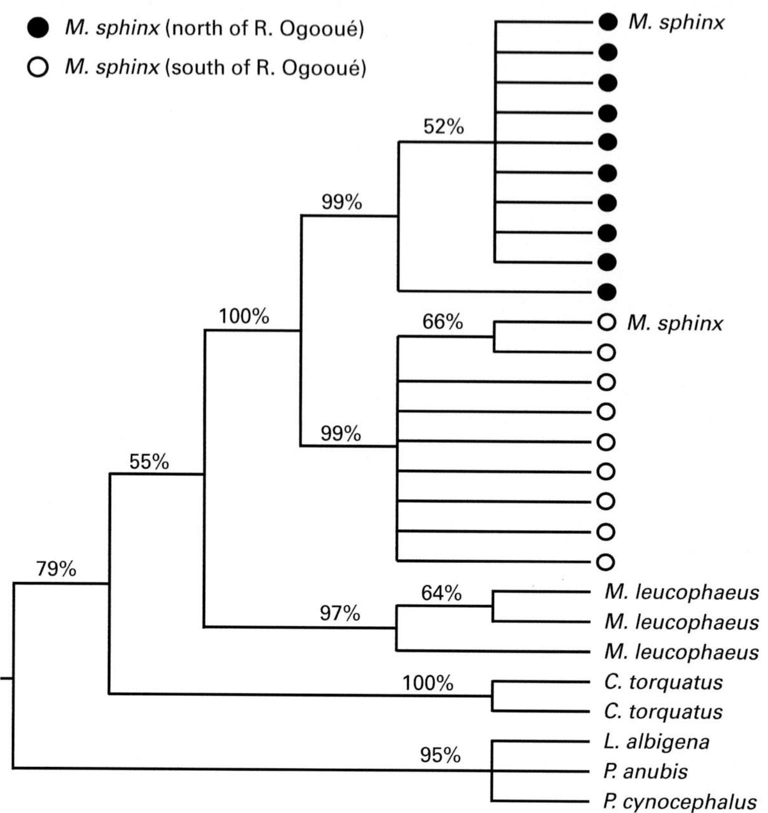

Figure 2.4. The *cytochrome-b* gene is highly conserved within primates. However, this (partial sequence) phylogeny shows that two distinct haplotypes of this gene occur in mandrill populations that live either to the north or to the south of the Ogooué River in Gabon. Bootstrap values are percentages of 1000 replications. (Redrawn and modified from Telfer *et al.*, 2003.)

Telfer *et al.* (2003) to the conclusion that two haplotypes occur, that the Ogooué River forms a distinct barrier between them, and that these haplotypes probably diverged as long as 800,000 years ago (Figure 2.4). Interestingly, mandrills belonging to the two haplotypes also differ in another important respect; each is subject to infection with a different subtype of the mandrill simian immunodeficiency virus (SIVmnd). Thus, mandrills in central and southern Gabon are carriers of SIVmnd-1, whereas those in northern and western Gabon are carriers of SIVmnd-2 (Souquière *et al.*, 2001).

At some time in the future, mandrills to the north and south of the Ogooué River might be recognized as belonging to separate subspecies. In order to resolve this question, detailed comparisons of their morphology and anatomy, as well as further genetic studies, will be required. It is interesting to reflect that the spectacular red and blue facial colouration of the male mandrill, which is fully developed in adults belonging to both haplotypes, must be at least 800,000 years old, given that the phylogenetic divergence demonstrated by Telfer *et al.* (2003) dates to that period. This leads us to wonder how long ago the mandrill and drill began to diverge as separate species. What factors might have led to their speciation? Unfortunately, there is no fossil evidence to guide us in exploring these interesting questions. In its absence, molecular studies are of special value in predicting likely divergence dates. Moreover, climatic changes, which have occurred in Africa over the past eight million years, have resulted in alternating periods of rainforest contraction and expansion. Hence, we must also consider whether these events might have played some role in determining the origins and speciation of the genus *Mandrillus*.

Rainforest distribution and evolution

The history of vegetational and climatic changes in tropical Africa during the last eight million years has been modelled with reference to measurements of pollen, and other fossils, contained in deep sea sediments (ocean cores), as well as to studies of plant fossils from various sites, dated using potassium–argon techniques (as reviewed by Hamilton, 1988, and Hamilton and Taylor, 1992). During the Late Miocene epoch, between 8.8 Mya and 6.4 Mya, equatorial Africa would have been warm and humid, much as it is today, with extensive lowland rainforest coverage. Grubb (1982, 1990) has posited that it was during this period of stable climate and major rainforest expansion that the ancestral mangabey stock increased its range, becoming widely distributed throughout the expanding rainforests.

However, during the late Miocene and the early Pliocene epochs, between 6.4 Mya and 4.6 Mya, major global climatic changes occurred. The Mediterranean Sea had become isolated and had gradually evaporated, leaving vast salt deposits (Denton, 1999; Hamilton, 1988) and creating a more arid climate throughout the region. This 'Mediterranean salinity crisis' also caused 'the remaining oceans to become less saline, contributing to a major increase in Antarctic ice' (Hamilton, 1988). In equatorial Africa, the drier climate caused the rainforests to recede, but forested areas continued to exist in widely separated refuges, where local conditions (such as the presence of major rivers or of mountains) favoured their survival. The putative distributions of these forest *refugia*, that persisted during dry climatic periods, are shown in Figure 2.5. They included the Ogooué River system in Gabon, the Sanaga River, the Cameroonian mountains, and the Niger delta (Hamilton, 1988; Kingdon, 1980, 1997; Morley and Kingdon, 2013).

Isolation events have had profound effects upon speciation in the fauna and flora of the African rainforest. Separation of ancestral mangabey populations during the arid phase between 6.4 Mya and 4.6 Mya may thus have affected their evolution. When the climate became wetter, and the forests expanded again during the later Pliocene (4.6–2.43 Mya), there had emerged a new, larger-bodied form of mangabey-like monkey, specialized for a primarily terrestrial existence. Molecular evidence places this split, between *Cercocebus* and *Mandrillus*, at somewhere between 4 Mya and 5 Mya (4 Mya: Goodman *et al.*, 1998; 4.32 Mya: Telfer, 2006; 5 Mya: Page *et al.*, 1999).

It will be argued during the course of this book that the common ancestral *Mandrillus* stock, which gave rise to the extant mandrill and drill, had already developed much of its marked sexual dimorphism, in body weight and various secondary sexual traits, as well

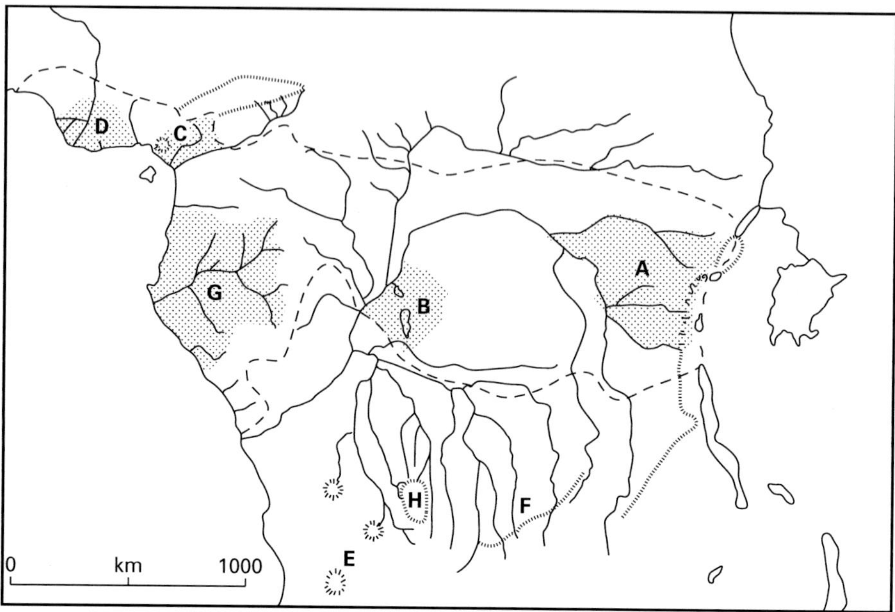

Figure 2.5. In the remote past, forest refuge areas that persisted during arid periods in Central and Western Africa ensured the survival and affected the speciation of primates such as the mandrill and drill. **A**: Central refuge. **B**: Southern Zaire basin refuge. **C**: Cameroon refuge. **D**: Niger refuge. **G**: Gabon or Ogooué basin refuge. **E**: North Angola refuge. **F**: Southern scarps of Zaire basin. **H**: Lunda Plateau. (From Hamilton, 1988.)

as its unusual social and mating system, prior to the divergence of *Mandrillus sphinx* and *M. leucophaeus* as two distinct species. Paul Telfer (2006) places this last divergence event at around 3.17 Mya, based upon his own research and a review of the available mitochondrial DNA evidence.

Between approximately 2.43 Mya and 1.0 Mya, the tropical African climate entered a phase of cyclical changes in warm/wet and cold/dry conditions, occasioned by ice age and interglacial fluctuations in the northern hemisphere (Hamilton, 1988). These climatic events caused cyclical changes in rainforest expansion and contraction, during which the drill's ancestors became confined to forests between the Sanaga and Cross Rivers, while the mandrill continued its separate development, in forests to the south of the Sanaga River. During the last one million years, cyclical changes in climate have continued, but have doubled in length (to approximately 100,000-year intervals) and have become more extreme. The phylogenetic divergence of mandrills to the north and the south of the River Ogooué occurred during this period (Telfer *et al.*, 2003) and, much more recently, the isolation of drills on Bioko Island, which became separated from the African mainland as sea levels rose. Mainland populations of the drill have undergone marked reductions in the more recent past (3000–5000 years ago), owing to climate change and forest contraction (Ting *et al.*, 2012). These populations subsequently made a remarkable recovery, as the forests expanded, to attain the vastness and grandeur that has endured until comparatively recently. Now, as human destruction of the rainforest

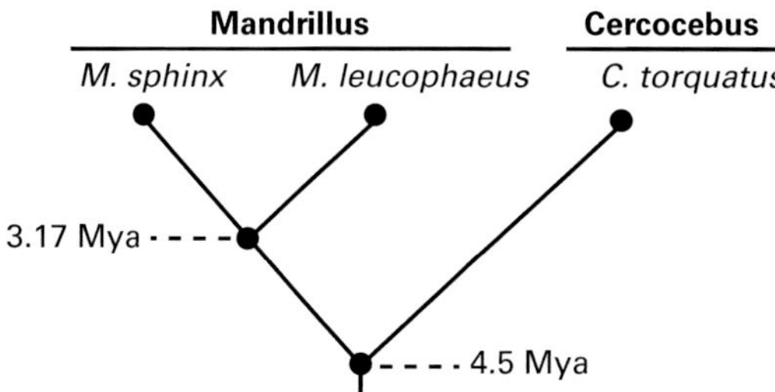

Figure 2.6. Diagram summarizing the time-scale of evolutionary relationships between *Mandrillus* and *Cercocebus*.

and its wildlife continues apace, drill numbers are again declining at an alarming rate (Morgan *et al.*, 2013).

Figure 2.6 summarizes what is currently known about the time spans and evolutionary relationships between *Mandrillus* and *Cercocebus*. It is chastening to reflect that, long before the genus *Homo* had emerged in Africa at around two million years ago, the genus *Mandrillus* had already been in existence for at least two million years. Speciation, giving rise to *M. sphinx* and *M. leucophaeus*, preceded the advent of *Homo ergaster* and *H. erectus* by more than a million years, and thus occurred long before anatomically modern *H. sapiens* arose in Africa, at around 200,000 years ago.

3 Morphology and functional anatomy

Anatomy is destiny.

Freud

Osman Hill (1970) has provided a detailed account of the comparative anatomy of both species of *Mandrillus*. My intent here is to focus on those facets of the mandrill's structure that will provide a useful background for further consideration of its behaviour, reproductive biology and evolution. Of special concern are those traits that are sexually dimorphic, including the secondary sexual traits that exhibit such spectacular development in adults of this species. The degree of similarity of these traits between the mandrill and drill must also be discussed, as well as some interesting differences between the two species.

External features

Mandrills are notable for their thickset bodies, large heads, long snouts and stumpy tails; Plate 1 shows an adult male and a female displaying the marked differences in body size and secondary sexual traits that are so characteristic of this species. Adult males weigh more than 30 kg and females 9–10 kg (Setchell *et al.*, 2001; Wickings and Dixson, 1992a). Data on the weights and crown–rump lengths of male and female mandrills, ranging in age between one year and 11 years, are provided in Table 3.1.

The mandrill's pelage is brown dorsally and on the flanks, and consists of long soft hairs, each of which is annulated with alternating light and dark pigmented bands. The hair on the abdomen is sparser and lacks pigment. In some animals (especially some adult males), a longitudinal fringe of white hair is visible on the ventrum (the epigastric fringe: Plate 3). In the adult male, longer hair forms a crest on the crown of the head and nape of the neck; this crest is erected during some facial displays. The adult male mandrill does not have a well-defined cape of long hair on the shoulders and upper body in the same way as do male hamadryas baboons or geladas. Nonetheless, some mature mandrill males have longer hair on the neck and upper body, so that it resembles a mane. The same is true of the adult male drill.

The mandrill has a beard, which is yellowish orange in colour and, although this is present in both sexes, it is much better developed in the adult male. Stiff, white, tactile hairs (vibrissae) are present around the upper and lower lips. A patch of white, hairless skin occurs behind each ear, and these white areas are prominent when the head is viewed from the rear, even in the shadows of the forest floor. Both sexes have long hairs in the

Table 3.1. Age-related changes in body weights and crown–rump lengths of semi-free ranging mandrills

	Body weight (kg)		Crown–rump length (cms)	
Age (y)	Male	Female	Male	Female
1.0–1.5	3.52±0.38	3.34±0.60	39.2±2.0	38.4±1.8
2.0–2.5	5.28±0.37	4.48±0.27	46.1±0.9	44.3±0.3
3.0–3.5	6.50±0.20	6.28±0.27	50.0±0.5	50.8±2.2
4.0–4.5	9.45±0.61	7.57±0.26	55.3±1.8	53.6±0.3
5.0–5.5	12.68±1.57	8.24±0.41	56.8±0.8	53.1±1.3
6.0–7.0	18.13±2.75	8.61±1.08	63.9±3.8	54.4±3.1
7.0–8.0	23.67±3.17	8.59±0.99	66.3±4.2	54.0±1.7
8.0–9.0	30.17±2.55	9.30±1.45	69.5±1.1	52.8±1.3
9.0–10.0	31.11±2.10	9.34±0.78	70.6±2.4	——
10.0–11.0	32.21±3.15	——	72.0±6.0	——

Data are means±SEM. Numbers of animals measured in each age band exhibited ranges of 6–18 (male body weight); 2–26 (female body weight) and 2–12 (crown–rump lengths in both sexes). (After Setchell *et al.*, 2001.)

mid-sternal region; these specialized hairs serve to retain the watery secretions produced by a dense layer of apocrine glands in the underlying skin (Hill, 1954, 1970). The sternal gland is much larger and more active in adult males; the hairs are also longer in males and brown in colour. The gland is used for scent-marking, as will be discussed later.

The skin of the face, rump and genitalia displays marked specializations in both sexes of the mandrill, and again these traits are highly sexually dimorphic. In the adult male, the skin of the mid-nasal strip and the fleshy tip of the nose is scarlet, especially so in the most dominant males. There is considerable individual variabilty in the distribution as well as the intensity of this red colouration, so that in some cases, the prominent brow ridge, the eyelids, and the lips also redden at maturity. Flanking the mid-nasal strip and running the length of the male's snout on each side are bony paranasal swellings; these are covered in bright blue skin, which is raised to form a series of six to seven longitudinal ridges, separated by grooves (Plate 6). One of the mandrill's older common names, 'ribnose baboon', refers to these distinctive facial features.

It should be noted here that the greatest differences in the secondary sexual traits of mandrills and drills concern the facial adornments of the adult males. Unlike the mandrill, the adult male drill lacks coloured sexual skin on the face, except for a transverse band of red skin just beneath the lower lip (Plate 2). The male drill's disc-shaped facial mask is black, fringed with white hair on the cheeks and around the jawline. There is also a whitish beard. The paranasal swellings are narrower and lack the longitudinal ridges as well as the blue colouration seen in mandrills (Figure 3.1 and Plate 8).

Returning to the male mandrill, the rump and external genitalia, like the face, are covered in sparsely haired and colourful sexual skin (see Plate 4). The shaft of the penis and the pubic, inguinal and peri-anal areas are red; the main, central portion of the rump

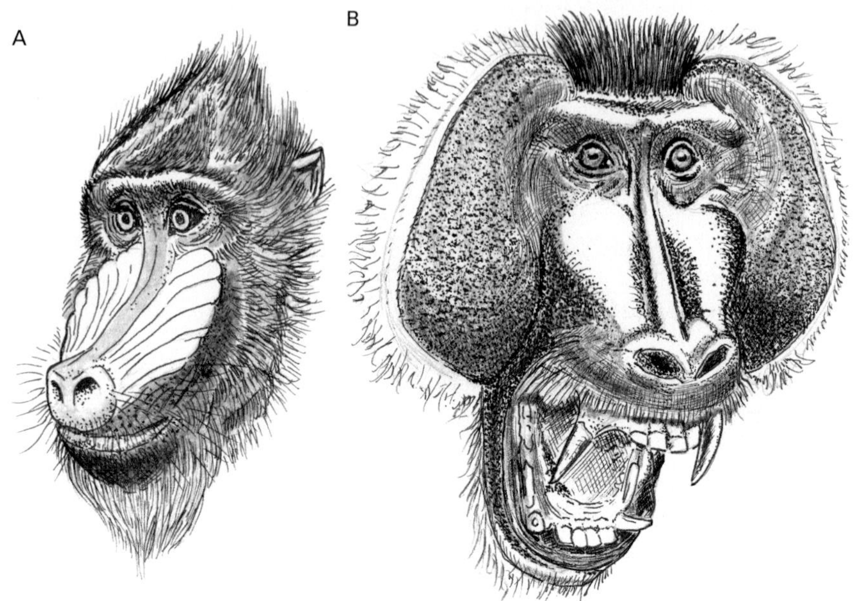

Figure 3.1. Portraits of (**A**) an adult male mandrill, and (**B**) an adult male drill. The drill is making an open-mouthed threat face. (Author's drawings.)

is bluish, with the skin grading to violet tints laterally and on the pendulous scrotum. In the adult male drill, sexual skin colouration on the rump and genitalia is similar to that of the mandrill.

The hindquarters and tail of the most dominant male mandrills are rendered larger and even more conspicuous by deposition of fat; hence the term 'fatted male' is applied to such high-ranking individuals, whereas some other adults are leaner in appearance, and are referred to as being 'non-fatted' (Wickings and Dixson, 1992b). It is important to stress that, in reality, a spectrum of morphotypes occurs between the fatted and non-fatted conditions. Differences between the fatted and non-fatted condition may be appreciated by comparing the male mandrills pictured in Plates 3, 4 and 5.

In the female mandrill, the facial colouration is much more muted than in the adult male. The female's muzzle is much smaller, lacks paranasal swellings, and the nose is either a light red colour or brown, owing to the presence of epidermal melanin. The skin of the paranasal region may also be dark in colour in some individuals, but bluish in others (Plates 10 and 11). Some of these individual differences probably have a genetic basis; thus, in some mandrill matrilines the females all develop stronger red and blue tints, while in others the face is very dark, being almost black, like that of the female drill. Endocrine factors affect the skin of the face, so that some females show stronger development of red colouration when they are pregnant. Setchell *et al.*, (2006b) measured fluctuations in the red facial colouration of female mandrills during menstrual cycles, and suggested that these changes might signal fertility. However, I think that this is unlikely, given that the changes reported were very small in comparison with the changes in female sexual skin swelling and colouration that occur during the menstrual cycle.

Female mandrills and drills both exhibit red sexual skin on their genitalia and perianal regions. The sexual skin becomes oedematous during the follicular phase of the menstrual cycle, to produce a marked swelling that gradually detumesces during the luteal phase of the cycle. In the mandrill and the drill, there is an ovoid area of paler skin surrounding the vulva, whereas the outer portion of the swelling is red or pinkish. In the pregnant female, by contrast, the sexual skin tends to be darker red and has a wrinkled appearance (Plates 12–14).

In both the mandrill and drill, female sexual skin swelling morphology is similar to that which occurs in the semi-terrestrial mangabeys (*Cercocebus* spp.), which are their closest phylogenetic relatives among the Papionini. Thus, the swelling affects the skin surrounding the vulva, clitoris and anal regions in *Mandrillus* and *Cercocebus* (Figure 3.2). The dorsal portion of the mandrill's swelling, which includes the anal field, becomes larger as females age, so that in multiparous females it is broader in posterior view (Huchard *et al.*, 2009). In the arboreal mangabeys (*Lophocebus*), the swelling does not include the skin surrounding the anal area. Figure 3.2 includes a diagram of sexual skin morphology in the grey-cheeked mangabey (*Lophocebus albigena*). In this example, the anal field is hidden from view behind the dorsal edge of the large vulval swelling.

Growth and development

Remarkable changes in the appearance of the pelage, sexual skin and other traits occur during growth and development in both sexes of the mandrill. Neonatal males and females weigh about 640 g; their hair is sparse and white, with a black cap on the scalp and a strip of blackish hair extending along the spine (Figure 3.3). The face is hairless and flesh-coloured. During the first two to three months, however, infants gradually transition to full development of brownish-green pelage on the body, limbs and head, while the face and snout become darkly pigmented, owing to deposition of melanin (see Plates 15 and 16).

Both sexes continue to grow at similar rates until two years of age, when males begin to put on weight significantly faster than females (Wickings and Dixson, 1992a). Female mandrills exhibit an earlier and less marked growth spurt than males. This begins from around 18 months onwards, and peaks at three years of age (Setchell *et al.*, 2001). By 3.6 years, on average, females develop their first sexual skin swellings, and they deliver their first infant at around 4.6 years of age, so that reproduction commences before maximum body weight is attained.

Males, by contrast, have a later but more pronounced growth spurt (from age four years until approximately eight or nine years), during which they attain weights (30–35 kg) that are three times those of mature females (Figure 3.4). It is during this period that males transition through puberty to adulthood. This is a gradual process that shows considerable individual variability, and which is affected by social factors (Setchell and Dixson, 2001a, 2002), as will be discussed in Part II of this book. The

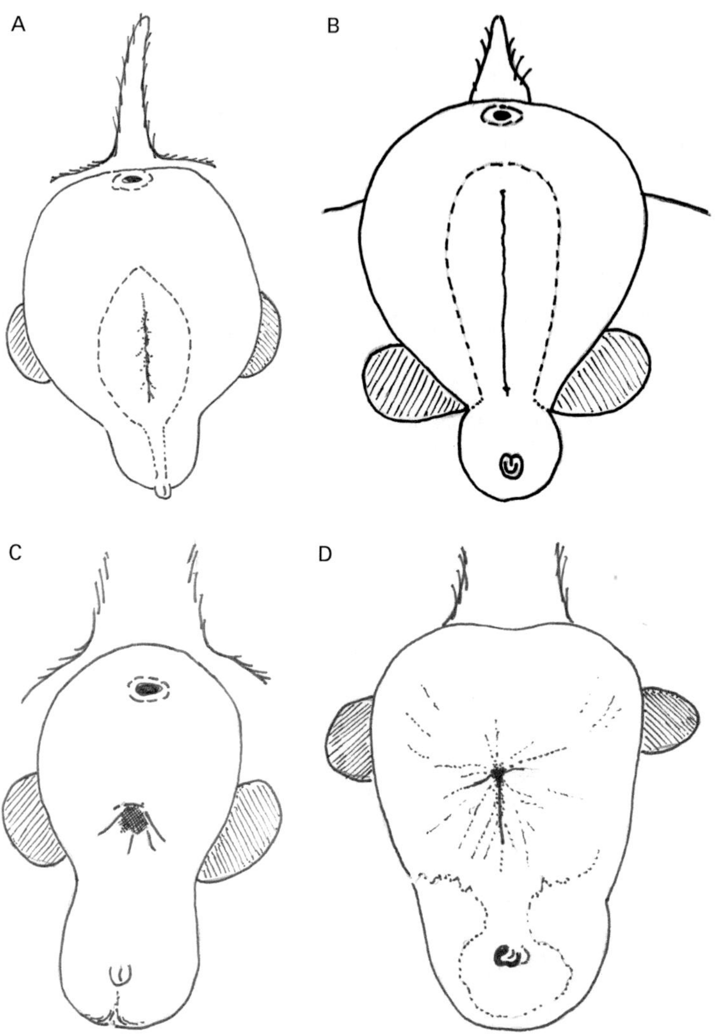

Figure 3.2. Sexual skin swellings of the following. **A:** *Mandrillus sphinx*. **B:** *M. leucophaeus*. **C:** *Cercocebus atys*. **D:** *Lophocebus albigena*. (Author's drawings.)

testes begin to increase markedly in volume at 5.5 years, circulating testosterone rises, and by six years of age, the florid masculine secondary sexual adornments begin to appear. By the time they are nine years old, most males exhibit full development of red and blue sexual skin on the face, rump and genitalia, as described previously. The testes have grown to adult size by this stage, the long upper canines are fully erupted, and the male's sternal gland is active. Figure 3.5 provides a summary of the timing, and the sequential nature, of these changes during puberty and adolescence in male mandrills.

A brief discussion concerning the physical basis of sexual skin in mandrills is necessary here, although more will be said about this subject in later chapters that deal

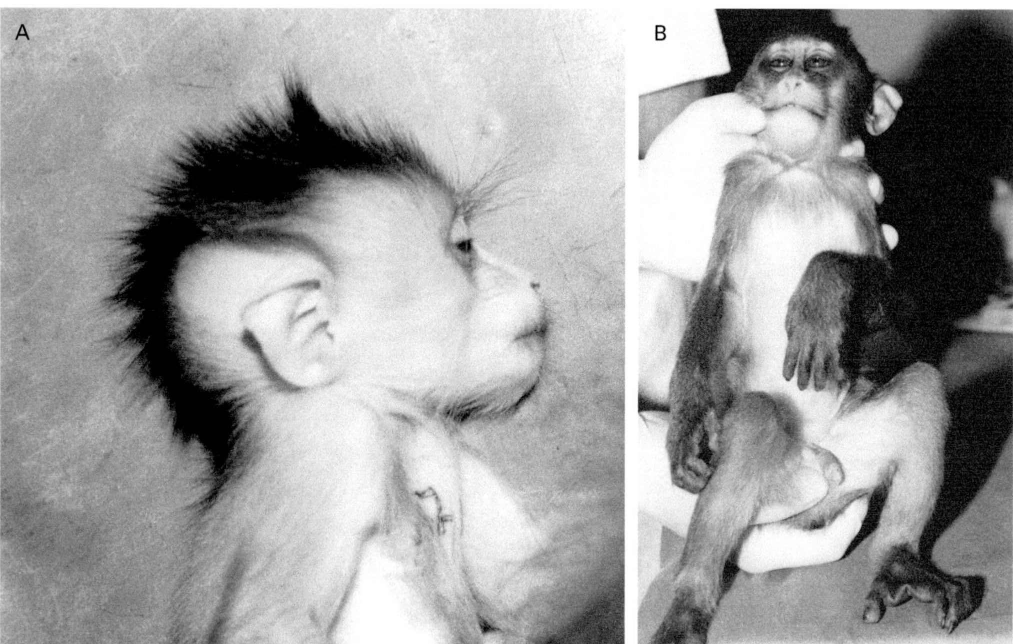

Figure 3.3. Infant mandrills. **A**: at two weeks of age. **B**: at approximately three months of age. (Author's photographs.)

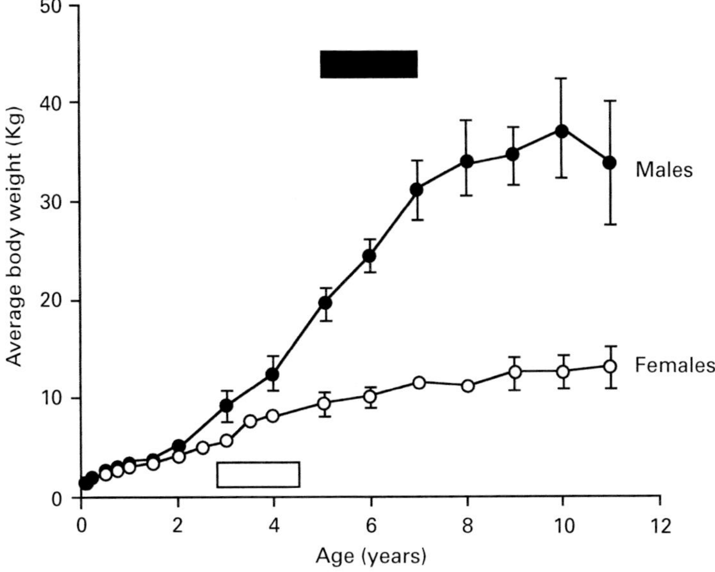

Figure 3.4. Growth in body weight from birth to adulthood in the mandrill. Data are means (±SEM) for males (filled circles) and females (open circles). Horizontal bars indicate the approximate age span for attainment of puberty (filled bars: males; open bars: females). (From Dixson, 2012, based upon data in Wickings and Dixson, 1992a.)

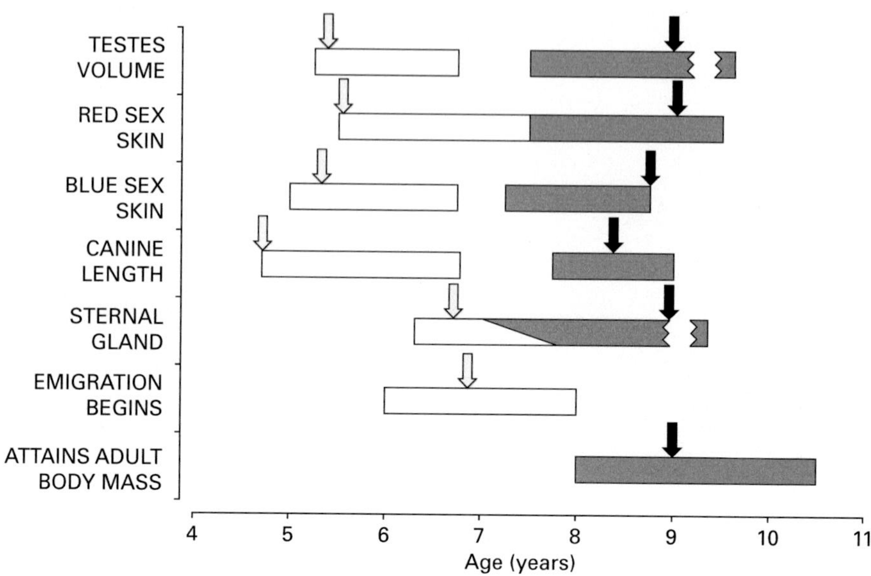

Figure 3.5. Sequence and timing of events that occur during puberty and adolescence in male mandrills. Open bars, age range during which development of each trait begins (mean ages are indicated by the arrows). Shaded bars, age range during which development of each trait is completed, with the black arrows marking the means. (Redrawn and modified from Setchell and Dixson, 2002.)

with reproductive biology. It is important to note that red sexual skin in anthropoids is a result of vascular mechanisms, and that it is hormone dependent in adults of both sexes. Female sexual skin colouration is stimulated by increases in oestrogen secretion during the follicular phase of the menstrual cycle; water retention and swelling are also oestrogen-dependent traits (Dixson, 1983a). In male monkeys, such as rhesus macaques, and hamadryas baboons, it was established long ago that the red sexual skin fades after castration, and is restored by testosterone treatment (Vandenbergh, 1965; Zuckerman and Parkes, 1939). However, in the male rhesus monkey, enzymatic conversion (aromatization) of testosterone to oestrogen by the sexual skin plays a significant role in promoting colour change (Rhodes *et al.*, 1997). Thus, it is possible that reddening of the face, rump and genitalia during puberty in male mandrills might also be effected by localized differences in the sensitivity and responsiveness of sexual skin to circulating testosterone (Setchell and Dixson, 2002).

Blue sexual skin is not dependent for its maintenance upon either testosterone or oestrogen, however, so that its colour is unaffected by castration in adult males of several species (e.g. scrotal colouration in the talapoin: Dixson and Herbert, 1974; and the patas monkey: Dixson, 1983a; see also Bercovitch, 1996). It is now known that the presence of collagen fibres in the dermis of the skin scatters and reflects light to produce the bluish tints. Previously, it was thought that light rays become scattered randomly, by the 'Tyndall effect', and are then reflected by dermal melanocytes to cause the blue colouration, as for example in the vervet monkey's scrotal skin (Price *et al.*, 1976).

However, more recent research by Prum and Torres (2004) on both vervets and mandrills demonstrates that the colour results from coherent (i.e. non-random) scattering of light by highly organized, dense arrays of collagen fibres. Moreover, Prum and Torres note that the same anatomical arrangement is found in some other mammals (e.g. *Marmosa mexicana*, a New World marsupial species), and has doubtless arisen a number of times during mammalian evolution. The striking blue facial adornments of the male mandrill provide another example of such convergent evolution. It will be argued later in this book that the common ancestor of the mandrill and drill would have lacked this blue paranasal colouration.

That vascular mechanisms might influence the blue colouration of the male mandrill's paranasal skin was suggested many years ago by Osman Hill (1955). Some recent work has implicated vascular mechanisms in controlling colour change in skin covering the head and wattles of male turkeys (Oh *et al.*, 2014). Alterations in blood flow affect the alignment of collagen bundles in the skin of these areas, so that its perceived colour shifts quickly from blue to white. Mandrills, however, do not display very rapid colour changes of this kind.

The mandrill's skeleton

When Hill (1970) wrote his account of the anatomy of the mandrill, he pointed out that its skeleton had 'never been fully described'. Indeed, he redrew an illustration, made in the nineteenth century by Martin (1841), of the skeleton of an adult male (Figure 3.6). This gives a general idea of the overall proportions of the adult male mandrill, including its large head, very short tail, broad scapula, and sturdy limbs.

Recent research on the skeleton of the limbs and limb girdles has produced some notable advances in our understanding of their functional anatomy. This newer work has also confirmed the results of the genetic studies, reviewed earlier, showing that a close phylogenetic relationship exists between *Cercocebus* and *Mandrillus* and that they represent sister taxa to *Lophocebus* and *Papio* in the Papionini (Fleagle and McGraw, 1999, 2002). Fleagle and McGraw showed that shared features of the scapula, upper arm and forearm of *Mandrillus* and *Cercocebus* reflect similarities of their musculature and locomotor abilities. These adaptations have evolved in relation to a mode of life that involves climbing in the trees, as well as vigorous manual foraging among the vegetation and leaf litter of the forest floor. Comparative measures of the pelvis, femur and tibia also unite *Mandrillus* and *Cercocebus* and distinguish them from *Papio* and *Lophocebus* (Figure 3.7). The hands and feet of mandrills are also markedly different from those of baboons. The thumb is longer than in *Papio*, and the toes are likewise much longer and stronger in mandrills and drills. The hallux ('big toe') is widely separated from the other toes, thus providing 'greater grasping power than in *Papio*, evidently in adaptation to a more arboreal environment' (Hill, 1970).

In addition to their studies of the limbs and limb girdles, Fleagle and McGraw (1999, 2002) examined dental traits, and found that the second (posterior) premolars of

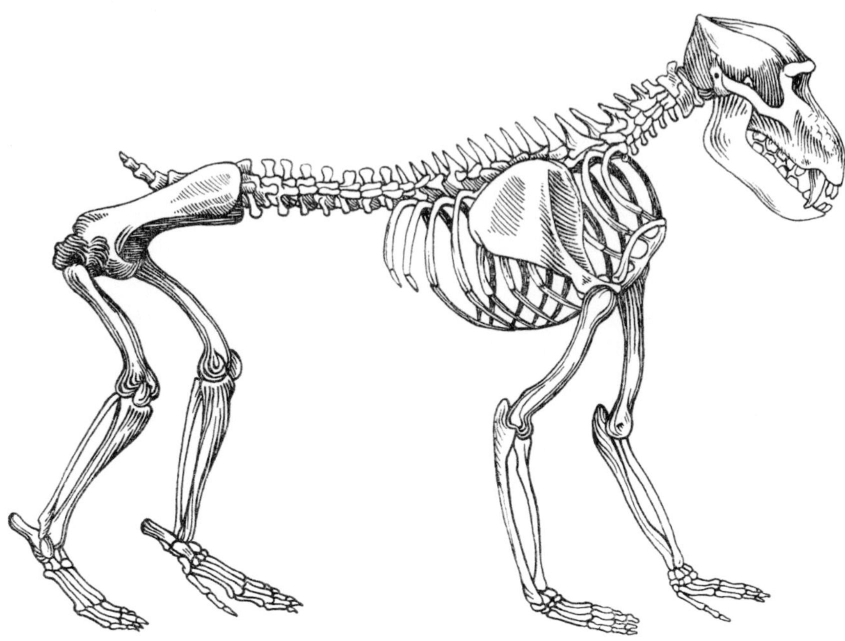

Figure 3.6. The skeleton of an adult male mandrill. (After Martin, 1841.)

Mandrillus are notably larger and better adapted as crushing teeth than in *Papio* and *Lophocebus*. Again, *Mandrillus* shares this condition with *Cercocebus* (Figure 3.7), as both are specialized *forest floor gleaners*, feeding on seeds and nuts, as well as using their teeth to break open rotting wood in search of insects and other invertebrates. Age-related tooth wear has been shown to be greater in mandrills than in savannah-living baboons. This is probably because of, in part, the greater importance of hard food items such as nuts and seeds in the mandrill's diet (Galbany *et al.*, 2014).

The adult dentition of the mandrill comprises, in each half of the upper and lower jaws, two incisors, one canine, two premolars, and three molar teeth; making 32 teeth in all, as is generally the case among Old World anthropoids. The upper (maxillary) canine teeth are especially large and dagger-like in adult male mandrills, and the male's jaws are huge in comparison to those of the adult female (Figure 3.8). A pronounced gap, or diastema, separates the upper canine from the second incisor. In the lower jaw, the first premolar tooth is much enlarged, and compressed laterally to form a cutting edge, which shears against the upper canine. This sectorial specialization of the lower first premolar occurs in other Old World monkeys, but it is especially pronounced in the mandrill (Hill, 1970).

The male's permanent canines begin to erupt at about 4.75 years of age, and reach their adult length (4–5 cms) by 8.75 years (Setchell and Dixson, 2002; Wickings and Dixson, 1992a). As with other sexually dimorphic traits, there are marked individual differences between males in the timing and degree of canine growth.

The bony paranasal (maxillary) swellings that develop along each side of the maturing male's snout can be seen in Figure 3.8, in which male and female skulls are shown in

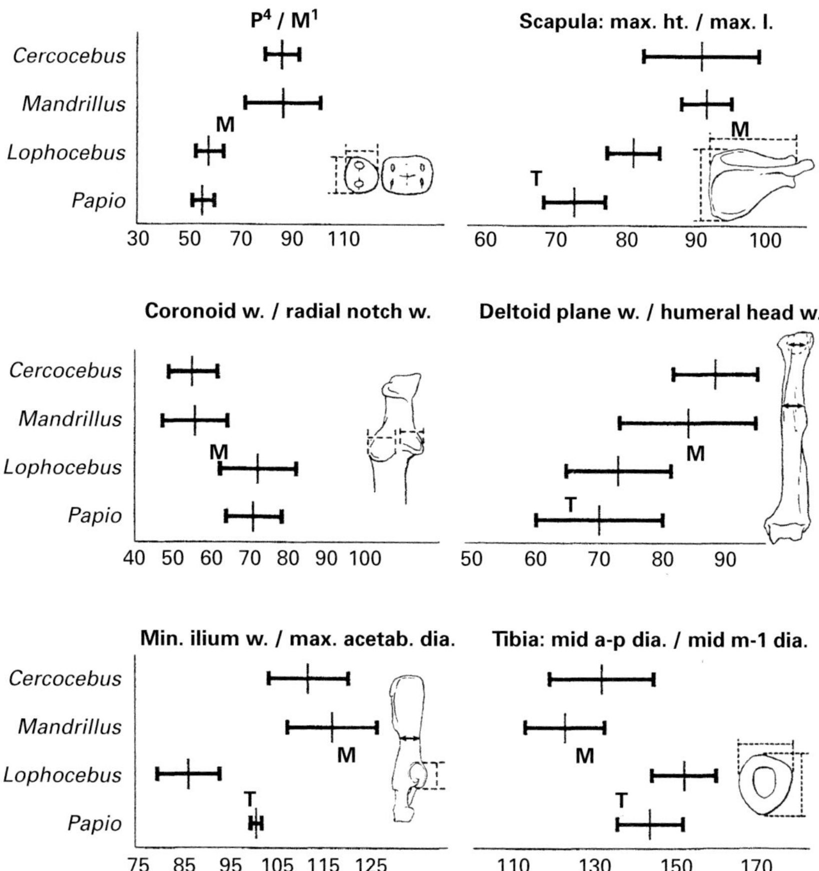

Figure 3.7. Morphological features of the limb bones, limb girdles and teeth that distinguish *Mandrillus, Cercocebus* and *Macaca nemestrina* (M) from *Lophocebus, Papio* and *Theropithecus* (T). (From Fleagle and McGraw, 1999. Copyright National Academy of Sciences, USA.)

lateral view. As noted previously, these swellings differ in size and shape in living adult males. Examination of mandrill skulls held in the collections of the Natural History Museum (London) and the Anthropological Institute of the University of Zurich showed that a continuum occurs in swelling sizes, rather than a marked division into well-developed versus poorly developed swellings (Martin and Dixson, unpublished observations). However, it transpired that the most extreme examples of enlarged paranasal swellings were present in some males that had grown to maturity in Zoos. The same phenomenon was identified in the skulls of captive-raised versus wild adult drills (Figure 3.9). Michelle Singleton (2012) has made a most valuable study of the unusual facial growth patterns of captive-raised male mandrills and drills. In some cases, their skulls exhibited 'increased robustness, pronounced cranial superstructures, lengthening of the anterior cranial base and face, increased facial height, and dorsal expansion of the paranasal ridges'. Singleton points out that captive females do not exhibit these

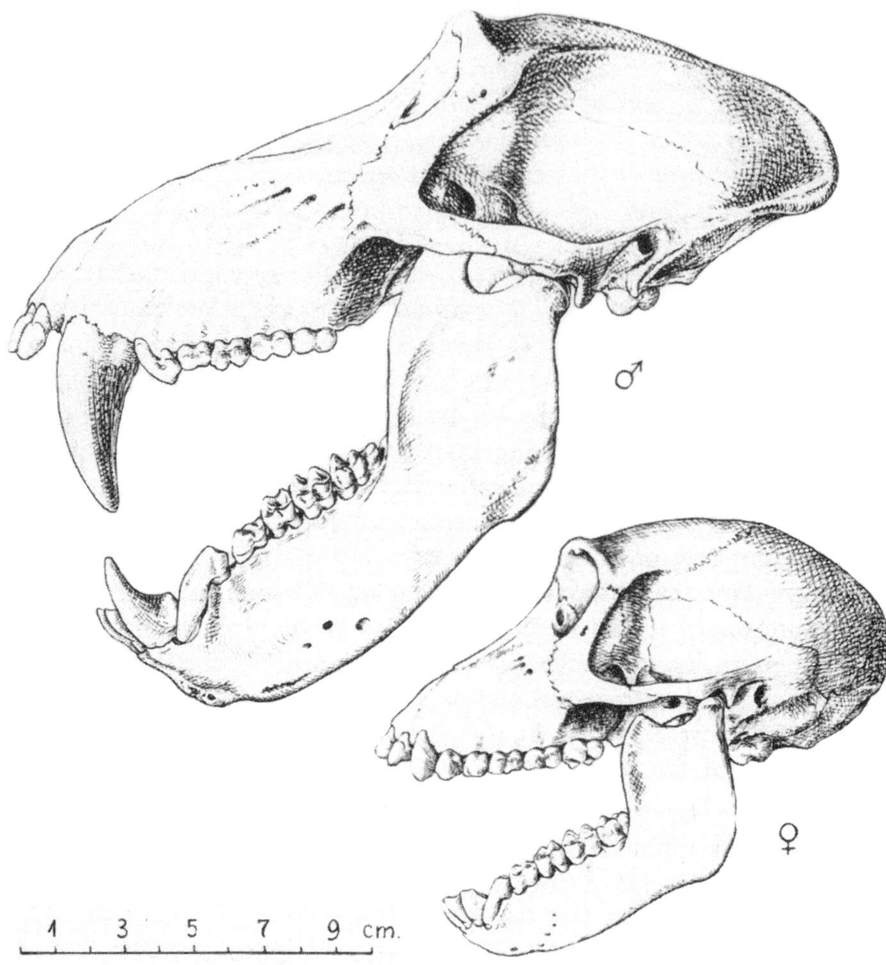

Figure 3.8. Skulls of adult male and adult female mandrills, drawn to the same scale. (From Schultz, 1969.)

cranial features, and she suggests that endocrine factors, rather than dietary ones, may influence cranial growth differently in the two sexes. Thus, increased secretion of testicular androgens, during puberty and adolescence, has major effects on craniofacial development (e.g in human males: Verdonck *et al.*, 1999). As will be discussed later in this book, adolescent male mandrills living in large social groups are subject to considerable social constraints upon the development of androgen-dependent secondary sexual traits, especially owing to the presence of more dominant males. The development of exaggerated masculine craniofacial traits may thus be linked to the relative absence of inhibitory social factors in small captive groups, lacking fully developed dominant males. I shall return to this interesting topic in Chapter 8 and in Chapter 10.

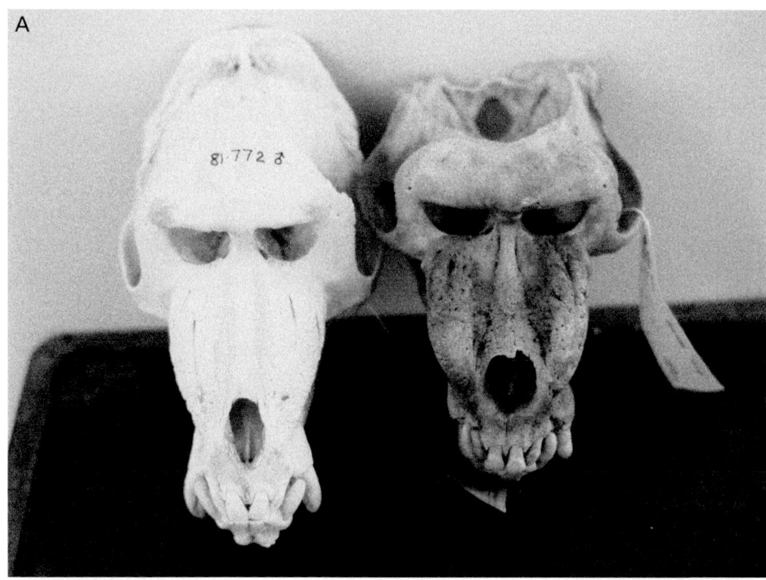

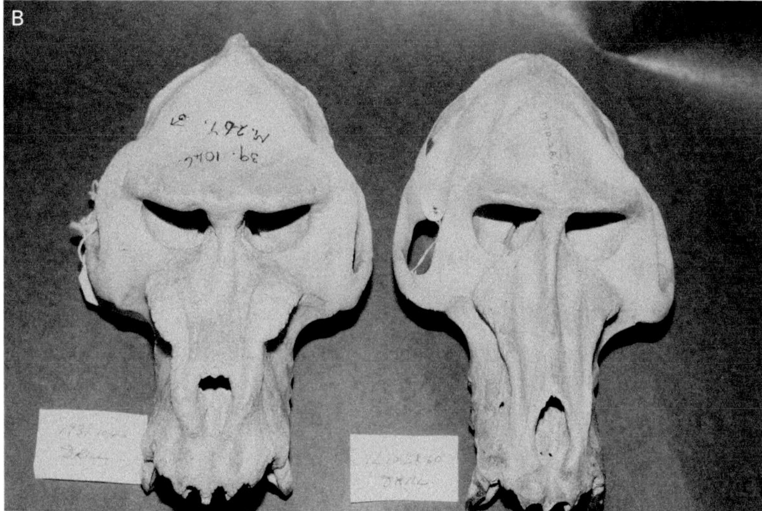

Figure 3.9. Skulls of adult male mandrills and drills to show the marked variations that occur in the shape and size of the muzzle and the prominent paranasal swellings that develop as males mature. **A**: Mandrill skulls. A robust example is shown on the right (in this specimen the posterior part of the cranium has been removed). **B**: Drill skulls. A robust example is shown on the left. (Author's photographs of specimens in the Natural History Museum, London.)

Reproductive anatomy

A discussion of the functional anatomy of the mandrill's genitalia provides important background information for later consideration of its unusual mating system, and the role played by sexual selection, via sperm competition (Parker, 1970) and cryptic female choice (Eberhard, 1985, 1996, 2009), during its evolution.

Hill's (1970) illustration of a sagittal section of the female genital tract of the drill is shown here, as well as my own drawing of a dissection of the male drill's reproductive organs (Figures 3.10 and 3.12). I have included these drawings of the drill's reproductive organs because so few illustrations are available of the mandrill's reproductive system, and especially of the internal genitalia of the male. However, it is known that the mandrill's reproductive system is very similar to that of the drill, as has been well-documented by Hill. Thus, the illustrations shown here should provide the reader with an accurate guide during the discussions of the mandrill's anatomy that follow.

Male genitalia

In the mandrill, as in all mammals, the male genitalia have evolved as an exquisitely complex, integrated system for the production, storage, and delivery of sperm (and accessory glandular secretions) to females during copulation. The reproductive systems of male mammals are structurally diverse, and this amazing diversity reflects, in part, the outcome of sexual selection, as well as natural selection, during evolution. Where the mandrill is concerned, we may start by noting that the testes are exceptionally large in relation to body weight in this species. Moreover, this is especially the case in high-

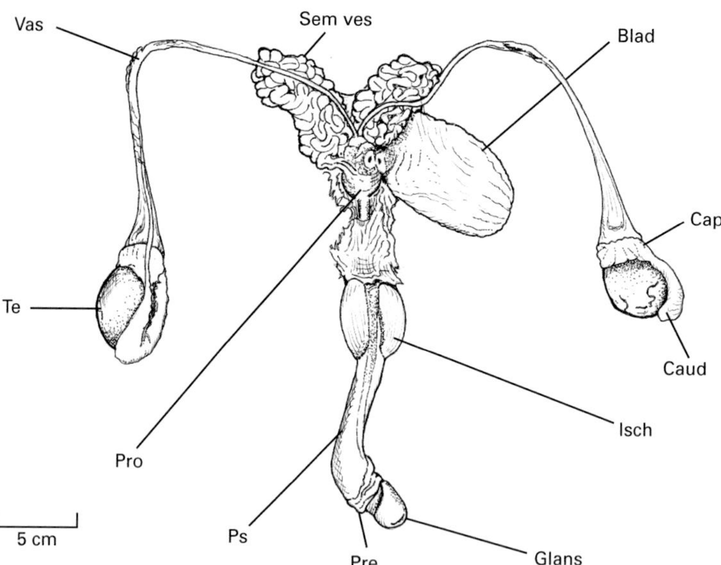

Figure 3.10. Dorsal view of a dissection of the reproductive tract of an adult male drill. The neck of the bladder has been transected, and the bladder reflected to one side. Blad, bladder; Cap, caput epididymis; Caud, cauda epididymis; Glans, glans penis; Isch, ischiocavernosus muscles; Pre, prepuce; Pro, prostate gland; Ps, shaft of penis; Sem ves, seminal vesicles; Te, testis; Vas, vas deferens. (Author's drawing.)

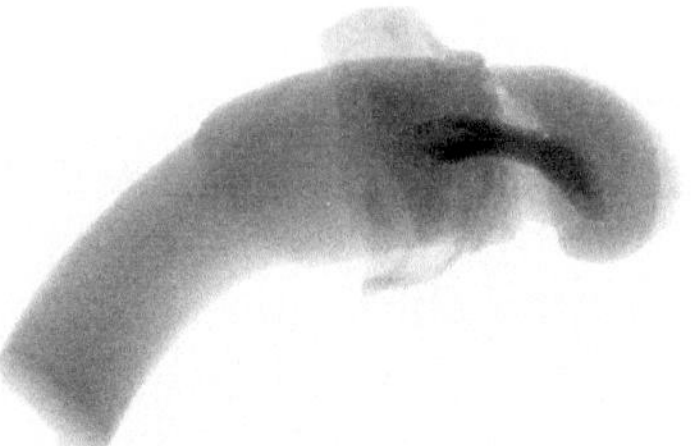

Figure 3.11. Radiograph of the penis of an adult male mandrill to show the position and shape of the penile bone (baculum). (From Dixson, 2012.)

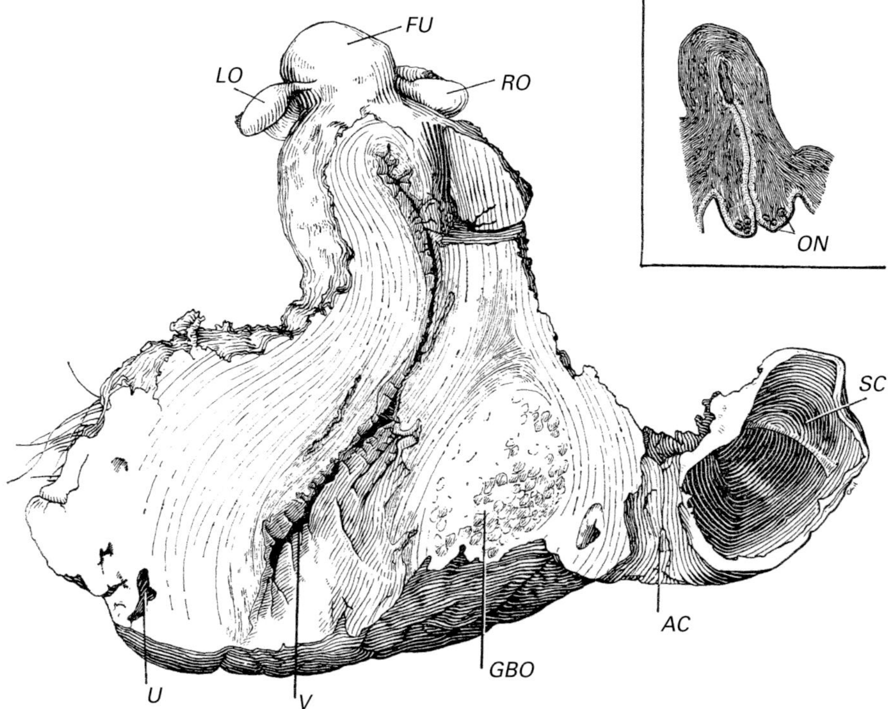

Figure 3.12. The reproductive organs of an adult female drill, in which the sexual skin is fully swollen. The swelling is shown in sagittal section, and a median sagittal section of the uterus is shown in the inset. AC, anus; FU, fundus of the uterus; GBO, bulbourethral glands; LO, left ovary; ON, ovules of Naboth (small cysts); RO, right ovary; SC, subcaudal lobe of sexual skin swelling; U, uterus; V, vagina. (After Hill, 1970.)

ranking males, where the average volume of each testis was initially reported to be 34.6 ml., as compared to 18.8 ml. in subordinate (non-fatted) males (Wickings and Dixson, 1992b). Relative testes sizes (i.e. testes size in relation to body weight) in dominant males are indeed as great, or greater than, those in some macaques and baboons, which have multimale–multifemale mating systems (Dixson, 1998a).

Females mate with multiple partners during the fertile period in these species, and sperm competition has selected for larger relative testes sizes so that males are better able to maintain high sperm counts (Harcourt *et al.*, 1981) during high frequencies of copulatory activity (Dixson, 1991).

Attached to the surface of each testis, and extending along its length, is the epididymis, which receives sperm as they leave the seminiferous tubules and issue from the excurrent ducts (rete testis). Sperm acquire the capacity to display motility, and undergo other physiological changes as they pass through the epididymis, which is large in both species of *Mandrillus* and has a well-developed cranial lobe (Figure 3.10). Sperm epididymal transit times are unknown for *Mandrillus*, and nor are there any data on numbers of sperm stored in the cauda epididymis prior to ejaculation. When sexual activity occurs, sperm are expelled from the cauda epididymis and are forced rapidly, by muscular contractions, along the length of the vasa deferentia. In Figure 3.10, each vas deferens is 27 cms long, as measured from its proximal (epididymal) end to the point where the two vasa deferentia unite. This occurs just dorsal to the neck of the bladder. Prior to ejaculation, sperm become mixed with secretions of the accessory reproductive glands. The seminal vesicles and prostate gland provide the bulk of the fluid portion of the ejaculate. The bulbourethral glands, which are not visible in the dissection shown in Figure 3.10, also make a small contribution to the seminal fluid.

The seminal vesicles are massive, lobulated structures in both the mandrill and the drill. The left and right lobes each measure 7.5 × 3.0 cms in the specimen illustrated here. The seminal vesicles averaged 16.2 gm in weight (at autopsy) in two adult male drills (Anderson and Dixson, 2009). Studies of seminal vesicular sizes in primates have shown that they are largest in males of those genera where large relative testes sizes and multimale–multifemale mating systems occur (Dixson, 1998b). The genus *Mandrillus* is an example of this phenomenon. I shall return to this subject in Chapter 10, where adaptations for sperm competition in the mandrill and drill are discussed in greater detail. Mention must also be made of the striated muscles, situated at the base of the penis, as one of their functions is to cause the forceful expulsion of semen during ejaculation. The bulbocavernosus muscles are especially important in this regard, whereas the smaller ischiocavernosus muscles serve primarily to elevate the shaft of the penis and to assist the male in attaining intromission. Only the ischiocavernosus muscles are visible in Figure 3.10, as the massive bulbocavernosus muscles have been removed by dissection. The great size of these muscles, in both the mandrill and drill, may also be linked to the nature of their mating systems, as will be discussed in Chapter 10.

The penis itself is very long, but quite thin, in both the mandrill and drill. The glans penis is helmet-shaped, and it is protected by a retractable prepuce. The penis contains a stout, rod-like penile bone (baculum), which arises from the distal end of the erectile bodies (corpora cavernosa) of the penile shaft, and extends distally into the left side of the glans (Figure 3.11). The baculum was 24.8 mm long in two adult mandrills that I measured, which is similar to the 21 mm reported for the drill (Hill, 1970). These measurements are comparable to those recorded for other papionins, including various macaques, baboons and mangabeys (Dixson, 1987, 2012).

Thus, elongation of the penis in *Mandrillus* has not been accompanied by increased length of the baculum. Bacular elongation occurs in primates that engage in prolonged patterns of intromission during mating and, as we shall see in Chapter 5, this is not the case in the mandrill.

There is some evidence that elongated penes are found in those species in which females have sexual skin swellings, so that co-evolution of the male and female genitalia may have occurred in this regard (as in *Pan troglodytes*: Dixson and Mundy, 1994). Hill (1974) commented on this possible correlation with respect to the mangabeys (*Cercocebus*), as follows:

> ...as observed by Pocock, the penis in *Cercocebus* is large and long in adaptation to the catamenial swelling of the female, thus ensuring penetration during copulation.

It will be recalled that the female mandrill and drill both exhibit sexual skin swelling, and that swelling morphology in *Mandrillus* is very similar to that of *Cercocebus*. This is likely to be a result of the close phylogenetic relationship that exists between these two genera, so that their common ancestor probably possessed the same type of swelling morphology. Although very few data are available for the mandrill, it is interesting that erect penile length (12.5 cms) exceeds the length of the vagina when the sexual skin is flat (9.2 cms), but correlates closely with vaginal length when sexual skin swelling is maximal (13.0 cms). Further research is needed to determine whether swelling size has driven the co-evolution of increased penile length via cryptic female choice (Eberhard, 1985, 1996) in *Mandrillus*, as is likely to be the case in the chimpanzee, for which much greater morphometric evidence is available (Dixson, 2012; Dixson and Mundy, 1994).

Female genitalia

Thus far, I have discussed some of the ways in which the male's reproductive organs act to produce and store sperm and accessory reproductive glandular secretions, and to transfer them to the female tract during copulation. It might be argued that the female's reproductive system has been honed during evolution only to receive and store the male gametes, prior to fertilization of her ova. However, this statement casts the female system as playing an entirely passive role, and this is not the case, for the female's reproductive anatomy and physiology play crucial roles in determining whether sperm survive, and how they are transported to the oviduct where fertilization takes place. Sperm must traverse the vagina, cervix, uterus, uterotubal junction and oviduct in order to gain access to the single ovum that is released during each menstrual cycle by a female mandrill. It is well known that although hundreds of millions of sperm enter the vagina during mating in various mammals, only a tiny fraction of these gametes (some hundreds or a few thousands) can be recovered from the ampulla of the oviduct at any given time (Harper, 1994; see Dixson, 2012 for a review). The female's reproductive system presents a set of anatomical and physiological sieves and barriers, which influence whether sperm gain access to ova. Thus, cryptic female choice (Eberhard, 1985, 1996, 2009) as well as sperm competition may occur within the female's reproductive tract when gametes of more than one male are present during her fertile period.

I preface my account of the anatomy of the female reproductive system in this way because, although most studies of cryptic female choice have involved invertebrates, there is increasing evidence that this type of sexual selection also occurs in mammals. Cryptic female choice may have favoured selection for complex penile morphologies, and specialized patterns of copulatory behaviour in males, as well as elongation of the oviducts in females (Anderson *et al.*, 2006; Dixson, 1987, 1991, 2012). Because mandrills engage in multiple partner matings, we should be mindful that sexual selection at the copulatory and post-copulatory levels has probably influenced the structure and functions of the reproductive systems of both sexes.

The sagittal section of the female reproductive tract shown in Figure 3.12 is unusual, because this female had died during the late follicular phase of her menstrual cycle. Heightened secretion of ovarian oestrogen during the follicular phase of the cycle causes water retention by the tissues of the sexual skin. This swelling had attained its maximum size. It is during this phase that ovulation is most likely to occur, although its exact timing is quite variable. Laparoscopic studies of female baboons have shown that 38.5% of ovulations occur on the last day of maximal swelling and 17.8% during the four preceding days (Wildt *et al.*, 1977). Sexual skin 'breakdown' (BD) occurs the day after the final day of maximal swelling, and then the sexual skin gradually undergoes detumescence during the second half (luteal phase) of the cycle. Almost 27% of ovulations occur during the initial day of detumescence (BD day) in the chacma baboon. Unfortunately, there have been no laparoscopic studies or measurements of hormonal changes during the menstrual cycle reported for either the mandrill or drill. However, data obtained from studies of other Old World species in which females have swellings (e.g. pig-tailed macaques and chimpanzees, as well as baboons) indicate that ovulations are most likely to occur during the final days of sexual skin swelling, or on BD day. The same temporal patterns are thus likely to occur in the mandrill.

Hill (1970) reported the presence of enlarged ovaries in the female drill illustrated here; the left ovary measuring 11 × 7 × 6 mm, and the right 12 × 7 × 6 mm. No histological studies were conducted, but given the presence of maximal sexual skin swelling, one would expect one ovary to contain an enlarged (dominant) follicle. It is the dominant follicle that secretes oestradiol and which produces the pre-ovulatory surge of this hormone. The oestradiol surge, in turn, triggers a surge of luteinizing hormone (LH) from the anterior pituitary gland. A single ovulation thus occurs at the end of the follicular phase; the ciliated fimbria that surround the entrance to each oviduct gather up the ovum and guide it towards the upper oviduct (ampulla). Hill noted that the oviducts are long and convoluted but, unfortunately, no measurements of oviductal length have been reported.

A simplex uterus is present, and this is quite small in the non-pregnant female; the body of the uterus being 41 mm long and 18 mm wide in the specimen shown here. Hill also provided drawings of the uterus of a mandrill that had recently given birth. In this case, the postpartum uterus measured 120 mm in length and was 68 mm wide.

The upper pole (fundus) of the uterus is rounded in shape. At its lower end, the muscular cervix separates the uterus from the vagina. The cervical canal is 'slightly sinuous', according to Hill, and 16–22 mm long in two specimens that he measured. The vagina is a very large and thick-walled passage; its operating depth is markedly increased when the female's sexual skin is swollen (Figure 3.12). This was discussed earlier in relation to possible co-evolution of vaginal and penile lengths in mandrills.

Finally, mention must be made of the clitoris, the smaller homologue of the male's penis, which is embedded in the pubic lobe of the female's sexual skin and situated ventrally, close to the entrance of the vagina.

4 Ecology and behaviour

> The woods are lovely, dark and deep,
> But I have promises to keep,
> And miles to go before I sleep,
> And miles to go before I sleep.
>
> Robert Frost

Social groups and the myth of the 'one-male unit'

Long before any detailed information on the social organization of wild mandrills or drills was available, a number of authors had commented on their relatively large group sizes as compared to those of other rainforest cercopithecines (Jeannin, 1936; Malbrant and Maclatchy, 1949; Sanderson, 1940). Malbrant and Maclatchy, for example, thought that mandrill groups might contain as many as 50 individuals, and that drills occur in *grandes hordes bruyantes*. Struhsaker (1969) recorded counts of nine to 55 (mean= 23.3) for 12 drill groups in Cameroon, but noted that 'most of these were incomplete'.

The earliest field study of any magnitude was conducted by Gartlan (1970) on drills of the Southern Bakundu Forest Reserve in western Cameroon. He counted drill groups ranging in size from 14 to 179 individuals; at least three adult males and 21 adult females were included in the largest group. Gartlan also encountered two solitary adult males. Subsequent fieldwork in Gabon (Jouventin, 1975a) and in Southern Cameroon (in the Campo Reserve: Hoshino *et al.*, 1984) also produced some useful data on mandrill group size and composition. Jouventin (1975a) proposed that the basic social unit in mandrills consists of one large adult male with five to ten females (with or without infants) and ten juveniles. He thought that a number of these small social units, or 'harems', sometimes unite to form large troops of up to 200 mandrills, while 'excess adult males live a solitary existence in the forest'. Hoshino *et al.* (1984) counted groups of between 15 and 95 mandrills at Campo Reserve. They concluded that two types of grouping occur in mandrills, 'one-male and multimale groups', but they observed solitary males as well on nine occasions. Hoshino *et al.* also regarded the one-male groups, which consisted of one large male and 14 other individuals on average, as permanent units that then coalesced from time to time to form larger multimale bands. The implication was, as proposed by Jouventin (1975a), that mandrill one-male units were, in general terms, equivalent to those that occur in polygynous species such as the hamadryas baboon or the gelada.

It is not difficult to understand why these earlier fieldworkers reached the conclusion that smaller groups of mandrills, or drills, might represent the one-male units of a polygynous mating system. One-male units of this kind were already known to occur in geladas (Crook, 1966) and in hamadryas baboons (Kummer, 1968). Fissioning within ancestral multimale–multifemale social groups is thought to have produced the one-male/multifemale units of both these species (Barton, 2000; Kummer, 1971, 1990). Both these African papionins are known to have complex, multilevel social organizations, however, so that their one-male units are the building blocks of larger social units (clans and bands) that constitute their large troops (Kummer, 1990). There is an Asiatic species, the golden snub-nosed monkey (*Rhinopithecus roxellana*), which also has a multilevel social organization. However, in this case, it is thought that troops have evolved by aggregation of spatially separate one-male/multifemale units, such as those that occur in many of the extant Asiatic colobines (Qi *et al.*, 2014).

Male geladas and hamadryas baboons are large and visually striking animals, with bright red areas of sexual skin, capes of hair and other secondary sexual traits indicative of a highly polygynous (one-male–multifemale) type of mating system. Given that mandrills and drills are so sexually dimorphic, and were traditionally considered by taxonomists to be forest baboons, it seemed logical to suggest that they might have similar social and mating systems to those of the much better-studied species. Hence, the existence of one-male units of mandrills and drills became established in the published literature (e.g. Barton, 2000; Harrison, 1988; Kingdon, 1997; Rowe, 1996; Stammbach, 1987).

Studies conducted at the Centre International de Recherches Médicales de Franceville (CIRMF), and at its Field Station in the Lopé National Park, have largely dispelled the myth of the mandrill one-male unit. A semi-free ranging mandrill group, living in a large enclosed area of natural rainforest at the CIRMF, was shown to have a multimale–multifemale social organization and mating system (Dixson *et al.*, 1993; Wickings *et al.*, 1993). Males did not form one-male units of females under these conditions; instead, the most dominant males mate-guarded individual females during periods of sexual skin swelling. Furthermore, the occurrence of less colourful non-fatted males was documented (Wickings and Dixson, 1992b). The existence of less colourful adult males had not been recognized by fieldworkers previously, and it is likely that many of the putative one-male units described in the field studies reviewed earlier actually contained multiple adult males. Indeed, subsequent research, conducted in the Lopé National Park on supergroups, indicates that mandrills do not form polygynous, one-male units (Abernethy *et al.*, 2002), nor do they form complex multilevel societies, like those described previously for hamadryas baboons, geladas, or golden snub-nosed monkeys.

Mandrill supergroups and subgroups

The Lopé studies revealed for the first time just how large mandrill groups can be. This is especially the case during the long dry season in Gabon, which lasts from mid-June to mid-September. It is during this period that the largest supergroups of mandrills have been encountered. These groups are sometimes referred to as ‘hordes’, a term that is evocative of their huge numbers, exceptional biomass and wide-ranging movements. Rogers *et al.* (1996) were able to observe one such supergroup over a 10-day period. It contained more than 625 animals, including 21 highly coloured mature males, 71 less colourful adult and sub-adult males, 247 adult and sub-adult females (at least 33 of which had sexual skin swellings) as well as 200 juveniles and 86 dependent infants.

This enormous group was by no means exceptional at the Lopé National Park, where the remoteness of the forest and lack of hunting pressures, coupled with the permanent presence of trained fieldworkers, has made it possible to study mandrill supergroups for the first time. Abernethy *et al.* (2002) were able to make 20 counts of at least four such supergroups over the course of three years. As well as direct observations, they also made filmed records of groups, as they crossed more open (savannah) areas in order to forage in forest fragments bordering the Ogooué River. Analyses of these films made it possible to obtain much more accurate data on group sizes and age/sex compositions. Supergroups varied in size from 338 to 845 individuals (mean, 620); Figure 4.1 shows data on the average age/sex compositions based upon 20 group counts for all months sampled by Abernethy *et al.* (2002). The authors stress that although numbers of adult females remained fairly constant throughout the year, numbers of mature males showed a distinct annual rhythm, with most males entering supergroups during the dry season months, which is when adult females develop their sexual skin swellings (Figure 4.2). Mature males (estimated to be >10 years old), as well as adolescent and sub-adult males (aged 6–9 years) emigrated from supergroups in large numbers at other times of the year. Thus, 'the mean number of mature males in the groups dropped from 9.4±4.7 in mating months to 1.5±0.8 in other months; a sixfold difference'. Figure 4.2 shows how few

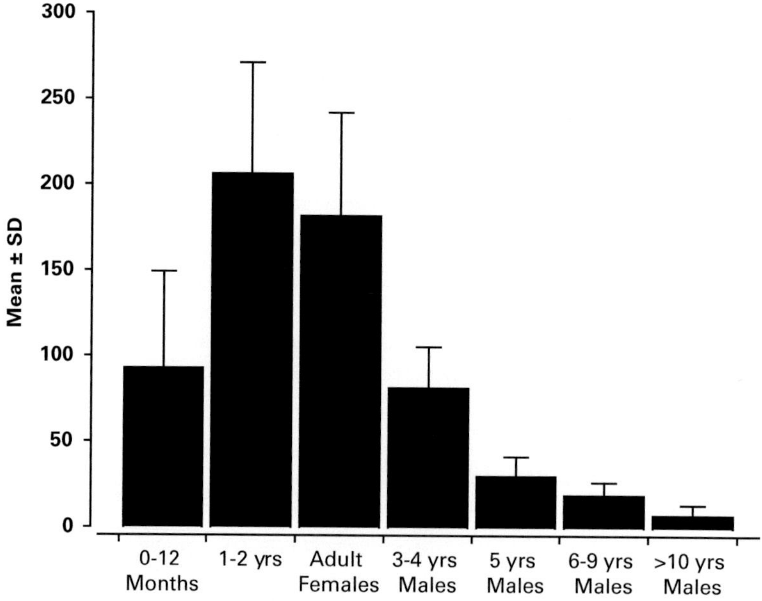

Figure 4.1. Age/sex composition of mandrill supergroups in the Lopé National Park, Gabon. (Drawn using data [20-group counts] in Abernethy *et al.*, 2002.)

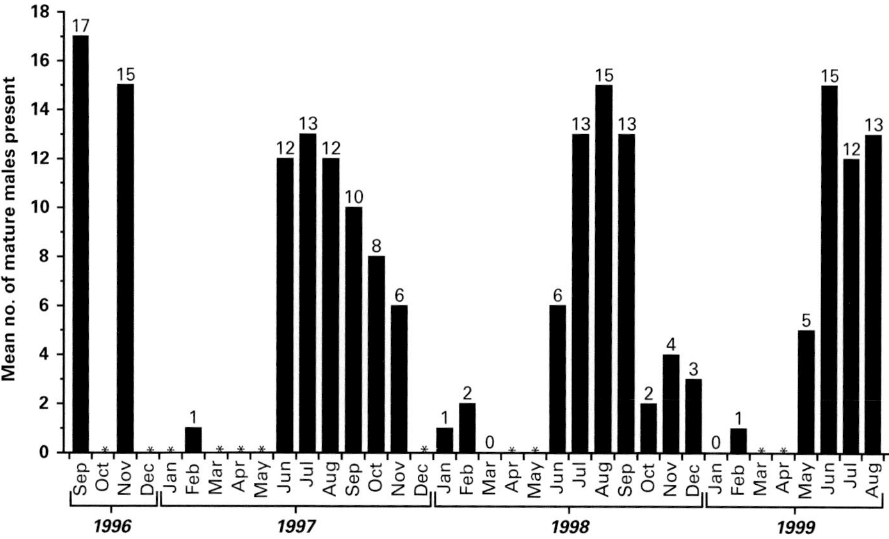

Figure 4.2. Seasonal changes in the numbers of adult males in mandrill supergroups during a three-year period. Data are monthly totals taken from 24 group counts. *, a month during which no data were recorded. (Redrawn from Abernethy *et al.*, 2002.)

mature males remained in the supergroups in December, January, February and March, with their numbers varying from zero to three during this period.

These results of fieldwork on mandrill supergroups have been augmented by more recent studies conducted at a second site, in the Moukalaba–Doudou National Park in southwest Gabon. Shun Hongo, of the Graduate School of Science at Kyoto University, spent 25 months studying mandrills in this National Park, and was able to film group progressions on three occasions (Hongo, 2014). Analysis of the films showed that two supergroups consisted of 350 and 442 individuals, while a third subgroup contained 169 mandrills. The two supergroups included only five or six adult males. These represented just 1.4% of total group numbers, as compared to the 36%–39% accounted for by adult females. Lone males were also seen on 11 occasions, as well as a possible 'bachelor group' consisting of just two adult males. Hongo could find no evidence that mandrills form one-male units; females and immature individuals were often closely associated during group progressions, but adult males did not herd or otherwise control harems of females.

I should emphasize that neither Abernethy *et al.* (2002) nor Hongo (2014) were able to observe the sexual behaviour of wild mandrills. The implication from their work is that sexual interactions and conceptions must occur in the multimale–multifemale supergroups, during the long dry season in Gabon, given that female mandrills develop their sexual skin swellings during this period. Females remain in their social groups all year round, whereas males are more mobile and often emigrate towards the end of the dry season. In Chapter 8, I shall describe in detail the sexual and associated behaviour of semi-free ranging mandrills. Females in such groups are philopatric, and they exhibit a strongly matrilineal social organization. Maturing males, in the age range of six to nine years, actively peripheralize and emigrate from their natal groups. They become 'lone males' but often re-enter the social group once they are fully adult. Some socially dominant adult males may remain as permanent residents in captive groups, and they compete to mate-guard individual females when they develop their swellings during the long dry season. However, females can also mate with multiple partners, including adolescent and lower-ranking adult males, so that sexual interactions are remarkably complex, even in groups that are only a fraction of the size of those that exist in the Lopé and Moukalaba–Doudou National Parks.

Before I address the feeding ecology and ranging behaviour of mandrills, it is germane to discussions of the evolution of the genus *Mandrillus* to consider whether the drill might share with the mandrill the propensity to form supergroups. Do male drills enter these groups for mating purposes, or emigrate and live as lone males for extended periods of time? It will be recalled that the largest group of drills recorded by Gartlan (1970) in the Bakundu Forest contained 179 individuals, including at least three adult males and 21 adult females. Sadly, most of the Bakundu Forest has been destroyed and drills are now extinct in that area. However, long-term data on drill group sizes are available from another site, Bakossiland in southwestern Cameroon (Wild *et al.*, 2005). Between the years 1970 and 2002, 105 direct visual counts were made of drill groups, and of solitary males in the Bakossi forests, especially on Mt Kupe. Group sizes ranged from five to 400 (mean±SEM, 93.1±8.4), and eight lone males were also recorded.

Unfortunately, it was not possible to make counts of the various age/sex classes, or to establish whether seasonal changes occurred as regards the presence of adult males in the drill groups of Bakossiland. Complete counts were almost impossible to achieve, and many must have been underestimates, as was shown to be the case when direct counts were compared to those made using filmed records of mandrill groups in Gabon (Abernethy *et al.*, 2002). Nonetheless, it is clear that, until recently, very large groups of drills still roamed the forests of Bakossiland; 17 groups contained between 101 and 200 individuals, five groups contained 200 to 300 individuals, and four groups had minimum counts of 400 drills (Figure 4.3). Thus, it seems highly probable that smaller groups and supergroups occur in drills, just as in mandrills. No data on numbers of females with sexual skin swellings are available and very little can be said about the mating system of free ranging drills. However, sexual behaviour and seasonal patterns of reproduction have been studied in captive drill groups. The results of this work will be discussed in Chapter 7, as there are some useful comparisons to be made with the mandrill.

Feeding ecology and ranging behaviour

There are three substantial published accounts concerning the feeding ecology of the mandrill (Hoshino, 1985; Lahm, 1986; Rogers *et al.*, 1996). These reports contain detailed lists of the species of plants and animals eaten by mandrills, but it is not my intention to repeat these listings here. Rather, my goal is to give some overview of the foraging and ranging behaviour of mandrills. The mandrill is a forest floor gleaner, as well as being a species that utilizes the lower and mid-levels of the rainforest in its quest for food. Some appreciation of the mandrill's feeding ecology is required, in order to

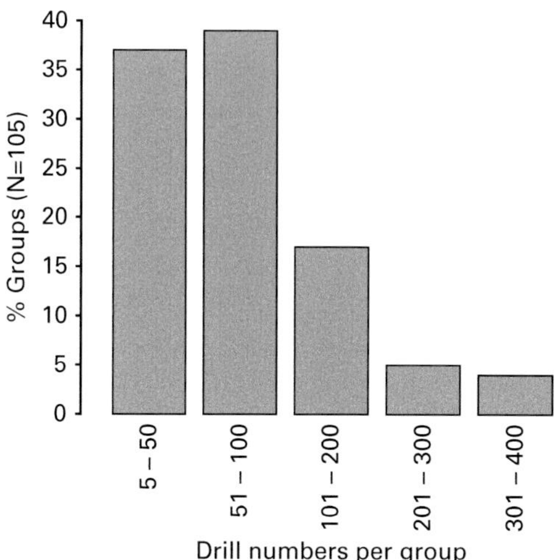

Figure 4.3. Group counts (N=105) of drills in Bakossiland, Cameroon. (Based upon data in Wild *et al.*, 2005.)

understand its fluid social groupings, subgroups and supergroups, as well as the propensity for adult males to spend long periods of time foraging alone.

Jiro Hoshino (1985) studied mandrills for 28 months in southwest Cameroon, in the Campo Reserve. Faecal analysis showed that just over 84% of their diet consisted of fruits (including seeds as well as pulp), with the remainder being made up of leaves and animal matter (mostly small arthropods). In all, 113 species of plants contributed to the diet, and Hoshino noted that in 101 of these species, fruits were consumed. Seeds, rather than other parts of the fruit, were often the major portions eaten (39 species). He concluded that 'seeds clearly have a special importance in the mandrill's diet'. Sally Lahm (1986) also stressed the importance of seeds as dietary items for mandrills in Gabon (at Makokou, and in the Lopé National Park), and noted that these monkeys may be important seed dispersers for many plant species. It will be recalled that mandrills display dental and other adaptations consistent with feeding on hard items, such as seeds, and for vigorous manual foraging on the forest floor. Hoshino often found the crushed remains of seeds in their faeces. Seed consumption was not confined to feeding in the trees during fruiting periods. Mandrills foraged very actively on the forest floor, and they often consumed fallen fruit, as well as the seeds that remained long after the soft parts of fruits had begun to decompose. Thus, 68% of plant food items at Campo Reserve were obtained at ground level.

Surveys of the tree species in this forest led Hoshino to identify two fruiting seasons, a major fruiting season extending from September to March and a minor season between April and August (Figure 4.4). Because trees in fruit were often widely dispersed in the rainforest, the mandrills ranged over long distances in order to exploit these patchy and clumped resources. At certain times, individual tree species supplied the major part of the diet. Thus, in March and April, mandrills at Campo Reserve fed mainly on the fruits of *Grewia coriacea* and, in September and October, on the fruits of *Sacoglottis*

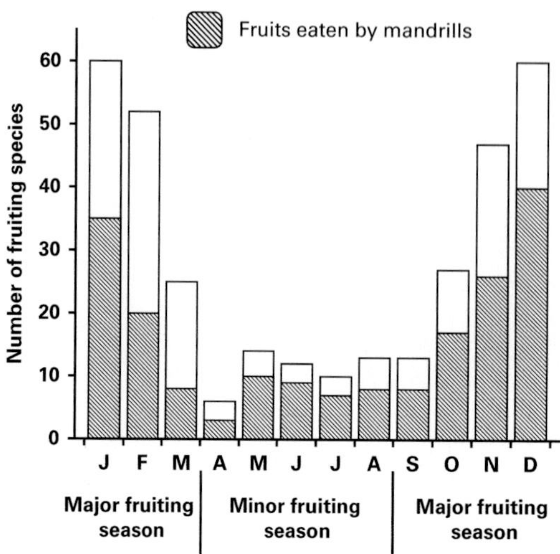

Figure 4.4. Seasonal changes in total numbers of fruiting species, and of those fruiting species eaten by mandrills in the Campo Reserve, Cameroon. (Redrawn and modified from Hoshino, 1985.)

gabonensis, which are especially plentiful during this period of the major fruiting season. Consumption then decreased during the minor fruiting season, and the mandrills included a greater proportion of leaves in their diet during this period. These were mainly the leaves of herbaceous plants growing in the forest understorey, and especially monocotyledonous species belonging to the families Graminiae, Marantaceae and Zingiberaceae.

Mandrills, however, are omnivorous, and they search through the leaf litter of the forest floor and examine rotten logs in order to feed upon ants, termites, crickets, spiders and snails. Small vertebrates are also taken, and there is evidence that mandrills include birds, and their eggs, as well as frogs and mice in their diet. One case has been reported of mandrills killing and eating a bay duiker (*Cephalophus dorsalis*); therefore, the possibility exists that they will take larger prey, should the opportunity arise (Kudo and Mitani, 1985).

In Gabon, Lahm (1986) recorded the 'eclectic feeding behaviour' of mandrills on 'a diverse array of invertebrates'. These included 14 species of ants and eight species of beetles, as well as large hunting spiders. The remains of tortoises, frogs, shrews, brush-tailed porcupines and swamp rats were also identified in the faeces of mandrills, or as part of their stomach contents.

Figure 4.5 shows the vertical distribution of mandrills observed by Hoshino (1985) in the various strata of the rainforest, in relation to the 86 types of plants eaten by the monkeys. Almost 67% of sightings were of mandrills in the lowest stratum, within five metres of the ground, or foraging on the forest floor itself. Most of their plant foods were available at this level, especially if one includes fallen fruits. The great majority of invertebrate species consumed by mandrills were also obtained on the forest floor. Hence, Hoshino concluded that 'mandrills are basically terrestrial'. As Figure 4.5

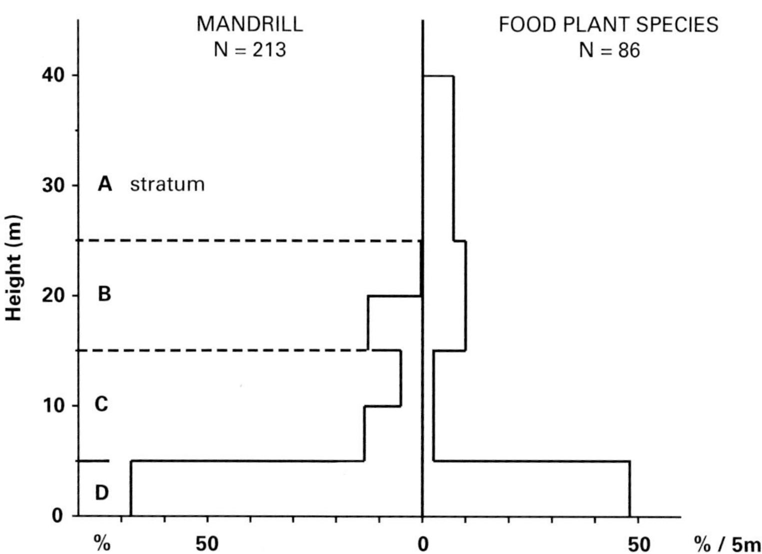

Figure 4.5. Vertical distribution of mandrills (left) and their food plant species (right) in a Cameroonian rainforest. (Redrawn from Hoshino, 1985.)

shows, they do climb, and a little over 10% of sightings were of monkeys at heights of 15 m or more. Observations of mandrills in the naturalistic enclosures at the CIRMF show that juveniles climb more frequently and are able to move from tree to tree by jumping across small gaps. Adult females also do this, but during group progressions, adults are more often seen at ground level. Lahm (1986) noted that adult mandrills of both sexes forage in the trees, but it is mainly the smaller group members that do this, and they are also more likely to climb trees in response to potential threats.

Mandrill groups are known to travel considerable distances each day in search of food. Hoshino calculated the distance travelled by the largest group he studied (95 animals) to be 4.5 km per day during the major fruiting season, and 2.5 km per day during the minor fruiting season. Longer travel distances during the major fruiting period probably occurred because the mandrills fed more on fresh fruits at this time, and fruiting trees are widely dispersed in the forest. During the minor fruiting season, animals foraged more intensively for the seeds of fallen fruits and for invertebrates on the forest floor. The animals left their sleeping trees by about 07.00 h, and moved steadily throughout the day, without any obvious midday rest period, until about 17.00 h, when the whole group climbed into the trees before nightfall.

Hoshino considered the total weights (biomass) of small and large mandrill groups in relation to the distances they travelled each day, and to food availability during the different fruiting seasons. He reasoned that a small group, containing 15 mandrills (one adult male, six adult females and eight juveniles, which Hoshino estimated to have a total mass of 111 kg) might obtain enough fruit from a single *S. gabonensis* tree to satisfy its daily needs during the height of the fruiting period (September/October). A large group (90 mandrills) would have to forage on four trees to achieve the same result, however. The longer distances (4.5 km/day) travelled by the large group at Campo Reserve probably brought it into contact with as many as seven trees of *S. gabonensis*, so that it 'could easily satisfy its need for this fruit from mid-September to the beginning of October'. Clearly, the needs of a supergroup, containing hundreds of mandrills must far exceed those of the groups observed by Hoshino.

The fact that mandrills are sometimes seen in small groups and sometimes in larger groups thus probably relates to their feeding ecology, and to the very complex seasonal changes in food availability that occur throughout the year. Let us consider for a moment the huge numbers of mandrills that have been counted in supergroups during the long dry season, in the Lopé National Park. As was discussed earlier, the largest supergroup studied by Abernethy *et al.* (2002) contained at least 845 individuals. Using data on the weights of animals of known ages in the CIRMF colony (Wickings and Dixson, 1992a), I estimate that the various age/sex classes of mandrills listed by Abernethy *et al.* in this supergroup would have had an approximate total biomass of 7400 kg! This huge biomass is equivalent to that of at least 116 adult humans; yet mandrill supergroups succeed in finding sufficient food to maintain their members for substantial periods during the long dry season. Rogers *et al.* (1996), who followed one such supergroup for ten days, estimated that it had moved approximately 14 km during this time. The group made use of the savannah–forest mosaic at the Lopé National Park, visiting isolated islands of forest to exploit fruit resources that had been depleted elsewhere, owing to greater interspecific competition in the main forest block. The dry season is also advantageous for foraging through the leaf litter of the forest floor, and Rogers *et al.* were able to show that mandrills were consuming large numbers of arthropods (especially ants and termites), as well as seeds and plant fibre. Stem, pith and leaves from monocotyledonous plants were especially important, and species belonging to the Marantacae and Zingiberaceae were preferred (as noted also by Hoshino, 1985).

How large are mandrill home ranges? Fieldworkers have only recently begun to obtain accurate answers to this question, although earlier authors had ventured to make estimates. Rogers *et al.* (1996), for example, noted that their mandrill supergroup (>600 individuals) made use of a total area of almost eight km^2 over a ten-day period. However, they also stressed that mandrills are seen only sporadically in forest near to the Lopé Field Station, except during the dry season, and that '8 km^2 is a small fraction of the total annual range required for a community of potentially associating and interbreeding sub-groups'. At Campo Reserve in Cameroon, Hoshino *et al.* (1984) thought that mandrill groups (numbering 15–95 animals) might utilize ranges of between five and 28 km^2. Still earlier, Jouventin (1975a) considered that a home range of 40–50 km^2 might be used by a large mandrill group. Harrison (1988) thought that mandrills in the Lopé National Park possibly ranged over areas of 36–45 km^2, during the course of nine months.

White *et al.* (2010) have conducted by far the most detailed study of ranging behaviour in mandrills. They used telemetry to track the movements of a supergroup containing at least 700 mandrills, in the northern part of the Lopé National Park. Radio transmitters were attached to a total of 27 individuals; 14 females and young males and 13 adult males. Attention focused upon the females and younger males, as these remained with the group all the time, whereas (as we have seen) adult males often

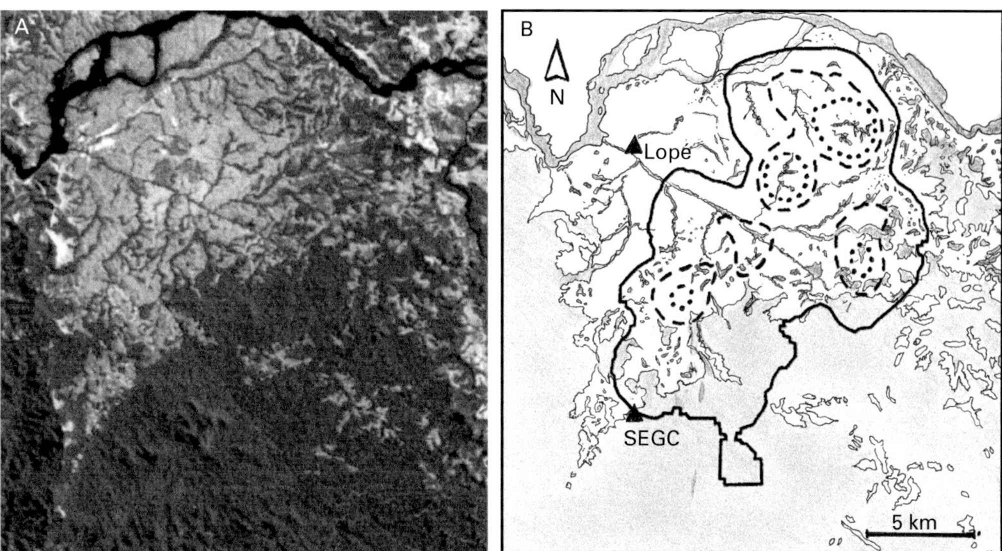

Figure 4.6. The home range of a mandrill supergroup. **A:** An aerial view of the northern part of the Lopé National Park, bordering the Ogooué River in Gabon, where the home range mapping study was conducted. **B:** The supergroup's home range, mapped using the fixed kernels contour (FKC) method. SEGC, Primate Research Station. (Aerial photograph from Abernethy *et al.*, 2002; home range map redrawn and modified from White *et al.*, 2010.)

leave groups for long periods. White *et al.*'s data are impressive, as they span six years (1999–2004) in the life of this mandrill supergroup. The group included extensive areas of gallery forests and isolated forest fragments within its home range, as well as the *Marantaceae* forests, which occur in the northern section of the Lopé National Park. Using the minimum convex polygon (MCP) technique to measure home range gave a total area of 182 km^2 (including open savannah not utilized by the animals), and 89 km^2 when only suitable forest habitat was considered.

Given the limitations of the MCP technique, White *et al.* also used other measures of home range size, as they provided a more accurate picture of how mandrills distribute their time within particular parts of their range. Thus, the fixed kernels contour (FKC) technique showed that the group spent more than half its time in less than 10% of its total range (Figure 4.6). The supergroup sometimes split into two to four subgroups, which reunited after varying periods. The total distance the group travelled each day was measured on 17 occasions, most of which were during the long dry season (range 1.96–10.2 km; mean= 4.94 km). As far as the authors could tell, only one supergroup occupied the range shown in Figure 4.6, but the area of the range varied significantly from year to year (Table 4.1).

The home range measures obtained by White *et al.* (2010) are consistent with the patchy distribution of the mandrill's major foods (i.e. fruits and seeds), seasonal changes in food availability and the overall biomass of its supergroups. White *et al.* calculated that their supergroup had a biomass of roughly 4852 kg. Previous work has shown that a

Table 4.1. Annual home range estimates made by White *et al.* (2010) for a mandrill supergroup, in the Lopé National Park, Gabon

Year	Home range (km^2) Total area	Home range (km^2) Excluding grass savannah
1999	161	74
2000	95	27
2001	149	66
2002	116	39
2003	77	22
2004	113	50

Home range estimates were made by using the 95% probability contour of fixed kernel contours (FKC) measures. (Data are from White *et al.*, 2010.)

positive correlation exists between home range area and group mass, for primates in general (Clutton-Brock and Harvey, 1977). White *et al.* noted that their home range data for the mandrill, which has the largest groups and the greatest group biomass of any primate species, nonetheless fit very well within the overall pattern reported by Clutton-Brock and Harvey.

Some comparative observations on drill ecology

Like the mandrill, the drill is a predominantly terrestrial species, a 'forest floor gleaner', and an omnivore feeding upon fruits, seeds and invertebrates (Astaras 2009; Gartlan, 1970). Christos Astaras observed that drills in the Korup National Park in Cameroon sometimes foraged in trees at medium heights (3–10 m), but were usually encountered at lower levels or on the floor of the forest. Group sizes at Korup averaged 43 drills (range 25–77), and Astaras recorded groups splitting or fusing together on several occasions. Whether these groups might have represented component parts of a larger supergroup is not known. Two solitary males were also observed during this study. Almost nothing is known about patterns of immigration or emigration by male drills, but clearly some males do leave their social groups. Astaras found that drill groups travelled an average of 2.9 km/day, foraging almost continuously, but sometimes with a period of rest or reduced activity between 12.00 h and 14.00 h. The animals usually climbed into the trees at about 17.30 h, well before dusk, and slept in adjacent trees connected by lianes and overlapping branches. He was unable to collect detailed information about drill home ranges, but he noted that 'a minimum of two groups utilized the 38 km^2 of the study area'.

Thus, it seems probable that mandrills and drills share many features of their feeding ecology and ranging behaviour. It has been suggested that drills are more exclusively animals of the rainforest, whereas mandrills will sometimes utilize open habitats, as in the savannah–forest mosaic areas of the Lopé National Park. However, this may simply underscore our lack of knowledge concerning drill populations in the wild, rather than reflecting a genuine interspecific difference. On Mt. Kupe in Cameroon, for example, drill groups quite commonly traverse open grassy areas near the summit of the mountain, and forage there before descending into the primary forest. We should keep in mind that both mandrills and drills have survived throughout long periods during their evolution when their rainforest habitats contracted owing to climate change. They survived in refuge areas where forests persisted on high ground, or close to major rivers. The ability to move between forest patches, as mandrills still do in the northern part of the Lopé National Park, may thus have been very advantageous during the long history of the genus *Mandrillus*.

5 Social communication

Mandrills employ a complex repertoire of visual, vocal and olfactory displays for social communication; they also make use of tactile cues in a variety of contexts, as during mother–infant interactions, when grooming and during copulatory behaviour. The elements of social communication have been well studied in captive and semi-free ranging groups of mandrills, and some useful comparative information is available for the drill. However, where wild mandrills are concerned, only their vocalizations have been studied in any detail, probably because it is much easier to obtain data on the sounds made by wild mandrills than it is to observe their social interactions at close quarters. I shall begin, therefore, by considering the mandrill's vocal repertoire and a seminal paper by Kudo (1987), in which he considered the likely social organization of mandrill groups in the light of his studies of their vocal communication.

Vocal communication

Hiroko Kudo (1987) spent 14 months studying free ranging mandrills in the Campo Reserve in Cameroon, during which time he was able to identify ten vocalizations used for communication. Details of these are listed in Table 5.1. Two further types of vocalizations were heard, but he was not able to record them or to identify their functions. In addition, he noted that all age/sex classes of the mandrill make 'tooth grinding' sounds. I have heard only adult males making these sounds, however. As they open and close their jaws, the upper canines rub against the lower teeth (whether the canines or first premolars, I cannot say for certain) in a honing action. Sometimes a male actively works his jaws in this way, and the teeth make a high-pitched squeaking sound as they grind against each other.

Kudo recorded several long-distance calls that were of particular importance for communication, as the mandrills spread out over a considerable area while foraging or when travelling rapidly through the forest in search of fruiting trees. A distinctive low-frequency sound, the 2-phase grunt (2PG), was made only by adult males. The 2PG consists of short, deep groans, repeated every two seconds or so. Males tend to remain near the rear of the group during progressions, and Kudo thought that only one, or a few adult males, gave 2PG vocalizations. He considered that this might indicate 'the existence of a leader male' in such groups.

Table 5.1. Vocal repertoire of the mandrill

Name of vocalization	Class of animal making the sound	Context
LONG DISTANCE CALLS		
2-Phase grunt	Adult males only	During group progressions and when mate-guarding individual females
Roar	Adult males only	During group movements. Acoustically similar to the 'wahoo' bark of adult male Baboons
Crowing	All animals except adult males	Contact calls during group movements or when animals are widely dispersed during foraging
SHORT DISTANCE CALLS		
Yak	All animals except adult males	Tense situations
Grunt	All age/sex classes	Aggressive situations, also a contact call
Growl	All age/sex classes	Mild alarm
K-alarm	All age/sex classes	Intense alarm
K-sound	Unknown	Unknown
Scream	All age/sex classes	Fear and submission
Girney	Infants, juveniles and adult females	Appeasement, frustration

(From Kudo (1987), plus author's observations at the CIRMF, Gabon.)

However, Kudo also thought that vocal communication within and between mandrill subgroups was much more consistent with the social organization of this species being basically multimale–multifemale, rather than representing a closed, one-male unit type of system. Later research would confirm Kudo's ideas, which were well in advance of their time.

Observations of semi-free ranging mandrills at the CIRMF confirm that male rank is correlated with emission of the 2PG. In the group that I studied most intensively, there were three fatted males, while three non-fatted adult males also lived outside the group. The fatted males were heard making 2PG vocalizations, as the group moved through the forest, and the alpha male vocalized most frequently. The 2PG was also given regularly whenever a male was following and mate-guarding females during the annual mating season. It was especially the alpha male in the group that mate-guarded and vocalized in this way. It sounded as if the two phases of each grunt corresponded to the male breathing in and out, to produce deep 'huffing and puffing' sounds. The alpha male's 2PG vocalizations may have served, in part, to ward off other males; some lower-ranking individuals moved away swiftly whenever they heard the alpha male following a female through the forest.

Adult males at Campo Reserve sometimes emitted 'roars' during group movements. These loud, low-frequency barking sounds were made singly, or repeated two to three times, rather than as a continuous series. At the CIRMF, roars were occasionally made by solitary adults, as well as by group-associated males, in response to disturbances (e.g. the sound of a car on the service road next to the enclosure). Other age/sex classes in wild

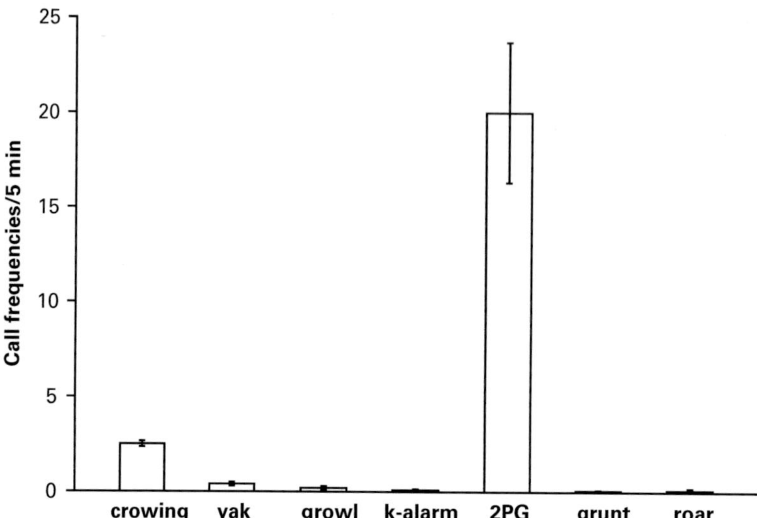

Figure 5.1. Frequencies with which free-ranging mandrills vocalize. 2-Phase grunting (2PG), by adult males, is the most frequent vocalization, followed by 'crowing', which other group members employ as a contact call. (Redrawn from Kudo, 1987).

mandrill groups did not make either roaring or 2PG vocalizations. Instead, 'crowing' vocalizations were given by adult females and younger animals of both sexes. These higher-pitched calls were often heard when groups were on the move, being taken up in chorus by large numbers of animals. Similarly, crowing calls were often heard in the CIRMF groups, and it appeared that they served to maintain contact between group members, and to coordinate their movements. Adult males did not make this crowing vocalization.

Kudo (1987) quantified the frequencies with which various mandrill calls were heard at Campo Reserve, the most frequent being the 2PG, followed by crowing. Other vocalizations occurred at much lower frequencies (Figure 5.1).

Short-distance calls included 'yaks', given in tense situations, sequences of rapid growls and the 'K-alarm', a sharp, two-syllable vocalization made by highly alarmed animals of all age/sex classes (Table 5.1). Mandrills of both sexes, and all ages, also utter screams and squeals when frightened and during flight; these vocalizations often accompany the fear grimaces given by subordinate monkeys. 'Girneying' or soft 'mewing' sounds were most often given by infants at the CIRMF. As an example, when infants were being weaned, their mothers sometimes actively rejected or refused to re-establish contact with them. Infants made persistent mewing sounds in response to such rejections.

Most of the vocalizations given by mandrills are very similar or identical to those given by drills. Knowledge of the drill's communicatory biology is less complete, but it is relevant to note that adult male drills also give 2-phase grunts and roaring vocalizations in the same contexts as adult male mandrills, and that other age/sex classes use crowing as a contact and group cohesion call (Astaras, 2009; Gartlan, 1970).

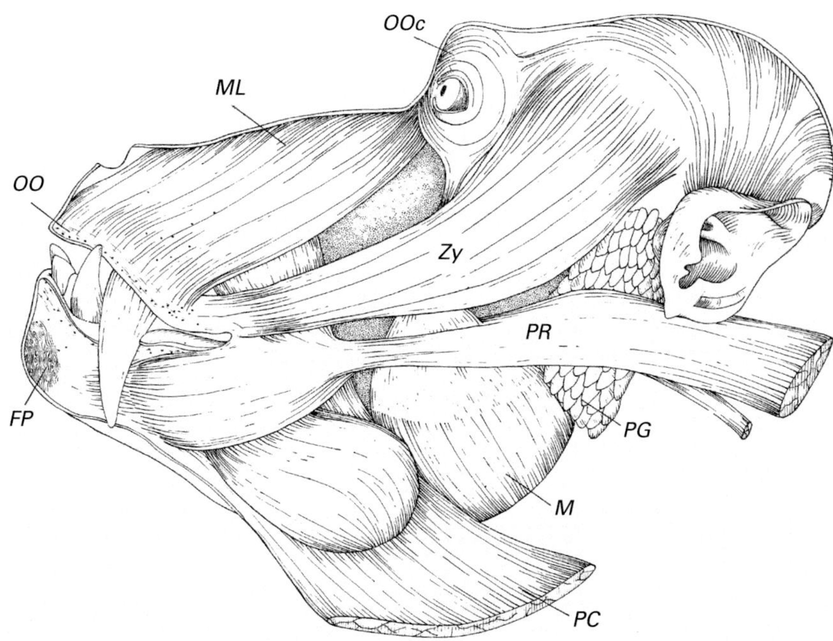

Figure 5.2. The superficial facial musculature of an adult male mandrill. FP, fibrofatty pad; M, masseter; ML, maxillolabialis; OOc, orbicularis oculi; OO, orbicularis oris; PC, cervicoplatysma; PG, parotid gland; PR, notoplatysma (risorius); Zy, zygomaticus. (From Hill, 1970.)

Facial expressions and other visual displays

Figure 5.2 shows a dissection of the superficial facial musculature of an adult male mandrill. Lip movements play important roles in facial expression, and the mandrill has well-developed maxillolabialis muscles, enabling it to raise its upper lip, while the orbicularis oris muscles make it possible to purse the lips and perform other movements that alter the shape of the mouth. The zygomaticus and notoplatysma (risorius) muscles serve to retract the mouth corners. The eyes and brows of the mandrill are also highly mobile and expressive. Eye closure is brought about by contractions of the orbicularis oculi, while the corrugator supercilii muscles (not shown in Figure 5.2) enable the mandrill to move its brows. All of these muscles fall under the control of the seventh cranial nerves (VII), as is also the case in other anthropoids. Note that in the account that follows, the major facial displays are described and treated as distinct entities, but they should not be regarded as completely separated from one another. Intergradations sometimes occur as, for example, when young mandrills display forms intermediate between the 'play face' and the 'grin' during bouts of rough-and-tumble play.

One cannot observe a group of mandrills for very long without realizing the important and subtle role that facial communication plays in their social and sexual lives. Van Hooff (1962, 1967), whose comparative work on the ethology of primate facial expression remains a classic in this field, was able to include some observations of captive

Table 5.2. Facial displays of the mandrill and Van Hooff's classification of homologous displays among the Old World monkeys and apes

Common name	Van Hooff (1967)	Context and function
Stare	Staring open-mouth face	Threat display, often accompanied by head bobs, ground slapping and piloerection
Fear grimace	Frowning bared-teeth scream face	During agonistic interactions to deter aggression by a dominant
Aggressive fear grimace	Staring bared-teeth scream face	Contains elements of aggression as well as fear. Fleeing and squealing may also occur
Duck face ☺	Pout face	Indicates submission or anxiety in young animals and low-ranking females
Play face	Relaxed open-mouth face	An invitation to play, and to continue a play bout
Grin	Silent bared-teeth face	Signals non-aggressive intent. A pre-copulatory display in males, often combined with head-shaking movements
Lip smack	Lip-smacking face	Facilitates approach and contact. Occurs as part of male's grinning display
Teeth chattering	Teeth-chattering face	Sometimes seen during male's grinning display

☺ It is unclear whether the duck face is homologous with the pout face.

mandrills and drills in his studies. Thus, it is clear that the genus *Mandrillus* shares many of its facial displays with other Old World monkeys, and especially with other members of the Papionini. Nonetheless, some unusual specializations of facial expression occur in the mandrill and drill; I shall discuss these in greater detail as well as providing an overview of the entire repertoire of facial displays.

Table 5.2 lists the various facial expressions used by mandrills, along with some notes on the contexts in which they occur, their common names, and their equivalents in Van Hooff's (1967) classification of primate facial displays.

Agonistic displays include the aggressive stare and head bob, two types of grimacing expressions, and the strange 'duck face'. The stare is probably equivalent to Van Hooff's 'staring open-mouth face'. During this aggressive expression, the brows are raised, the eyes wide open, and the mouth is also opened to varying degrees. However, the orbicularis oris muscles are sometimes contracted, and the mouth corners pulled forward. When making this expression, mandrills and drills do not always open their mouths (for an example of the drill's open-mouth threat display, see Figure 3.1).

Both *Mandrillus* species often 'head bob' during aggressive stares, jerking the head slightly upwards, then strongly down and forwards towards the other monkey (Figure 5.3). In adult males, the crest of hair on the scalp and nape of the neck is usually erected, as are the long hairs of the shoulder mane. Sometimes the animal slaps the ground with one hand, and this may carry over into a forward lungeing movement, or active chasing by the aggressor. I analysed 338 instances of staring and head bobbing in a mandrill group at the CIRMF. In each case, the sequence of events that followed the display was recorded in detail. The staring display was given most often (but not

Figure 5.3. Aggressive staring and head bobbing in the mandrill (**A**) and drill (**B**). Note the forward thrust of the head, and piloerection of the crest and mane in the male drill. (Author's drawings.)

exclusively) by dominant animals towards lower-ranking individuals, the most common responses by submissive animals being avoidance, presentation postures, or grinning. In 66 cases, staring and head bobbing rapidly escalated into lungeing and chasing sequences; half of these resulted in flight and 42% evoked other types of submissive

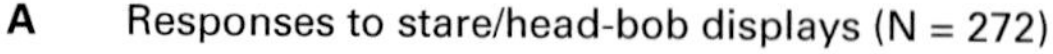

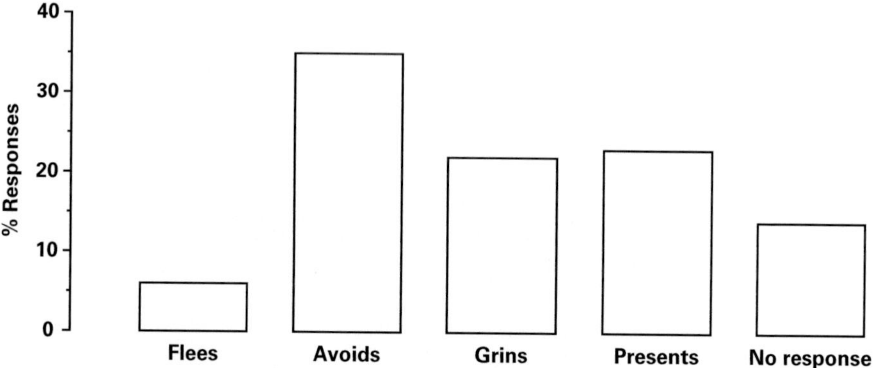

B Responses to stares/head-bobs accompanied by a lunge/chase (N = 66)

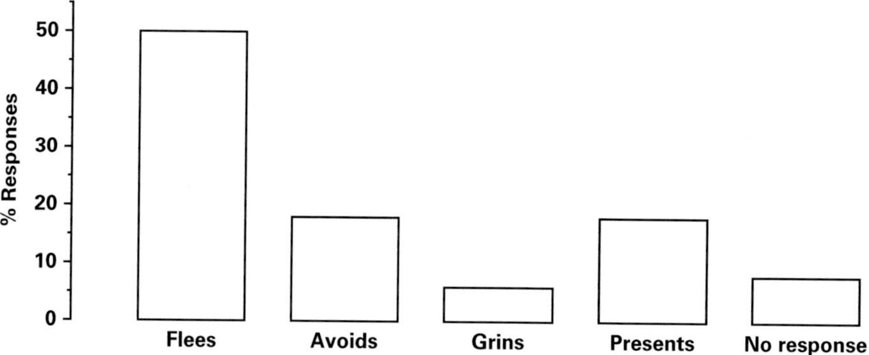

Figure 5.4. Responses shown by mandrills to the following. **A:** Facial threats alone. **B:** Facial threats combined with lunges and chases by more dominant members of their social group. (Author's data.)

responses by the recipients of aggressive displays (Figure 5.4). Confident and potentially aggressive male mandrills may also grasp a branch and bounce up and down, kicking vigorously with their feet. These branch-shaking displays were given almost exclusively by adult and adolescent males.

Another expression that is likely to be significant in agonistic contexts is yawning. It is not included in Table 5.2 because its role as a well-defined facial display is far from clear. Yawning occurs in a wide variety of vertebrates (Walusinski, 2010), including many primate species (Anderson, 2010), and often in situations involving aggressive or sexual interactions. Such 'tension yawns' have been observed in many Old World monkeys, including talapoins and macaques, as examples (Dixson 1977; Dixson *et al.*, 1975). There is a sex difference in the frequency of yawning, and this is likely to have a neuroendocrine basis. Thus, adult male rhesus monkeys tend to yawn more often than females; castration leads to a decrease in yawning frequencies, which is reversible by

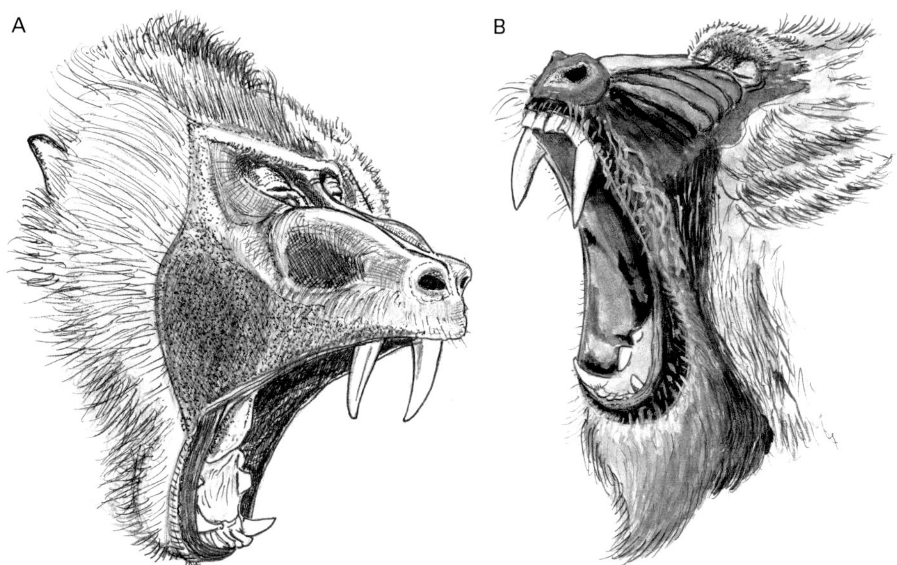

Figure 5.5. 'Tension' yawns. **A**: Adult male drill. **B**: Adult male mandrill. (Author's drawings.)

testosterone treatment. Adult male mandrills yawn, on average, five times more frequently than females, according to my own observations. Tension yawns of the adult male mandrill and drill are shown in Figure 5.5 (see also Plate 9). Such behaviour probably serves to display the enormous canines, and thus to deter potential challenges from other males in agonistic and sexual contexts. It is difficult to support this conclusion with data, however, as males often appear to yawn 'spontaneously' in a variety of situations. Yet, in my opinion, these yawns should not be dismissed as epiphenomena, as they are likely to be significant in social contexts.

Submissive mandrills sometimes squeal and make a fear grimace, which is the equivalent of the 'frowning bared-teeth scream face' in Van Hooff's (1967) descriptive system. When making this expression, the monkey's brows are lowered protectively, and the eyes partly closed, while the mouth is opened wide with lips strongly retracted to expose the teeth. The fear grimace is shown in response to threats or actual attacks, and it has the same form and functions in the mandrill as in many other monkeys, including macaques and baboons, as examples. The mandrill also displays a variant of the grimace, which van Hooff calls the 'staring bared-teeth scream face'. This is similar in form to the fear grimace, except that the eyes are wide open and the animal stares at the dominant individual. It is an intermediate type of display, which carries elements of aggression as well as fear, and it may be used, along with loud vocalizations, to solicit support from other group members. Thus, I have seen female mandrills do this when soliciting support from an adult male, in order to be defended against an aggressor.

The final submissive facial expression that requires consideration is the 'duck face', which occurs in both the mandrill and drill. This is a subtle facial expression that involves pursing the lips, puffing out the mouth at its corners, and lifting the brows slightly while the gaze is directed towards another animal (Figure 5.6). Davis (1976) noted this

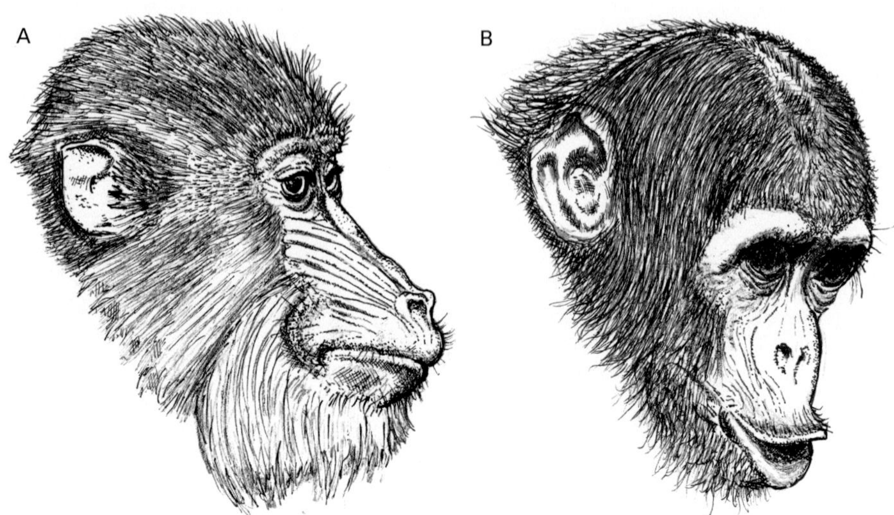

Figure 5.6. **A**: The mandrill's 'duck face' display. **B**: A juvenile chimpanzee making a 'pout face'. (Author's drawings.)

expression in a captive-raised infant mandrill in situations where it appeared to be anxious. Wood (2007) also noted its occurrence in semi-free ranging drills, as a submissive facial display. In the CIRMF mandrill group, I saw the duck face only rarely, but the expression is subtle, and difficult to identify unless viewed at close range. It probably occurred more often than I realized. In those instances where details of the social context were evident, the display was given primarily in response to aggression (76% of cases), and in association with avoidance, flight, or presentations by lower-ranking monkeys (82% of cases). It occurred in adult females, as well as in youngsters of both sexes, but I never saw it exhibited by an adult male. The duck face of *Mandrillus* shows some similarities to the pout face described by Van Hooff in infant macaques, baboons and chimpanzees when they are anxious and seeking physical contact with their mothers. However, these species part their lips, at the centre of the mouth, to create a small funnel during the pout (Figure 5.6), whereas mandrills and drills purse their lips and puff out the mouth corners. Thus, it is by no means clear whether the mandrill's duck face is truly homologous with the pouting facial expressions of other Old World anthropoids.

Affiliative (distance reducing) facial displays exhibited by mandrills include the play face, lip smacking and the distinctive grinning display. The play face, or 'relaxed open-mouth face' of Van Hooff, is often seen in mandrills, as in other Old World monkeys, when individuals are playing, and especially during rough-and-tumble and mock-biting episodes. The eyes are open and the brows may be slightly raised, but the mandrill does not stare fixedly when making a play face. The mouth is opened wide, but the teeth are usually covered by the lips, although they may be visible during the wrestling and mock-biting episodes that accompany play, especially in younger animals.

Lip smacking involves seeking eye contact with a conspecific, while rapidly opening and closing the lips and protruding and retracting the tongue. As Van Hooff (1967) has pointed out, this display is less commonly observed between familiar group members but tends to occur when 'animals meet for the first time, or after a period of separation'. In the mandrill, the most frequent use of lip smacking is by adult males, as they follow and court females. Lip smacking is often combined with the male's grinning display, and it is the rhythmic tongue movements (rather than lip movements) that are most obvious under these circumstances.

Of all its visual displays, it is the mandrill's unusual grinning facial expression that has most attracted the attention of ethologists and primatologists. Some earlier studies asserted that the grin is an aggressive expression, perhaps because of its superficial resemblance to a snarl (Andrew, 1963). It is the case that male mandrills sometimes grin at each other during tense and potentially aggressive interactions. These are formidable animals, in which threat may be juxtaposed with appeasement; the occurrence of grinning actually reduces the likelihood that a fight will break out. It is now well established that the grin mainly occurs in a variety of non-aggressive contexts (e.g. Bout and Thierry, 2005; Dixson, 1998a; Otovic *et al.*, 2014). Indeed, Van Hooff (1967) was well aware of this fact, as he interpreted the mandrill's grin as being a variant of the non-aggressive 'silent bared-teeth face', which has been described for many Old World monkeys, as well as in chimpanzees. The lips are retracted and the teeth bared during this display, but mandrills are unusual because their lips are often kept closed at the centre of the mouth, and raised at its corners, so that the lips assume a ' ∞ ' shape. This is not always the case, however, so that the lips sometimes remain parted at the centre of the mouth, as is more typical of other species. Variants of the grin face, in both the mandrill and drill, are shown in Figure 5.7 (see also Plate 8).

An important difference between the mandrill and other papionins, such as macaques and baboons, is that the mandrill's grin often occurs as part of a complex and highly sexually dimorphic visual display. Adult male mandrills often shake the head from side to side and make rhythmic tongue movements while grinning. Sometimes they also move their jaws in the manner of the 'teeth-chattering' displays that occur in some other Old World monkeys, and especially in the macaques (Van Hooff, 1967). The hair of the crest on the scalp and nape of the male's neck is sometimes erected. Males also exaggerate these various movements, which presumably serve to enhance the visual impact of their displays. Sometimes, a seated male may grasp his knee or thigh as he grins and engages in head shakes. Adult females, by contrast, tend to grin quite rapidly, and they rarely exhibit head shakes, tongue movements or piloerection.

Captive infant mandrills first grin at about three weeks of age (Mellen *et al.*, 1981). In the CIRMF mandrill groups, infants younger than one year of age often grinned at one another. Infants and juveniles grinned as an invitation to play or during extended play bouts. As male mandrills transition from adolescence to adulthood, head-shaking, tongue movements and piloerection displays become more frequent, as accompaniments to the grinning facial expression.

Table 5.3 shows the various combinations of the display elements used by adult male mandrills during their displays. On average, these males exhibited the grin alone or in combination with just one other element (i.e. with tongue movements, head shakes or scalp piloerection) in 70% of their grinning displays. The remaining displays included

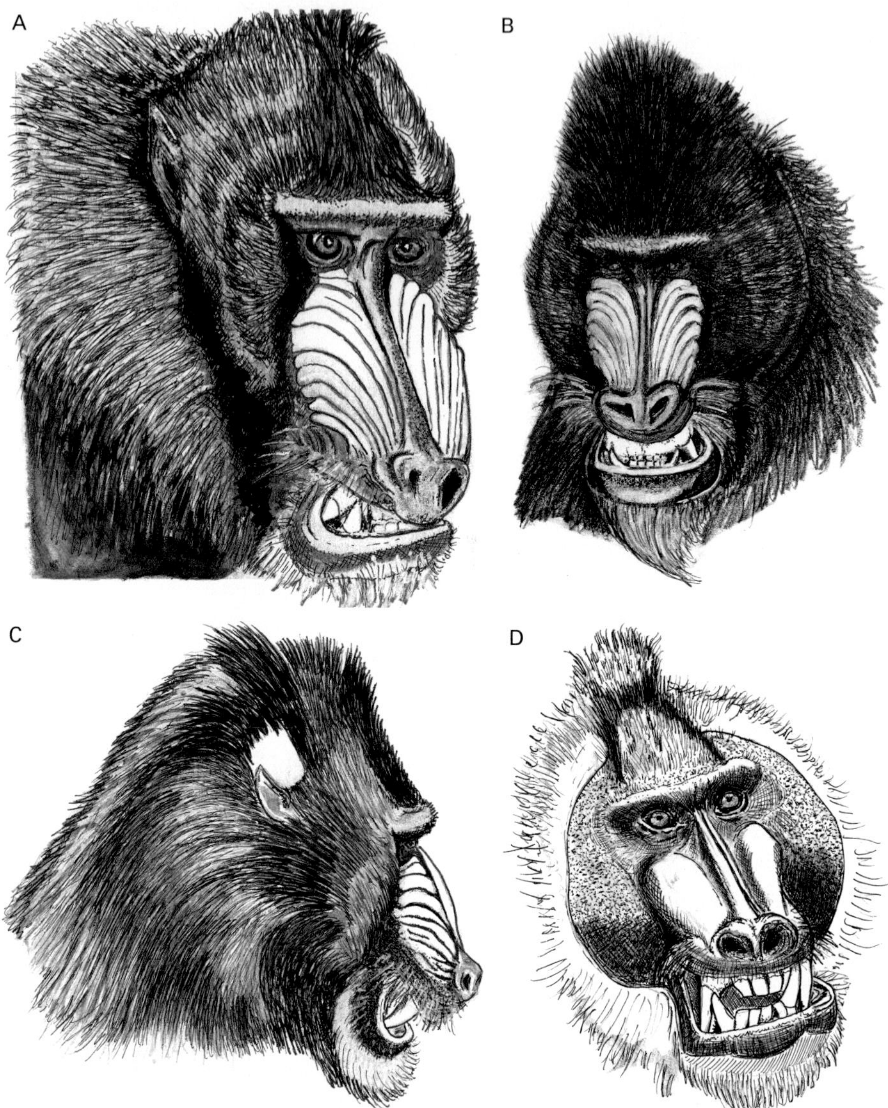

Figure 5.7. The 'grin face', or 'silent bared-teeth display' of the mandrill and drill. **A:** Adult male mandrill showing the typical ' ∞ ' shape assumed by the lips during the grin. **B:** Here the adult male's lips are parted at the centre of the mouth, as is more typical of other Old World monkeys during the silent bared-teeth display. **C:** Adult male mandrill, in lateral view, to show piloerection of the crest on the scalp and neck during the grin. **D:** The grin face of an adult male drill. (Author's drawings.)

three elements or all four elements combined. I was unable to discern any obvious pattern of display combinations, as regards the social or sexual contexts in which they occurred. It was clear, however, that males grinned at one another in situations where aggression might occur, and that their displays served to decrease the likelihood of such aggressive escalation. Thus, both dominant and subordinate male mandrills may grin at

Table 5.3. An analysis of the grinning displays (N = 324) made by adult male mandrills, to show the frequencies of the various display elements (grinning, head shaking, piloerection of the scalp crest and lip smacking)

	Male no. 14		Male no. 7		Male no. 15	
Display elements	N	%	N	%	N	%
G alone	**40**	25	**21**	21.5	**11**	16
G + SH	**20**	13	**6**	6	**27**	40
G + CR	**27**	17	**40**	41	**6**	9
G + LS	**18**	12	**4**	4	**5**	7.5
Sub-total	**105**	67	**71**	72.5	**49**	72.5
G + SH + CR	**8**	5	**11**	11.2	**7**	10
G + CR + LS	**17**	11	**14**	14.3	**3**	4
G + SH + LS	**15**	9	**1**	1	**4**	6
G + SH + CR + LS	**13**	8	**1**	1	**5**	7.5
Grand total	**158**	100	**98**	100	**68**	100

G, grin face; SH, head shaking; CR, scalp crest piloerection; LS, lip smacking. (Author's data.)

one another, but the frequencies of displays by subordinate males tend to be greater. For example, the alpha male in CIRMF Group 1 (no. 14) directed only nine grin displays at the second-ranking male (no. 7) over a two-month period, but received 53 grinning displays from this individual. The grin may thus, on occasions, function as a submissive display; indeed, as we have seen, it is sometimes given by subordinates in response to aggressive staring and head bobbing, or when being chased by higher-ranking monkeys.

Although the alpha male rarely grinned at other males in his group (a total of 14 times in two months), he frequently directed his displays towards females (143 displays during the same time period). Here the grins and associated display elements clearly fulfilled a distance-reducing function, as during the play bouts of young animals. Males exhibited grinning displays when seeking or maintaining proximity to females, and immediately prior to mounting them. It is likely that the male's exaggerated grinning and head-shaking displays evolved in this context to 'reassure' the much smaller females, and to lessen the likelihood that they might flee, or refuse his mount attempts. Figure 5.8 compares frequencies of grinning displays given by adult and subadult males towards females in the CIRMF group during the annual mating season. The higher frequencies of the most dominant (fatted) individuals are consistent with their greater involvement in courtship and copulatory behaviour.

The role of the male mandrill's grinning and head-shaking displays in sexual contexts will be discussed further in Part II of this book. However, at this point it is relevant to consider the phylogenetic relationships among *Mandrillus, Cercocebus* and *Lophocebus* in relation to the possible origin of head-shaking and grinning displays. Wallis (1981) described the occurrence a 'slow head shake' display given by grey-cheeked mangabeys (*Lophocebus aligena*) as one male approached, and supplanted another. Males of this species also head-shake (the 'head flag') when inviting females to copulate. Ally Mwamende (personal communication) has observed similar pre-copulatory head flagging

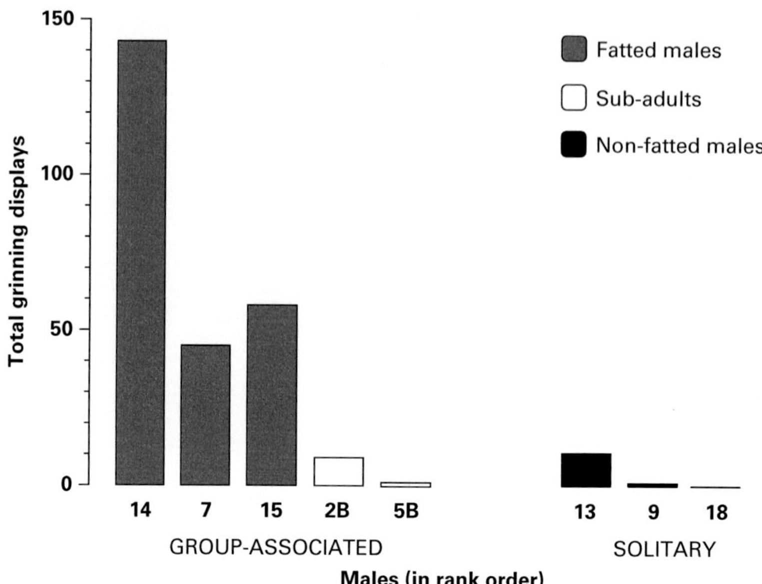

Figure 5.8. Total numbers of grinning displays made by adult and adolescent male mandrills throughout a single mating season. Group-associated and solitary males are arranged in rank order (the highest-ranking male is on the left, in each case). (Author's data.)

by free ranging male Sanje mangabeys (*Cercocebus sanjei*). Therefore, it is possible that the stereotyped head-shaking displays of male mandrills and drills might derive from less complex forms of such behaviour, which were present in their common ancestor with the mangabeys. The occurrence of head shaking in both *Cercocebus* and *Lophocebus* indicates that it is an ancient pattern. The mandrill's grin, as already discussed, derives from the 'silent bared-teeth face', which is widespread in Old World monkeys, and thus would have been present in the common ancestors of *Mandrillus* and *Cercocebus*.

Tactile and olfactory communication

Grooming

Mandrills groom the pelage of other group members (allogrooming) as well as themselves (autogrooming), and this behaviour doubtless fulfils important social as well as hygienic functions. Groomers typically use their fingers to search those parts of a conspecific's body that it cannot easily reach itself (see Plate 11). Groomees often alter position during a grooming bout in order to focus the partner's attention on areas such as the rump and the back.

Table 5.4 shows an analysis of 432 allogrooming bouts observed in the semi-free ranging Group 1 at the CIRMF, throughout the course of two annual mating seasons. It should be noted that most of these data were collected during the late morning and early afternoon, when this group was visible on the fringes of the forest. The mandrills received

Table 5.4. Distribution of grooming activity (N= 432 grooming bouts) in a semi-free ranging mandrill group

Groomers	Recipients	N	% of total grooming
1. Mothers	Infants	159	36.8
2. Adult females except infants	Matriline members	98	22.68
3. Adult females	Other matrilines *	15	3.47
4. Adult females	Adult females **	79	18.3
5. Adult/sub-adult males	Adult/sub-adult males	1	0.23
6. Adult/sub-adult males	Adult females	41	9.5
7. Adult females	Adult/sub-adult males	79	18.28

*, excluding any adult/sub-adult males in other matrilines; **, female recipients also contributed to scores 2 + 3 categories, hence totals for all seven categories exceed 100%. (Data were collected by the author at the CIRMF in Gabon, during two annual mating seasons.)

extra food every day at this time. It was my strong impression that at other times, as when the group entered its sleeping site in the late afternoon, or lingered there in the early morning, that grooming bouts were more frequent, and prolonged. I was also unable to collect any data during long periods during the day when the group retreated deep into the forest. However, given the limitations of the information presented in Table 5.4, some patterns are worthy of comment. Thus, adult female mandrills groomed the adult and juvenile members of their matrilines significantly more frequently than other group members (Wilcoxon Test: $N = 14$, $T = 17$, $P<0.05$). This finding was more significant if the data on mothers grooming infants were included in the analysis ($N = 14$, $T = 1.5$, $P<0.01$). Mothers, daughters and grandchildren were often seen in close proximity, foraging together, resting and occasionally engaging in mutual grooming. Males, and especially sub-adults and adult males, by contrast, groomed others relatively infrequently. They received roughly twice as much grooming from the group's adult females as they gave ($N = 12$, $T = 9.5$, $P < 0.05$). I never observed adult males grooming one another; indeed, they rarely stayed in proximity long enough for grooming to occur. One advantage of the otherwise limited data set summarized in Table 5.4 is that it was collected during periods when mate-guarding was in progress. As we shall see in Chapter 8, there is no evidence that allogrooming forms a significant part of courtship or mate-guarding interactions in mandrills. Grooming occurred very infrequently in mating contexts.

Scent-marking

Among its many unusual anatomical and behavioural traits, the genus *Mandrillus* is unique among the Old World monkeys in possessing a specialized sternal gland, which it employs for scent-marking. Although scent-marking has been mooted to occur in several other Old World monkeys (e.g. in some guenons: Loireau and Gautier-Hion, 1988), the evidence is tenuous at best, and histological confirmation for the existence of

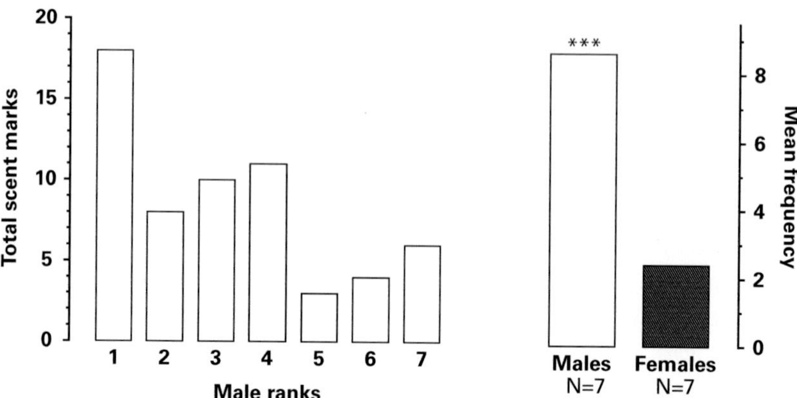

Figure 5.9. Frequencies of scent-marking by mandrills (using the sternal gland). ***, P<0.001. (Drawn using data in Feistner, 1991.)

scent glands is lacking. In the drill and mandrill, however, it has long been known that a dense concentration of apocrine glands occurs in the skin of the sternal region of adult monkeys, where specialized hairs serve to retain the glandular secretions (Hill, 1954, 1970). The sternal gland develops at puberty and is markedly sexually dimorphic, being much larger and more active in adult males (Setchell and Dixson, 2002).

Research on other primate species that scent-mark using specialized cutaneous glands shows that the secretions provide a kind of 'chemical fingerprint', concerning an individual's identity, age, sex and social status (e.g. in marmosets and tamarins: Epple *et al.*, 1993). However, at the behavioural level, very little is known about why mandrills scent-mark, and what role the sternal gland might play in their social or sexual communication. The only quantititative information on patterns of scent-marking behaviour in mandrills has been reported by Anna Feistner (1991), who studied the semi-free ranging mandrill group at the CIRMF during the early years after its establishment. She noted that sub-adult and adult founder males in the group scent-marked significantly more frequently than the adult females. Immature mandrills of both sexes also made sternal rubbing movements, long before their cutaneous glands had developed, but they did so comparatively rarely, and there was no sex difference in marking frequencies (Figure 5.9). The earliest incidence of sternal marking observed was by a male aged just seven months.

Feistner noted that high-ranking males in the group marked more frequently than lower-ranking males; later research showed that gain and loss of alpha status in male mandrills is associated with increases, or decreases, in the secretory activity of their sternal glands (Setchell and Dixson, 2001b). Mandrills mark the trunks of trees, and the majority of sites selected for marking are near ground level. The animal sniffs a tree trunk and rubs its lower lip on the bark, before clasping the trunk with its arms, and rubbing the chest upwards and downwards several times. It is interesting that certain trees in the rainforest at the CIRMF were marked and over-marked repeatedly by mandrills (Feistner, 1991); it seems that the choice of scent-marking sites is not simply made at random. Unfortunately, nothing is known about scent-marking behaviour in free-ranging mandrill groups, or about the role that olfaction might play in communication between groups and solitary males in the population.

Analyses of mandrill sternal gland secretions have revealed that they contain complex mixtures of volatile chemical compounds, especially in adult males (Setchell *et al.*, 2010c). Attempts have been made to correlate scent composition with major histocompatibility complex (MHC) genotypes, and patterns of mate choice in mandrills (Setchell *et al.*, 2010a; Setchell *et al.*, 2011b). A discussion of mate choice in male and female mandrills, and its role in the evolution of their secondary sexual adornments, will be found in Chapter 10.

Sexual behaviour

Mandrills are seasonal breeders, in the sense that females develop their sexual skin swellings during the long dry season (i.e. from June to September in Gabon), and males mate-guard and copulate during these months. In the semi-free ranging groups studied in detail at the CIRMF, dominant (fatted) males mate-guarded individual females during periods of sexual skin swelling, whereas lower-ranking (non-fatted) adults or sub-adult males adopted alternative, opportunistic mating strategies.

The female mandrill may invite copulation (i.e. behave proceptively: Beach, 1976) by presenting her rump to the male (Figure 5.10). She adopts a quadrupedal presentation posture, as is typical of many monkeys, and faces away from the male, but sometimes turns her head to make eye contact with him. She may also wiggle her tail rapidly back and forth, in the sagittal plane. Females sometimes pace to-and-fro in close proximity to a male ('parading behaviour', as has also been described in *Macaca nigra*: Dixson,

Figure 5.10. Sexual presentation posture of a female mandrill. The male inspects her sexual skin swelling. (Author's drawing.)

1977). Such displays occur most often during those periods when females have well-developed sexual skin swellings.

Although proceptive approaches and presentations are important aspects of female sexuality, it is the case that males are generally more sexually assertive, and that the majority of mount attempts with maximally attractive (i.e. swollen) females are male initiated. Males follow such females, look at their sexual skin swellings, and sometimes sniff the female's vulva prior to mounting.

With regard to possible olfactory signals of female sexual attractiveness, there has been one report (Charpentier *et al.*, 2013) that mandrills exhibit 'flehmen'. Flehmen occurs in a variety of mammals, such as antelopes and goats, and especially as part of masculine pre-copulatory behaviour. After sniffing and licking the female's genitalia or urine, the male raises the head, and wrinkles the upper lip; these responses are associated with transport of sexually arousing chemical cues, via the nasopalatine ducts in the roof of the mouth, to the vomeronasal organ (Estes, 1972; Evans, 2003). I have not observed such behaviour in mandrills, however, despite making detailed records of sexual interactions in this species. Although nasopalatine ducts persist in adults of some Old World monkeys (including the mandrill), anatomical studies have failed to find any evidence of a vomeronasal organ, although vestiges of it may be present during foetal development (Evans, 2003; Maier 1997, 2000). Only five out of the 21 mandrills tested by Charpentier *et al.* exhibited flehmen, with one male accounting for 93% of observations. Thus, I think that the behaviour recorded by Charpentier *et al.* in the mandrill is rare and idiosyncratic; it does not represent a homologue of flehmen in other mammals.

As was noted previously, the male mandrill frequently grins at the female, as he seeks to gain proximity to her, and his displays may serve to lessen the likelihood that she might flee. The male may grasp the female with one hand while inspecting her genitalia, or place both hands on her hips (the 'hip touch') without progressing to a full mounting posture. A mount attempt, if it ensues, is not always accepted, so that the female may refuse by moving away rapidly. The male mandrill will sometimes run after the female, so that bouts of following develop into active chasing prior to the mount. If mounting does occur, then the male grasps the female's hips with his hands, and may attempt to clasp her calves with his feet. This double foot clasp position is a phylogenetically ancient trait, as it occurs in most of the Old World monkeys. However, the male mandrill's size renders the double foot clasp position impractical, and his feet usually remain on the ground (Figure 5.11; Plate 19). Smaller, pubescent males sometimes attain the full double foot clasp position, however. So large are adult male mandrills that females often gradually sink into a crouching position during copulation, as the male's weight presses down on the female's hindquarters.

The copulatory pattern of the male mandrill consists of a single mount, with intromission and pelvic thrusting, prior to the attainment of ejaculation (Table 5.5). Copulation is therefore relatively brief, and intromission lasts for less than 60 seconds; the male's final pelvic thrust tends to be more rapid and deeper than those that occur earlier in the sequence. This is the most phylogenetically widespread of the four copulatory patterns defined for primates (Dixson, 2012). It occurs, for instance, in species as diverse as

Table 5.5. Data on the copulatory patterns of adult male mandrills. In all cases, ejaculation occurred during a single mount with intromission and pelvic thrusting

		Number of pelvic thrusts	
Male ID	N	Pre-intromission	During intromission
CIRMF 14	85	0–4 mean = 0.74	12–32 mean = 19.74
CIRMF 7	19	0–2 mean = 0.44	12–17 mean = 13.95
CIRMF 15	11	0–6 mean = 1.3	11–26 mean = 17.73
CIRMF 13	5	0–3 mean = 1.4	13–24 mean = 16.8
ZSL 1	4	3–17 mean = 8.0	14–18 mean = 15.75

CIRMF, Centre International de Recherches Médicales, Gabon; ZSL, Zoological Society of London; N, total mounts scored for each male. (Author's data [ranges and means].)

Figure 5.11. Copulatory posture of the mandrill. (Author's drawing.)

capuchins, mangabeys and gorillas. During the mount, the female mandrill sometimes turns her head to look at the male. As the male ejaculates, he ceases thrusting and muscle spasms occur, especially in his legs. Females, by contrast, do not show any overt signs of orgasm. Neither sex vocalizes during or after copulation, and as the male dismounts, the female sometimes runs a short distance away from him. In some anthropoids, female orgasm occurs during mating, as for example in the stump-tail macaque, in which the female makes a distinctive 'climax face', (Chevalier-Skolnikoff, 1974), and exhibits uterine contractions and other physiological responses (Slob *et al.*, 1986). The absence of female orgasm and of female copulatory or post-copulatory vocalizations in *Mandrillus* is interesting, as such responses do occur in some other papionins and have been implicated in post-copulatory female choice (for examples, see Maestripieri *et al.*, 2005, and a review by Dixson, 2012).

Successful copulations result in the male depositing a white seminal coagulum, which is often visible as it protrudes from the female's vulva. Females sometimes reach back to pick at this, and remove portions of the coagulum. However, mandrills do not engage in post-copulatory genital grooming; indeed, such behaviour is rare among the anthropoids, with the exception of some of the New World marmosets and tamarins (Nunn and Altizer, 2006).

Among the highest-ranking males in the CIRMF mandrill colony, copulations often took place during extended mate-guarding episodes, especially when females were at the height of sexual skin swelling, and were more likely to ovulate. Other males engaged in opportunistic, alternative mating strategies. It is notable that adult male mandrills only mate-guarded individual females and never attempted to sequester small units or 'harems' in the manner of male hamadryas baboons or geladas.

Mate-guarding in mandrills is a complex phenomenon, involving the maintenance of mutual proximity, sometimes for some days (see Plates 21–23). A female may attempt to avoid a guarding male, for example by climbing into the trees where he cannot pursue her so easily. If, as is often the case, more than one female has a large swelling, then males have the problem of partitioning their courtship efforts between potential partners. How the web of sexual strategies deployed by male and female mandrills plays out in their enormous supergroups in the wild is unknown. However, we now have very good information about patterns of sexual activity, and resulting reproductive success, in semi-free ranging mandrills at the CIRMF (e.g. Dixson *et al.*, 1993; Setchell *et al.*, 2005a; Wickings *et al.*, 1993). These topics are explored in detail in Chapter 8.

From a comparative and evolutionary perspective, it is interesting that high-ranking males of at least one mangabey species (*Cercocebus sanjei*) have been observed to sometimes mate-guard females that are at full swelling (Ally Mwamende, 2009). Males follow individual females, and emit grunting vocalizations. Such behaviour is reminiscent of (although less pronounced and stereotypical than) the persistent following, and deep two-phase grunting displays of male mandrills during their mate-guarding episodes.

Socio-sexual behaviour

In the mandrill, as in many Old World monkeys and apes, elements of sexual behaviour, such as presentations and mounting postures, have become ritualized and incorporated into the broader sphere of social communication. Indeed, mandrills of the same sex, as well as the opposite sex, often present to one another in a variety of social contexts (Plate 20). Socio-sexual presentations occur, for example, when one animal approaches another, and then sits with it or engages in grooming. The behaviour has an affiliative or distance-reducing function in such circumstances. Presentations often also occur in mandrills as a response to aggression, or the threat of aggression, from higher-ranking group members. This point was made earlier, in relation to presentations given in response to staring and head-bobbing threat displays (Figure 5.4). In a separate analysis involving 656 aggressive acts of all types (Frei, 1991), it was shown that recipients responded by presenting to aggressors in 24% of cases, or took evasive action (fleeing or avoidance) in 46% of cases. Socio-sexual presentations in mandrills thus serve as part of reconciliation behaviour (DeWaal, 1989), which occurs between individuals after an agonistic encounter. Indeed, in 90% of those cases where mandrills presented in response to threats, chases or attacks by conspecifics, the aggressive episode was promptly terminated.

It is important to note that socio-sexual presentations are often qualitatively different from the presentation postures used by females to invite copulation. Animals that present when approached by a more dominant individual will often do so very rapidly, and then move away. Monkeys that present in response to overt aggression, by contrast, may hold the posture for longer. Moreover, submissive or fearful presentations typically involve flexing of the limbs, and lateral bending of the spine, so that the monkey appears to 'cower' as it looks back at the higher-ranking aggressor. The tail is often flexed forwards, and the mandrill may reach back with one hand to grasp its rump. This differs from the proceptive presentation posture displayed by females as they invite copulation, as this usually involves an upright stance, an absence of limb flexure, and sagittal tail movements. Examples of these two types of presentations are shown in Figure 5.12. In practice, a continuum of intermediate forms occurs. For example, a low-ranking female may sometimes partially flex her legs when inviting copulation from a high-ranking or aggressive male. These variations provide good examples of what Darwin identified as the principle of 'antithesis' in the form and function of visual displays. In *The Expression of the Emotions in Man and Animals*, Darwin (1872) noted that the postures and gestures employed by a submissive or frightened mammal, such as a dog, are often the reverse of those displayed by a confident or aggressive conspecific. The presentation postures of

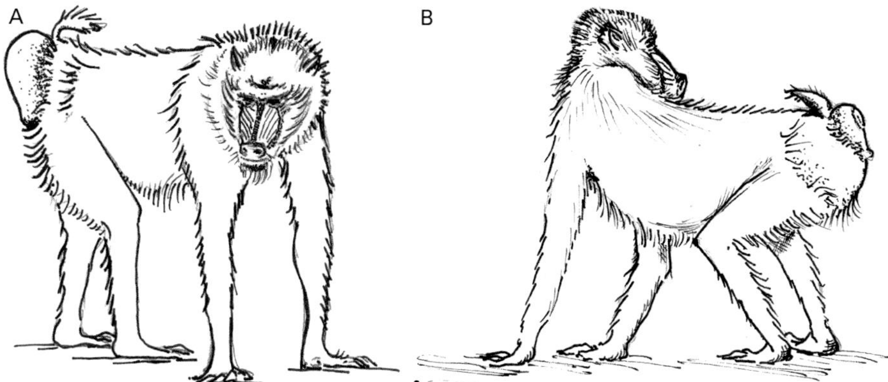

Figure 5.12. Sexual and socio-sexual presentation postures of female mandrills. **A**: a typical sexual invitational posture. **B**: Socio-sexual presentation by a low-ranking female. Note the limb flexure, and lateral curvature of the spine. (Author's drawings.)

mandrills follow this principle (Figure 5.12), as is also the case in yellow baboons (Hausfater and Takacs, 1987) and in other Old World monkeys, many of which display such postures in socio-sexual contexts (Dixson, 2010, 2012).

Adult male mandrills only occasionally present to one another, but juvenile and adolescent males frequently present to higher-ranking group members. When receiving such presentations, adult and sub-adult males often touch the genitalia or handle the penis of the lower-ranking individual. In hamadryas baboons, adult males employ presentation and genital manipulation as part of reciprocal notifying displays, which occur between the leader males of harems (Kummer, 1968). Such displays serve to resolve conflicts and inhibit aggression between leader males (Colmenares, 1990, 1991). The absence of such reciprocal notifying displays between adult male mandrills is consistent with the fact that they do not form harems.

Observations of socio-sexual behaviour in mandrills do not provide support for Wickler's (1967) ideas concerning the evolution of socio-sexual mimicry in this species. The brightly coloured rump and genitalia of the male mandrill are not associated with the use of presentations or other overt forms of genital display, presentations being much more prevalent in younger males that usually lack full development of sexual skin. Wickler posited that the red and blue facial colours of the male mandrill might have evolved to mimic those of the penis and scrotum (see Plate 18). The male mandrill's grinning display certainly draws attention to its face, including the red and blue muzzle. However, these colours are lacking in adult male drills, and yet they use exactly the same facial displays as male mandrills. In both the mandrill and drill, males have evolved to be massive and visually striking in adulthood, but the reasons for this probably have nothing to do with socio-sexual mimicry. Grubb (1973) pointed out that the superficial resemblances in colouration between the face and the genitalia in male mandrills may have arisen for physiological reasons. He commented that 'not fortuitously, but because the physiological mechanisms for the production of sematic pigments are limited, the face has convergently come to acquire bright colours similar to those of the genitalia'. Pierre Jouventin (1975b) agreed with this explanation, based upon his studies of captive and free-ranging mandrills.

Mandrills also exhibit same-sex mounting as a socio-sexual behaviour, although I have never observed adult or sub-adult males being mounted in this context. Some mounts or incipient mounts (hip-touches) by adult males upon females are probably of a socio-sexual nature, as they occur in response to crouching presentations given by females in agonistic contexts. Not all mounts by male mandrills are attempts at copulation, therefore. However, most socio-sexual mounts are made by adult females, or by juveniles of both sexes, with lower-ranking individuals typically being the recipients. Mounting also occurs as part of play bouts among infants and juveniles. Although no quantitative data are available on frequencies of play mounting in mandrills in the CIRMF groups, it appeared that male infants engaged in this more often than their female peers. Such a sex-difference is well documented for a number of Old World monkeys, including macaques and baboons, so that its occurrence in mandrills is to be expected.

In Japanese macaques (*Macaca fuscata*), some adult females form consortships with female partners; they mount one another and employ distinctive pelvic thrusting patterns to stimulate the vulval, perineal and anal (VPA) areas (Vasey, 2004; Vasey and Duckworth, 2006). In this case, the behaviour is unrelated to female social rank, and does not serve for reconciliation after agonistic encounters between partners (Vasey *et al.*, 1998). Instead, it appears to represent a pleasurable (hedonic) aspect of sexual relationships between female Japanese macaques. While same-sex mounting may have a partially hedonic basis in some other Old World anthropoids (e.g. female stump-tail macaques and bonobos), it is not equivalent to homosexuality. Rather, it is indicative of the strong bisexual potential that is shared by a significant number of Old World anthropoid species (Dixson, 2010). However, where the mandrill is concerned, I have only rarely observed same-sex mounting of this kind. Females never mate-guarded one another, vocalized or exhibited specialized patterns of VPA stimulation during iso-sexual mounts. Just occasionally, females did make pelvic thrusting movements while mounting one another; thus, it is possible that clitoral stimulation occurred during such interactions. In the great majority of cases, however, female–female mounts, like socio-sexual presentations, occurred as part of social communication between group members, and especially in relation to matters of social rank.

Social communication and social rank

The complex social groups in which mandrills, and especially female mandrills, spend their lives, constitute a forum for competition, as well as for cooperation in the struggle to survive and to reproduce. Male mandrills emigrate from their natal groups as they mature, returning to group life at intervals as adults, in order to compete for access to females during successive annual mating seasons. Solitary foraging probably has nutritional advantages, and it also reduces potential conflicts between males during non-mating periods. Females, by contrast, are philopatric, and they associate in order to exploit the resources (primarily food) that they require to survive and to rear their dependent offspring. Females thus form the core of mandrill society, and long-term relationships develop between female relatives within the social group. This type of matrilineal social organization is found in many other monkeys, including the macaques, baboons and mangabeys. For a female, therefore, social rank is affected by her position as a member of her matriline, so that daughters of high-ranking females experience social advantages as they mature, and are more likely to attain higher ranks in the group (relationships between rank and reproductive success are discussed in Chapter 8). Male mandrills do not inherit maternal rank; intra-sexual competition is intense and a male's rank appears to be largely a matter of individual prowess. Although male baboons or chimpanzees sometimes form 'alliances' with other males, working together to out-compete a high-ranking male, I never saw any instance of this among male mandrills. Perhaps because they spend large periods alone, male mandrills do not establish these kinds of cooperative social relationships with one another. The magnificent secondary sexual adornments, huge size and long canines of the adult male mandrill are thus highly adaptive in terms of advertising his status, and in making it possible for him to assert priority of access to females rapidly whenever he enters a group for mating purposes.

Long-term observations have revealed many clues concerning social rank and rank orders within mandrill groups. Measurements of agonistic behaviour have traditionally been used to assess rank in groups of primate species. In mandrills, analyses of the winners and losers of aggressive encounters allow one to assign dominance ranks to most group members (Bossi, 1991; Frei, 1991; Wickings and Dixson, 1992b). However, overt aggression may be comparatively infrequent between some individuals, including males that emigrate at puberty, and some adult males that live a solitary existence in the rainforest enclosures at the CIRMF. More subtle measures, of displacement and approach–avoidance behaviour, can also reveal rank relationships, however. Displacement involves one animal approaching and supplanting another, for example, in order to gain access to food or proximity to another conspecific. A lower-ranking individual will also sometimes actively move away from (avoid) another mandrill if it comes too close. Rank orders constructed by measuring displacement and avoidance relationships correlate well with measures of agonistic rank in mandrills (Wickings and Dixson, 1992b), as is also the case in baboons (Rowell, 1967a).

Edi Frei (1991), who studied socio-sexual relationships between mandrill group members at the CIRMF, found that a strong correlation exists between social rank (based on measurements of agonistic behaviour) and rank as reflected by presentation behaviour. He concluded that 'presenting stabilized the social structure by affirming the actual rank order, but presenting was also influenced by age-sex class, and the relationship between presenter and recipient'. Table 5.6 shows rank orders constructed by measuring same-sex presentations in mandrills, and those constructed by measuring agonistic interactions; the two rank orders are almost identical. Many years ago, Irwen Bernstein (1970) reported that status hierarchies based upon measures of socio-sexual behaviour do not correlate with aggressive rank orders in captive groups of some papionins (macaques, mangabeys and geladas). However, the mandrill clearly differs in this respect as its socio-sexual presentations are indicative of dominance rank, and dominant animals are rarely mounted by subordinates.

It will be noted in Table 5.6 that mothers and their daughters often occupy adjacent positions in the rank orders (e.g. female no. 2 and her offspring nos. 2C and 2D; female no. 12 and her offspring nos. 12A, 12C and 12D). Relationships between matrilines also affect priority of access to resources. In the CIRMF group, I frequently observed that

Table 5.6. Ranks of 14 adult female mandrills in a social group, based upon measures of their agonistic behaviour and socio-sexual presentations

Female no	2	2D	2C	5	10	12	12D	12C	17	12A	17B	17A	16	6
Agonistic rank	1	2.5	2.5	4	5	6	7	8	9	10	11	12	13	14
Sociosexual rank	1	2	3	4	5	6	7	9	10	8	12	11	13	14

Spearman rank correlation coefficient for agonistic vs socio-sexual rankings = 0.71; P<0.01. Females with capital letters after their ID numbers are offspring (e.g. 2C and 2D are the daughters of female no. 2). Frei and Dixson, unpublished observations.

members of each matriline tended to move and forage together, and that higher-ranking matrilines had priority of access to the additional food with which the group was provisioned each day.

Are females always lower-ranking than males in mandrill groups? The answer to this question is 'yes, at the level of individual relationships'. By the time male mandrills reach five years of age, they out-rank all the female members of the group. However, there are occasions when females form coalitions, and may turn upon a male and chase him off. I rarely observed this, but it sometimes occurred if males made facial threats at infants. The third-ranking fatted male in the CIRMF group did this occasionally, and I saw females chasing him as he fled into the forest to escape them.

Only one group attack by female mandrills upon an adult male has been recorded (Setchell et al., 2006a) and the circumstances under which this occurred were most unusual. Some time after a second semi-free ranging group had become well established at the CIRMF, seven adult males were added to its numbers, thus bringing about a marked destabilization of the group's social structure. A few weeks later, the highest-ranking introduced male attacked and severely injured the group's alpha male. The alpha male, too badly injured to escape, was then attacked by eight adult females; they pulled his hair and bit him repeatedly. This male was never seen again and was presumed to have died. The abnormal circumstances that precipitated this attack by females are unlikely to find a parallel in wild mandrill groups.

6 Matters of life and death

Sex differences in longevity

Nobody knows for certain how long mandrills might live in the wild. Records show that some captive mandrills live to be 30–40 years of age in Zoological Gardens (Hill, 1970). This is most unlikely to be the case in the wild, where predators, parasites, diseases, and the struggle to find sufficient food and to reproduce must exert a heavy toll. Even in the semi-free ranging mandrill groups at the CIRMF, where the animals enjoy a comparatively sheltered existence, females typically live until they reach their early twenties, while the median lifespan for males is 17 years (Setchell *et al.*, 2005a). These authors recorded a significant sex difference in mortality, with more males than females 'disappearing' (and presumed dead) in the large rainforested enclosures from the age of six years onwards. The oldest (post-reproductive) female in the CIRMF colony was still alive at 37.3 years, whereas the oldest surviving male was 21.2 years of age (Charpentier *et al.*, 2013).

The only estimates of longevity in free-ranging mandrills are those provided by Abernethy and White (2013), based upon their observations of radio-collared animals (15 males: three females). These authors reported that wild mandrills might have a lifespan of 12–14 years, but as the ages of the mandrills they radio-collared were unknown at the outset, these estimates are approximate at best. Under completely natural conditions, sex differences in mandrill mortality are pronounced. I say this because, although the percentage of males:females at birth is approximately 50:50, wild mandrill supergroups contain six times more adult females than adult/sub-mature males during the annual mating season (Abernethy *et al.*, 2002). Even allowing for the fact that some males probably remain on the fringes of groups during the annual mating period, and were thus not counted by Abernethy *et al.*, the mandrill clearly has a very high socionomic sex ratio (i.e. the ratio of adult females:adult males in the population).

Effects of predation

Male mandrills probably have a higher risk of mortality than females for a variety of reasons. Adolescent males emigrate from their natal groups, and it is likely that a significant proportion of these inexperienced young males might perish before they reach maturity. Predation by leopards may be a significant factor, as examination of leopard scats collected in the Lopé National Park has shown that mandrills form part of their diet (Henschel *et al.*, 2005). Given the formidable size and large canines of mature male mandrills, however, it seems unlikely that a leopard would risk attacking even a lone male, although adolescent emigrants, or the younger members of social groups, might be easily preyed upon.

The African crowned eagle (*Stephanoaetus coronatus*) is a formidable predator that specializes in killing forest monkeys. This eagle is certainly capable of preying upon infant and juvenile mandrills. It is also likely that pythons might take mandrills from time to time. Unfortunately, there are no data on predation rates, but the African rock python (*Python sebae*) is quite common throughout the geographical range of both *Mandrillus* species. At an average length of 4.8 m, and at 50 kgs in weight, adult rock pythons are certainly large enough to prey upon mandrills.

Do mandrills ever die because of accidental factors, such as snakebites? Given their foraging strategies, as forest floor gleaners, mandrill groups must encounter venomous snakes from time to time. A prime candidate for such accidental encounters is the gaboon viper (*Bitis gabonica)*. This viper hides under leaf litter on the forest floor during the day; it is exceptionally sluggish until roused, but has the largest fangs of any snake in the world. Its venom is neurotoxic, and its bite is often fatal to humans who have the misfortune to tread on one. Whether some mandrills might suffer a similar fate owing to bites from these vipers, or from other snakes, is unknown.

Parasites and diseases

Mandrills are hosts to a large array of parasites and can be afflicted by diseases of many kinds. Yet, in the great majority of cases, almost nothing is known about parasite loads or the impact that diseases might have upon the health or survival of free-ranging monkeys. Answers to these crucial questions must await the results of surveys on wild mandrill groups.

At the CIRMF, semi-free-ranging mandrills are hosts to various gastrointestinal nematode worms and protozoa (*Entamoeba* and *Endolimax* spp., and *Balantidium coli)*. However, the monkeys generally remain in good health as they receive regular veterinary care, so that their parasite loads are not associated with chronic illness, or correlated with the host monkey's sex or social rank (Setchell *et al.*, 2007). One study did indicate that some relationship might exist between higher faecal glucocorticoid levels in male mandrills and a 'higher diversity of gastrointestinal parasite infection'(Setchell *et al.*, 2010b). From an intellectually selfish (but purely scientific) point of view, it is unfortunate that the excellent health care afforded to the CIRMF mandrills has made it very difficult to conduct meaningful studies on the impact that parasites and diseases might have upon their reproductive biology.

One fundamental question concerning the mandrill's brightly coloured secondary sexual adornments involves an influential hypothesis that was first advanced by William Hamilton and Marlene Zuk (1982). These authors posited that colourful male adornments evolved, in part, as indicators of successful resistance to parasitic infection. Hamilton and Zuk based their ideas upon studies of passerine birds, and there have been few attempts to test the Hamilton–Zuk hypothesis in any primate species. Setchell *et al.*'s (2009) studies of the CIRMF mandrills failed to establish any correlation between parasite load and red facial colouration in either male or female mandrills. I venture to suggest that, even if a complete catalogue of the diseases and parasites of wild mandrills could be obtained, it would have little to tell us about the evolution of colourful secondary sexual adornments in males or females of this exceptional species. The Hamilton–Zuk hypothesis might have some, as yet untested, subsidiary role in this respect, but the mandrill's adornments probably evolved as a result of other selective forces, as will be discussed in Chapters 8 and 10.

One insight into the evolution of disease resistance in mandrills concerns the finding that basal white blood cell (WBC) counts are significantly higher in primate species where females commonly engage in multiple partner matings, as compared to single partner matings (Anderson *et al.*, 2004; Nunn *et al.*, 2000). This is thought to be a result of selection against sexually transmitted infections (STIs); the transmission risk being greater in primates that commonly engage in multiple partner copulations. The mandrill is one such species, and its WBC counts are higher than those of many polygynous and monogamous forms. However, we know very little about STIs in the vast majority of primates (Nunn and Altizer, 2006), although concerted efforts are being made to improve knowledge of these and other diseases in free-ranging populations (Nunn, 2012). It should not be assumed, for example, that because mandrills harbour two types of simian immunodeficiency lentiviruses (SIVmnd-1, and SIVmnd-2: Souquière *et al.*, 2001), that these are necessarily passed on by sexual contact. In the CIRMF colony, all the evidence points to horizontal transmission being a result of fights between male mandrills, with only one female being infected. This female had transmitted the virus to one of her offspring, probably during pregnancy (Estaquier *et al.*, 1991; Nerrienet *et al.*, 1998).

Mandrills, like many other primates, act as hosts to malaria (*Plasmodium*) parasites (Escalante *et al.*, 1998, 2005) and other blood parasites, including the microfilaria of various nematode worms. One of these infections is *loa loa*, which is transmitted to the host via the bite of a fly (*Chrysops*). Some of the microfilaria ultimately metamorphose into adult worms and migrate underneath the skin. Loaisis is endemic throughout the distribution ranges of both *Mandrillus* species, and it commonly afflicts humans (Pinder, 1988). Mandrills are probably also parasitized by the *tumbu* fly (*Cordylobia anthropophaga*), as its larvae burrow into the skin and feed upon many mammals, including humans. Wood (2007) records that infestations of *tumbu* fly larvae have caused the deaths of infant drills at the Pandrillus Rescue Centre in Nigeria. What might occur in wild drill and mandrill groups is not known, but it is likely that maternal grooming of infants helps to reduce parasite risk. Likewise, ticks are extremely common in the savannah/forest mosaic of the Lopé National Park, and mandrills frequently traverse the grassland to reach isolated pockets of forest during the dry season. Anybody who walks through these areas is attacked by ticks, and one presumes that the same must apply to the forest buffalo, mandrills and other species that venture into the open. Earlier it was noted that when mandrills groom one another, they often present otherwise inaccessible parts of their bodies to be groomed by conspecifics. Quantitative studies have shown that such behaviour occurs in many monkeys, as for example in talapoins (Dixson *et al.*, 1975), and among primates in general (Barton, 1985), ostensibly because it evolved to encourage removal of ectoparasites from areas that cannot easily be cleaned during self-grooming.

Part II

Reproduction

7 Seasonal patterns of reproduction

> The annual reproductive strategies of tropical mammals usually reflect three-way interactions between rainfall patterns, feeding strategies, and taxonomic limitations. As a result, almost every conceivable annual pattern of reproduction can be found in this region.
>
> Bronson, 1989

Rainfall patterns and reproductive strategies

The vast majority of studies dealing with seasonal breeding in mammals has involved species that live at higher latitudes, where photoperiodic cues are of crucial importance in determining the timing of reproductive events. The genus *Mandrillus*, by contrast, has existed for millions of years in the equatorial rainforests of Western Central Africa, where variations in rainfall, rather than day length, are the main markers of the seasons. Thus, the terms 'wet season' and 'dry season' have much more relevance to environmental conditions in countries such as Gabon and Cameroon than the terms 'spring, summer, autumn and winter', which are used to denote seasonal changes at higher latitudes.

In many rainforest monkey species, such as the guenons (*Cercopithecus* spp.), females give birth during the wettest periods of the year, whereas mating takes place during the dry season (Butynski, 1988). In southeast Gabon, where the CIRMF mandrill colony is situated, the long dry season occurs between June and September; these four months account for only 11% of the annual rainfall (Figure 7.1). Female mandrills develop their sexual skin swellings, mate and conceive during this drier period; on average 79% of putative ovulations occur between June and September (Figure 7.1). Thus, although some matings take place as early as May and as late as November in this semi-free ranging mandrill colony, the vast majority of conceptions occur during the peak months of the long dry season. It is therefore appropriate to regard the mandrill as being a species that has a mating season.

The gestation period of the mandrill lasts for 175 days on average (Setchell *et al.*, 2002). On that basis, we should expect a female that conceives early in June to give birth during November of the same year. A female that conceives towards the end of the mating season, however, at the end of September, would give birth early the following March. In fact, 83% of births occur between November and March (Figure 7.1). The mandrill birth season, therefore, coincides with the wet season in Gabon; 56% of annual

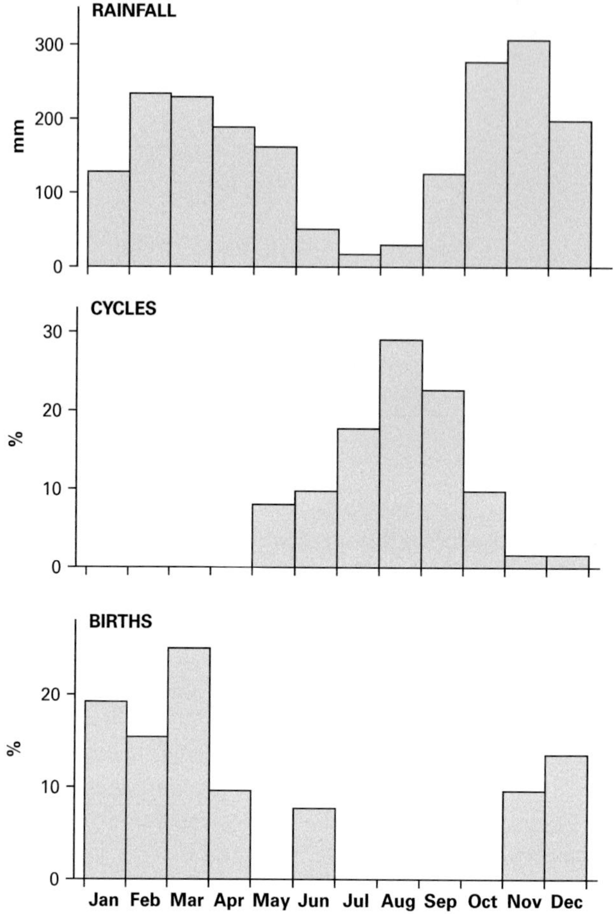

Figure 7.1. Seasonal patterns of rainfall, and reproduction in a semi-free ranging group of mandrills. Rainfall data are monthly averages over ten years for the Haute Ogooué region of Gabon (where the CIRMF mandrill colony is situated). Data on menstrual cycles refer to monthly occurrences of maximum sexual skin swellings during three years in Group 1. Dates of the resulting births (N = 52) were usually known to within a few days. (Author's data.)

rainfall occurs during the birth season. Rainfall also continues at high levels during the pre-mating season months of April and May when females are still suckling their infants.

Why do births occur mainly during wet seasons?

Why do mandrills, and so many other tropical mammalian species, exhibit these dry season/wet season patterns of matings and births? This question requires answers at both ultimate (evolutionary) and proximate (physiological) levels. At the ultimate level, it is likely that giving birth during wetter periods of the year has undergone positive selection because food availability, and hence female body condition and offspring survival, is greater during wet seasons. This is thought to be the case among the forest guenons, for example, as births coincide with greater availability of fruits, seeds, arthropods, young leaves and flowers during the rainy season (Butynski, 1988; Enstam and Isbell, 2007). The same might be true of the mandrill, which relies heavily upon a diet of fruits, seeds and arthropods, as was described in Chapter 4. Mothers in mandrill supergroups, and subgroups, must suckle and carry their offspring, as well as foraging and ranging over long distances each day. Greater resource availability during the wet season may be important for survival of mothers and infants.

Persuasive as this argument may be, there could be other reasons why females tend to give birth at the same time of year. Synchrony of births means that slower moving females, burdened with young offspring, may be less likely to be left behind during group movements. They may be less at risk of becoming isolated from other group members, and of being taken by leopards, crowned eagles and other predators. The matrilineal organization of mandrill groups, coupled with the tendency to give birth during the rainy season, means that females are moving, feeding and resting as members of cohesive social units. Moreover, as their infants begin to move independently, they encounter a cohort of similarly aged peers, and begin to engage in play and other activities that are important for normal social development.

At the proximate level, it can be argued that the physiological mechanisms that govern the onset of parturition are activated at the end of the 175-day gestation period in mandrills, irrespective of the time of year. Thus, the birth season is actually determined by the timing of the mating season, six months earlier.

What factors might control the timing of the mating season?

If we accept that selection has favoured seasonal breeding in the mandrill primarily because it is advantageous to concentrate births during periods of greater rainfall, then it is necessary to ask what mechanisms might coordinate sexual activity, so that conceptions occur at the appropriate time of year, during the dry season.

The first point to establish is that adult females, rather than adult males, are most responsive to seasonal factors, as regards changes in their underlying reproductive physiology. Figure 7.2 shows changes in female sexual skin swellings, and occurrences of births in the CIRMF mandrill group over a 21-month period, during 1996–1997 (Setchell and Dixson, 2001c). Two annual mating seasons and one birth season occurred during the period covered by these data, which refer to the longest-established and most stable of the two semi-free ranging groups (Group 1). The members of Group 1 included the descendants of the original founders that had occupied the same enclosure since 1984. Female mandrills in this group showed a well-defined annual rhythm of sexual skin swelling, followed by a birth season later in the year, as described earlier. Males, by contrast, did not exhibit marked seasonal changes in the red, testosterone-dependent sexual skin on the face, rump and genitalia

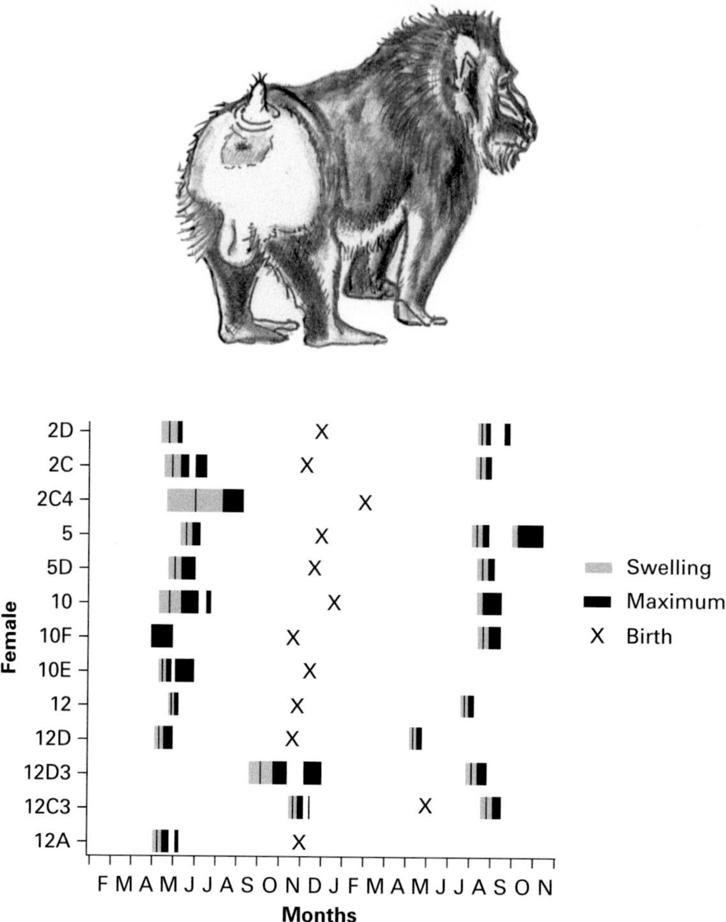

Figure 7.2. Patterns of sexual skin tumescence and births in twelve female members of a semi-free ranging mandrill group, in Gabon. (From Dixson, 2009).

(Setchell and Dixson, 2001c). Subtle individual changes did occur, however, as will be discussed later.

There have been no studies of spermatogenesis in mandrills, but a marked annual rhythm of spermatogenesis, such as occurs in some macaques (e.g. *Macaca mulatta*: Gordon *et al.*, 1976; Sade, 1964), is most unlikely to exist in mandrills. Male rhesus macaques, in which photoperiodic cues play a crucial role in determining the onset of the mating season, exhibit pronounced elevations in circulating testosterone, reddening of the sexual skin, increases in testicular volume and sperm production during the annual mating season. Cues from females, as well as changes in day length, stimulate these increases in gonadal function in males. Thus, if female rhesus monkeys are treated with oestrogen during the non-mating season, males show a resurgence of spermatogenesis, and activation of their sexual behaviour (Vandenbergh, 1969). We should not discount the possibility that more subtle changes in gonadal function and sexual skin colouration might occur in male mandrills during the mating

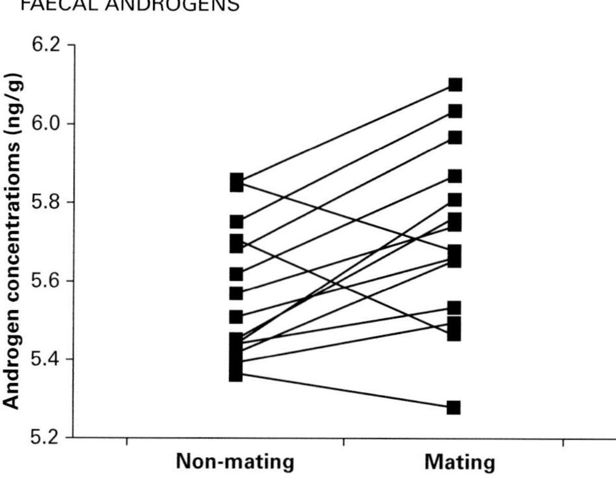

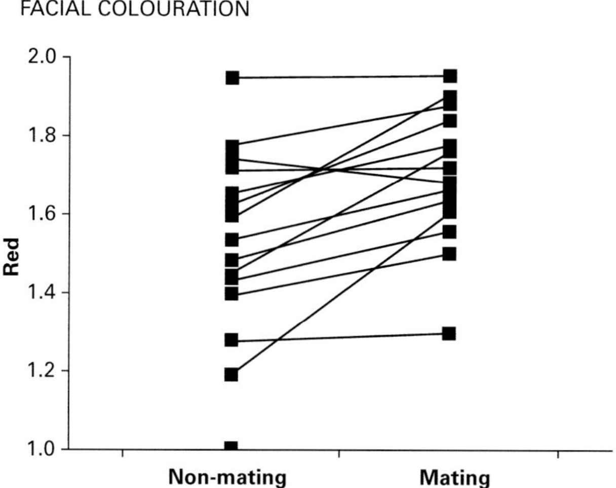

Figure 7.3. Small but statistically significant increases occur in concentrations of faecal androgens ($P = 0.012$) and in red facial colouration ($P = 0.002$) in male mandrills during those periods when females develop sexual skin swellings. (Redrawn from Setchell *et al.*, 2008).

season. Thus, measurements of faecal androgens in male mandrills have shown that higher concentrations are excreted during the mating season, and that small but statistically significant increases in facial redness are also measurable in some males during this time (Figure 7.3).

It will be recalled that, in the wild, many (but not necessarily all) adult male mandrills spend much of the time ranging alone, and then enter supergroups during the dry season for mating purposes (Abernethy *et al.*, 2002). Abernethy *et al.* showed that a positive correlation exists between the numbers of males in mandrill supergroups and the numbers of females that have developed their sexual skin swellings. It seems most likely that females are the primary respondents to whatever factors (rainfall, nutrition, social cues) determine the onset of the mating season, whereas males are more affected by the increases in female attractiveness, and by the inter-male competition that results when they join supergroups in order to mate.

In the CIRMF colony, some fatted males tend to remain with the social group throughout the year, while some others (mostly the non-fatted males) live a more solitary life, but haunt the fringes of the group during the mating season, and interact opportunistically with females. In the wild, it could be possible that some males are long-term residents that continue to forage with supergroups during non-mating periods. The published field data are not yet sufficient to be certain about this question. However, Dr Kate Abernethy has told me that 'most adult males do not have any contact or interaction at all with the supergroup outside the breeding season'. Rare exceptions occur, so that 'sometimes a male will enter the group for a few hours, days or even weeks outside the breeding season, and these sojourns may be in response to the odd female that cycles asynchronously, which we also see. But we did not always document a visibly fertile female when a male was visiting a group.' (Kate Abernethy, personal communication to the author). On current evidence, therefore, it appears that the great majority of adult males emigrate and remain outside supergroups once the annual mating season has ended.

The second point to establish concerning the mating season is that its timing does vary from year to year, and there are individual differences between females in the temporal patterning of sexual skin tumescence. In Figure 7.2, most females in 1996 developed full swellings earlier than usual, between the months of May and July, but the following year the usual peak swelling period (July–August) was re-established. Two young nulliparous females (Nos. 12D3 and 12C3) exhibited marked delays in attaining swellings in 1996 (until October–December), but had settled into the July–August pattern by the following year. Age is certainly one factor influencing these variations, and young females develop their first swellings at an average age of 3.6 years (range, 3.2–4.6 years: Setchell and Wickings, 2004a), and begin to reproduce two years before they attain adult body weight (Wickings and Dixson, 1992a). Additionally, as will be discussed in the next chapter, female social rank has significant effects upon fertility, so that some subordinate females may require multiple cycles in order to conceive, and they fail to do so during some years.

That social factors can disrupt the normal course of the mating season and menstrual cycle onset in mandrills is well illustrated with reference to some unusual events that occurred in the colony at CIRMF. First, I will discuss anecdotal evidence, and then provide some hard data relating to this subject. In 1991, several animals escaped from the enclosure occupied by Group 1 and visited the nearby suburbs of Franceville and the forests adjacent to the medical research centre. Included among the escapees was the group's lowest-ranking female (female no. 6); however, two weeks later she climbed back into the enclosure and rejoined the group. Despite the fact that these events occurred during the wet season when the majority of females were still pregnant, female no. 6 (not pregnant at the time of her escape) began to cycle once more and to copulate with the alpha male.

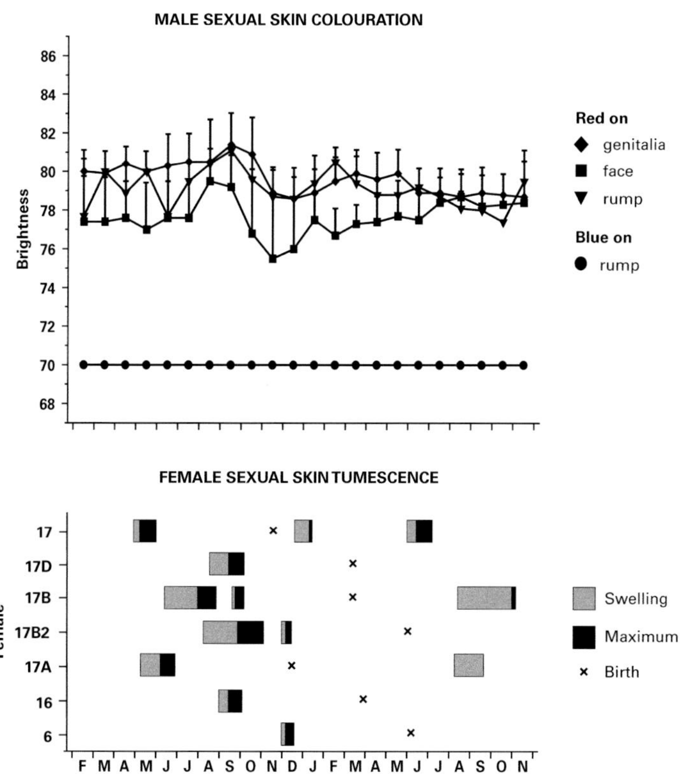

Figure 7.4. Effects of social disruption upon the timing of matings and births in a semi-free ranging mandrill group (Group 2) following its formation. Male sexual skin colouration during the same period remained relatively stable. Further details are provided in the text. (Redrawn from Setchell and Dixson, 2001c.)

The second and much better documented example concerns events following the splitting of Group 1, in 1994, when 17 mandrills (including six adult females and four adult males) were transferred to an adjacent enclosure to establish Group 2. What occurred in subsequent years was recorded by Joanna Setchell, and Figure 7.4 shows the timing of swelling cycles and births among Group 2 females during 1996 and 1997. Unlike the regular timing of mating and birth season events in Group 1, housed in full view of the adjacent rainforested enclosure, females in Group 2 showed very poor coordination of reproductive events. Their mating season extended from May 1996 until early January 1997, with births distributed between November and the following June. Meanwhile, the hormone-dependent, red facial colouration of the males in Group 2 remained relatively stable throughout the same period (Figure 7.4).

The exact nature of the proximate cues that cause the onset of sexual activity in seasonally breeding tropical monkey species, such as the mandrill, talapoin and many guenons, remains poorly understood. Indeed, the problem has received very little attention from a physiological perspective. In some way, the transition from wetter to dry season conditions is implicated in triggering the onset of the mandrill's mating season. Might this

Figure 7.5. In addition to foraging in their rainforested enclosures, mandrills at the CIRMF are provisioned daily with fruits and vegetables. (Author's photograph.)

involve changes in the nature of the food supply, associated with dryer conditions? Dietary and metabolic cues have been implicated in the control of hypothalamic–pituitary–gonadal function in various domesticated mammals (I'Anson *et al.*, 1991; Krasnow and Steiner, 2006; Sadleir, 1969; Schneider, 2004). Might seasonal changes in food availability act via these pathways to regulate reproduction in the mandrill? Logical as this suggestion may be, it is necessary to consider that mandrills in the semi-free ranging groups at the CIRMF have access to year-round provisioning (with fruits, vegetables and commercially produced 'monkey biscuits'), in addition to the natural foods in their rainforested enclosures (Figure 7.5). This makes it unlikely that the mating season could be triggered solely by dietary cues in this captive setting. Whether the decline in rainfall per se and the pronounced decrease in atmospheric humidity that occurs between June and September might have some more direct significance in affecting the hypothalamic–pituitary–ovarian axis remains a matter for conjecture.

Although we do not yet understand the precise amalgam of cues that controls the timing of the mating season in mandrills, a great deal is known about the underlying neuroendocrine substrates that govern menstrual cycles, and which must respond to the relevant environmental cues if the cycle is to be activated, whether at puberty or periodically throughout adulthood. Much of this knowledge derives from work on the rhesus monkey, but is applicable to other Old World anthropoids, including the mandrill. Central to this discussion is the gonadotrophin releasing hormone (GnRH) pulse generator, a collection of peptidergic neurons situated in the brain, in the arcuate nucleus of the mediobasal hypothalamus (Hotchkiss and Knobil, 1994; Knobil, 1974;). These neurons release hourly pulses of a decapeptide (GnRH), which is conveyed in the

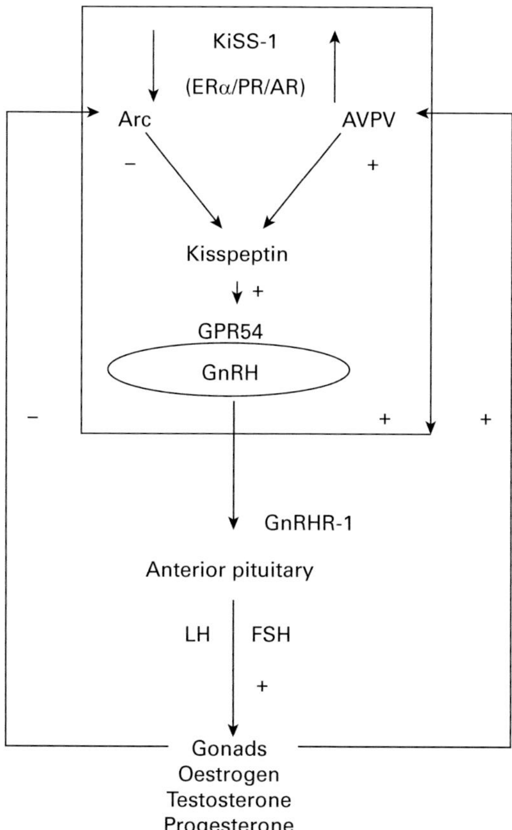

Figure 7.6. Diagram showing the positive and negative feedback exerted at the hypothalamic level by gonadal steroids, which bind to receptors on KiSS-1 neurons in the arcuate nucleus (Arc) in primates and sheep, and to both the Arc and the anteroventral periventricular nucleus (AVPV) in rodents. AR, androgen receptor; ER, oestrogen receptor; PR, progesterone receptor. (From Dixson, 2012, modified from Roseweir and Millar, 2009.)

bloodstream, via the portal vessels of the pituitary stalk, to the anterior pituitary gland. There, GnRH stimulates specialized cells (gonadotrophs) to secrete pulses of the gonadotrophins, luteinizing hormone (LH) and follicle stimulating hormone (FSH). The gonadotrophins, in turn, control ovarian function, and are essential to bring about cyclical changes in ovarian hormone secretion, and ovulation, during the menstrual cycle (Figure 7.6).

It should be noted, in Figure 7.6, that the GnRH neurons do not respond directly to the many influences (external and internal) that impinge upon the brain to regulate the pulse generator. Thus, they do not have receptors for the gonadal hormones. To better understand the control of GnRH secretion, it is necessary, as Iain Clarke (2011) has pointed out, to take 'one step back' and to consider 'neuronal pathways that converge on the GnRH cells'. These convergent pathways are currently thought to regulate the GnRH pulse generator via two types of RF-amide peptides: the kisspeptins, and gonadotrophin inhibitory hormones (GnIH).

Kisspeptins are widely distributed throughout the reproductive organs and the brain, including areas of the hypothalamus that contain GnRH neurons. Kisspeptins are the products of the *KiSS-1* gene and, in the rhesus monkey, *KiSS-1* occurs in the mediobasal hypothalalmus, closely associated with GnRH axons (Ramaswamy *et al.*, 2008). Metabolic and photoperiodic signals, as well as the positive and negative feedback effects of gonadal steroids, are funnelled via KiSS-1 neurons, which, in turn, coordinate the functions of the GnRH system and the hypothalamic–pituitary–gonadal axis (Roseweir and Millar, 2009).

Gonadotrophin inhibitory hormones (GnIH) were first discovered in birds (Tsutsui, 2009; Tsutsui *et al.*, 2000;), but homologues are now known to occur in mammals, including the rhesus monkey, where GnIH cells are situated in the dorsomedial and paraventicular hypothalamic nuclei (Smith *et al.*, 2010). Smith *et al.* showed that GnIH and kisspeptin neurons project to the cells of the GnRH pulse generator. It is thought that GnIH exerts its actions on reproductive function both at the level of the pulse generator and at the pituitary level to inhibit gonadotrophin secretion (Clarke, 2011; Parhar *et al.*, 2012).

Returning to the discussion of the mandrill's reproductive biology, when the activity of the GnRH pulse generator is inhibited, then gonadotrophin secretion by the pituitary gland is likewise blocked, and menstrual cycles cease (amenorrhoea). One physiologically important example of such inhibition concerns the amenorrhoea that occurs when mandrill mothers are lactating and suckling their young infants. We have seen that most births occur between November and March; females are thus lactating, suckling and transporting their growing infants in the months just before the mating season begins (i.e. June). Suckling itself inhibits the GnRH pulse generator, especially during the early phases of infant development; later on, the heavy metabolic demands upon the mother of sustaining her milk production also contribute to maintenance of lactational amenorrhoea. This generalization applies to the mandrill, just as it does to the other Old World monkeys and apes, as well as to humans (McNeilly, 2006).

There are, however, marked individual differences in the duration of lactational amenorrhoea in mandrills, so that some females in the CIRMF colony exhibit resumption of the ovarian cycle and ovulate 6.5 months after giving birth. The interval between parturition and resumption of ovulation may be much longer; in extreme cases, more than two years may elapse between successive births (Setchell *et al.*, 2002). A number of factors, including age and social rank, as well as previous reproductive history and (in all probability) as yet poorly understood aspects of body condition, influence how often and how successfully females produce their infants across the lifespan. These questions will be explored further in the next chapter, as part of discussions of the effects of social rank upon behaviour and reproduction in both sexes of the mandrill. In the context of seasonal breeding, it should be noted that although the combined duration of gestation and lactational amenorrhoea determines, to some extent, when a female is likely to begin cycling again, there must be other factors that ensure that such activity peaks during the long dry season months. Although analysis of 14 years of records of the CIRMF colony led us to the conclusion that 'mandrill births occurred in every month except August' (Setchell *et al.*, 2002), some seasonally atypical births were the result of periods of social disruption, and occurred under conditions that must be very different from those experienced by wild mandrills. In the wild, mandrills probably exhibit tighter coordination of their seasonal mating and birth patterns than has been described here. However, as was noted earlier in this chapter, Dr Kate Abernethy has observed occasional instances of female mandrills in the Lopé National Park cycling outside the mating season.

Comparative observations on seasonal breeding in the drill

Compelling evidence for the existence of seasonal breeding in the drill derives from work conducted by Dr Kathy Wood (2007) at the Pandrillus Rescue Centre in Nigeria. At Pandrillus Rescue Centre, drills are housed under naturalistic conditions in large, multi-male–multifemale groups, as is the case for the CIRMF mandrills in Gabon. However, in this region of Nigeria, the timing of wet and dry seasons is, broadly speaking, the reverse of that described previously. The wettest period in Nigeria is between the months of May and August. The majority of drill births occur during this wet season; Wood recorded that 71% of 153 births took place between May and September. The marked correlation that exists between drill birth peaks and rainy seasons in Nigeria is demonstrated more clearly by examining Kathy Wood's graph of monthly rainfall and birth frequencies, between the years 2000 and 2005 (Figure 7.7). The gestation period of the drill is the same as that of the mandrill (mean gestation length, 175 days), so that by counting

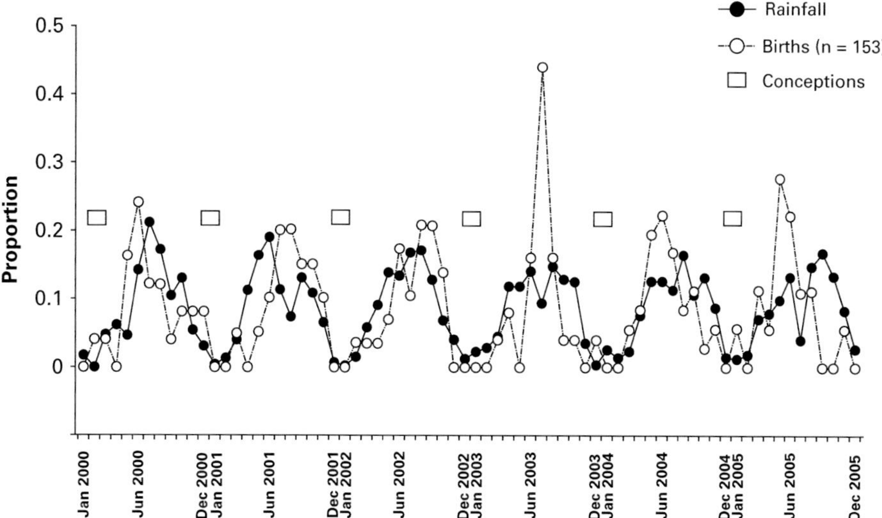

Figure 7.7. Circannual rhythms of rainfall and reproduction in large social groups of drills, held at the Pandrillus Reserch Centre, in Nigeria. (Redrawn and modified from Wood, 2007, with her permission.)

backwards from the annual birth peaks, it becomes apparent that most of the conceptions must have occurred during the dry season, and especially between the months of December/January and March. The putative mating seasons are indicated by the horizontal bars in Figure 7.7.

Unfortunately very little detailed information has been published concerning sexual behaviour in drills (Marty *et al.*, 2009), but it is clear that their mating seasons and birth seasons are closely aligned to changes in rainfall, just as in mandrills. It is probable, therefore, that the common ancestor of both *Mandrillus* species would also have exhibited such a pattern of seasonal breeding.

8 Behaviour and reproductive success

Semi-free ranging mandrill groups in Gabon

Between the years 1990 and 1992, at the Centre International de Recherches Medicales de Franceville (CIRMF) in Gabon, I conducted a longitudinal study of the behaviour of the semi-free ranging mandrill group (Group 1), the seasonal patterns of reproduction of which were discussed in the last chapter. This group was founded in 1983, when 14 young mandrills (six males and eight females) were released into a six-hectare enclosed area of secondary rainforest. By the time the observations described here began, the group contained 45 individuals, and it increased in size to 68 mandrills during the course of the project. All the animals except infants were fitted with numbered ear-tags; thus, it was possible to identify individuals quite easily when conducting behavioural observations.

The behavioural work benefitted from the valuable assistance of two students from the University of Zurich (Thomas Bossi and Edi Frei). Our detailed observations of the menstrual cycle and of sexual and associated patterns of behaviour extended throughout two annual mating and birth seasons. Blood samples were collected from all the infants born during the study period, as well as from their mothers and the various males in the enclosure (see Plate 24). Subsequently, Dr Jean Wickings conducted a DNA fingerprinting paternity analysis of these infants, so that, for the first time, it was possible to examine how the behaviour of male mandrills translated into reproductive success, in terms of infants sired with individual females. The dominance ranks of all the animals were known, as well as many details of their visual displays, vocalizations and grooming interactions (these were described in Chapter 5). Thus, it was possible to analyse how rank and other social variables might influence patterns of sexual behaviour and reproductive success in mandrills living under relatively natural conditions.

I shall discuss the results of this initial study in some detail, because fine-grained behavioural data were collected at a time when Group 1 remained relatively undisturbed and its social organization was stable. As was described in the previous chapter, mandrills were subsequently removed from Group1 (in 1994), in order to set up a second group in an adjacent enclosure. Subsequent transfers of additional adult males from Group 1, to enlarge Group 2, inevitably caused further social disruption. The following account focuses on the initial studies of Group 1, and then moves to a wider discussion of the results of more recent work involving both Groups 1 and 2, as they have gradually stabilized and progressively enlarged as a result of further breeding over the years.

Some comments on methodology

Behavioural observations of Group 1 were made from an observation tower, situated adjacent to the enclosure and commanding an excellent view over the valley, including a large open grassy area adjacent to the forest. The mandrills were accustomed to enter this area for about two to three hours every morning, and to receive additional food. Behaviour was recorded using a coded 'shorthand' system (interested readers will find a description of this scoring system in the Appendix). All sexual and associated behavioural interactions were recorded whenever they occurred rather than by a focal animal technique, as many copulations would have gone unrecorded using such an approach. The shorthand system also made it possible to record sequences of behaviour, and thus to determine which behavioural elements preceded or followed others during sexual or agonistic encounters. Female sexual skin swelling stages were recorded daily, as described later, as well as the amounts of time that the group remained in view. In this way, it was possible to calculate hourly frequencies of behaviour during each stage of the menstrual cycle.

The female mandrill's sexual skin undergoes rhythmic changes in swelling and detumescence during the menstrual cycle, as was discussed in Chapters 3 and 7. Initially the skin is 'flat' (for quantitative purposes this stage was assigned a score of zero, as shown in Figure 8.1), and lacks any signs of oedema or pink colouration. During the follicular phase of the cycle, the sexual skin swells and becomes pinkish, initially being a small region (+1), affecting the vulval area and pubic lobe, and then of medium size (+2) to incorporate the peri-anal field. Finally, the swelling reaches maximum size (large = +3). At this stage, it becomes progressively more turgid and shiny, pink or reddish in colour, and with a central area of paler skin surrounding the vulva (Plate 12). The swelling then begins to detumesce, and the first day of detumescence ('breakdown day') is easily recognized, as its turgor and shiny appearance

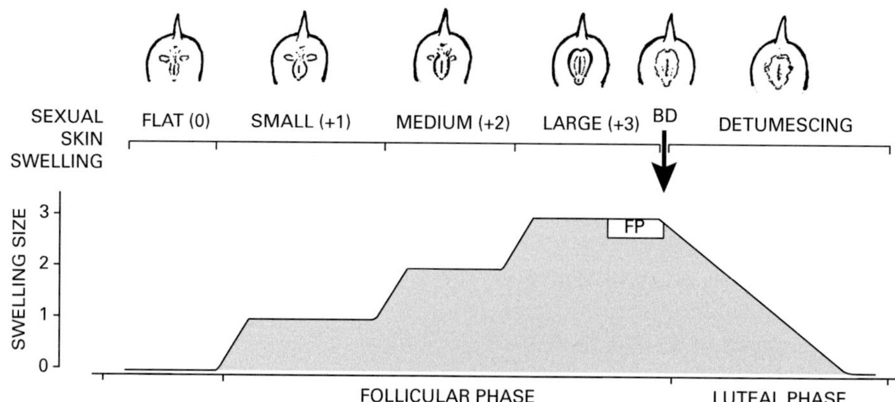

Figure 8.1. Diagram to show changes in sexual skin swelling during the various stages of the mandrill's menstrual cycle, including the timing of the putative fertile period (FP) during the final days of maximum tumescence, and the day of sexual skin breakdown (BD).

diminish very quickly. Detumescence then continues more slowly during the luteal phase of the cycle (approximately 15 days), or during the early post-conception phase (for about seven days) if the female has become pregnant. Menstruations were only rarely visible (Plate 13), and ten of the 14 females conceived on their first cycle during the mating season.

The stages of sexual skin swelling and detumescence were recorded for all females, using the system described here, and shown in diagrammatic form in Figure 8.1. As we shall see, the duration of the follicular phase of the cycle (swelling phases +1, +2 and +3) is highly variable in mandrills, so that some females may have very long follicular phases.

Webs of sexual encounters

Firstly, let us consider the complex web of sexual relationships that emerged when behaviour was recorded, across all stages of the menstrual cycle and throughout an entire mating season (Figure 8.2). Female mandrills presented sexually to most of the males in the group. The highest-ranking, fatted males (i.e. male numbers 14, 7 and 15 in Figure 8.2) received most of the females' presentations. Thus, the alpha male (no. 14) received 49% of all females' presentations, and accounted for 69% of all mounts that were observed during the 1990 mating season (Figures 8.2A and 8.2B). It was notable, however, that some females solicited copulations from, and were mounted by, low-ranking individuals, such as the adolescent males 2B and 5B. One further adolescent male (no. 6A) had begun to emigrate from the group (as occurs in all males aged six to seven years) and he was rarely seen interacting with other group members.

Three non-fatted adult males also occupied the enclosure (nos. 9, 13 and 18); these males spent most of the time alone and were only occasionally seen associating with females. However, male no. 13 was exceptional in being both non-fatted and of high rank, so that he was able to threaten and displace all other males except the alpha individual. Male no. 13 succeeded in copulating opportunistically with six of the 14 females.

Only fatted adult males were ever observed to engage in mate-guarding behaviour (Figure 8.2C). Mate-guarding episodes often lasted for some days, during which time a male followed and copulated with a single female, usually when her sexual skin was at maximal swelling. The dynamics of mate-guarding are considered in detail later. At this point, it should be noted in Figure 8.2C that the alpha (no.14) and the beta, group-associated, fatted male (no.7) often attempted to mate-guard the same females. Usually, the alpha male gained priority of access to females during the putative fertile period, but this was not always the case. It is also worth emphasizing that mate-guarding is probably physically very costly for male mandrills. The alpha male guarded on 126 days, which represented 82% of mating season days, in 1990. Declines in weight, and loss of general body condition were evident in both males (nos. 14 and 7) by the

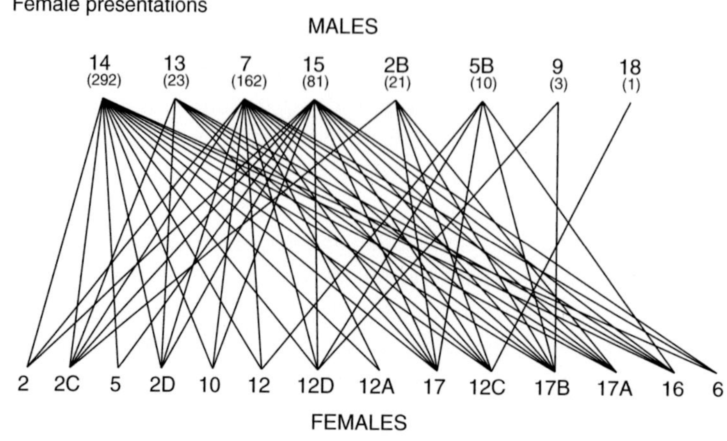

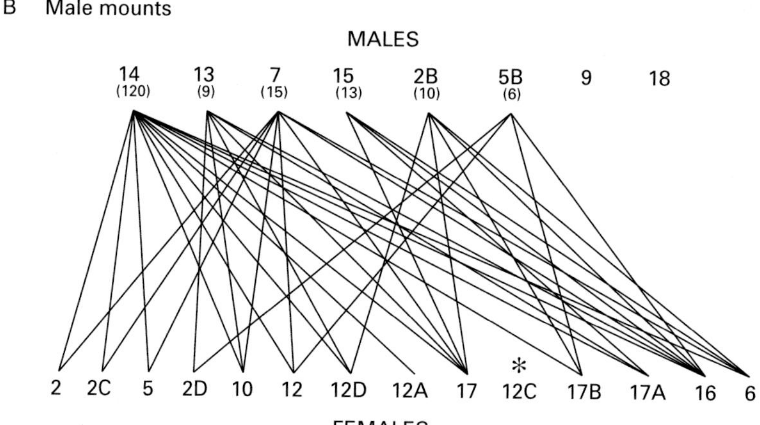

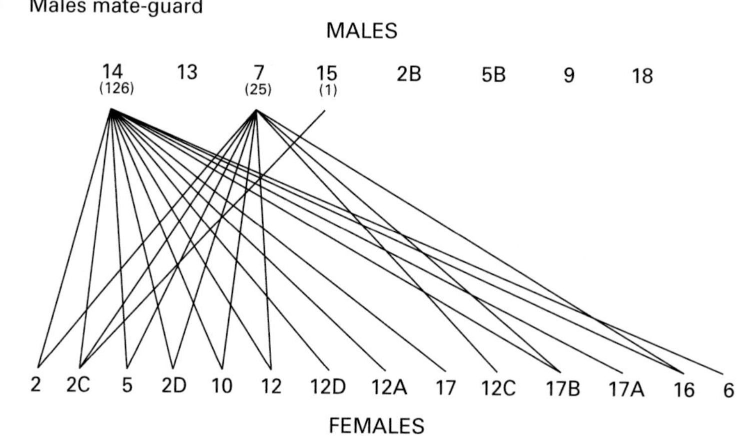

Figure 8.2. Webs of sexual interactions during an entire mating season in mandrill Group 1. **A:** Females' sexual presentations to each male. **B:** Males' mounting behaviour with individual females. **C:** Mate-guarding episodes. Identity numbers of members of both sexes are arranged in rank order (highest-ranking on the left). The total numbers of presentations received, mounts and mate-guarding episodes by each male are given in parentheses, below his ID number. (Author's data.)

end of the annual mating season, whereas the third-ranking fatted male (no. 15) appeared unchanged in his overall physical condition. It is probably significant that this particular male had only mate-guarded one female, on a single day, during the whole season.

The webs of sexual interactions depicted in Figure 8.2 also provide evidence for the existence of a multimale–multifemale type of mating system in this mandrill group (Dixson *et al.*, 1993). I say this because the adult males never made any attempt to assemble or control small groups or 'harems' of females. There were no 'one-male units' in this mandrill social group, and females invited copulation with a number of potential partners. As we shall see, mate-guarding is a transitory association between a male mandrill and a single female; the relationship ends as soon as the female's swelling begins to detumesce. Once the mating season is over, even these relatively short-term associations between males and individual females no longer occur. Mandrill society is thus clearly very different to the multilevel societies and their component one-male units that have been extensively researched in polygynous species, such as the hamadryas baboon (Kummer, 1990; Swedell, 2006) or the gelada (Dunbar and Dunbar, 1975).

Male rank and mating success

Figure 8.3 shows the marked relationships that existed between male rank and male mating success during the 1990 mating season.The second-ranking male (no.7) accounted for only 15% of ejaculations (with six females: Figure 8.3). By contrast, the highest-ranking male (no.14) accounted for just over 70% of all ejaculatory mounts observed directly, and these involved at least 12 of the 14 females that exhibited swelling cycles. I say at least because it was only possible to obtain a limited sample of the sexual activity of these male mandrills. Behavioural observations spanned, at most, two hours per day. Male no.14 was not seen to mate with female 12C, for example, and yet subsequent DNA analysis showed that he had sired her offspring.

Table 8.1 summarizes data collected throughout both mating seasons, during 1990 and 1991. The marked mating success of the most dominant, fatted male in the group is evident; he accounted for 63% of all mount attempts during the two mating seasons, and 74% of all ejaculations with females having large sexual skin swellings. His sexual activity was, therefore, focused within periods when females were most likely to conceive, and he was highly successful at mate-guarding such females, so that 80% of his ejaculatory mounts occurred during mate-guarding episodes. Although individual mating success was much less pronounced among the other males, there were overall statistically significant correlations between male rank, total mount attempts, numbers of ejaculations with females at full swelling, and hourly frequencies of ejaculation, as detailed in Table 8.1. Males initiated the majority of all mounts, but females sometimes refused their attempts, or terminated mounts prior to ejaculation. There was no significant correlation between female refusals/terminations and male

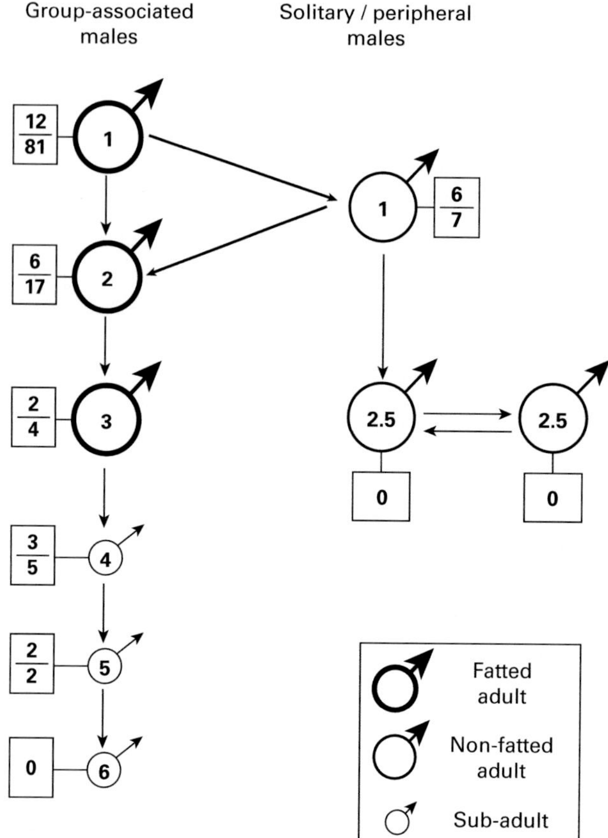

Figure 8.3. Male dominance and overall mating success in the semi-free ranging mandrills of Group 1 during a single mating season. Group-associated (fatted and sub-adult) males have a linear hierarchy, in which male rank is positively correlated with mating success. Boxes indicate the numbers of females mated (upper) and numbers of observed ejaculatory mounts (lower) for each male. Solitary/peripheral, non-fatted adult males mate opportunistically; only the most dominant peripheral male was observed mounting females. (From Dixson, 2012.)

dominance rank. However, the two lowest-ranking individuals (male nos. 9 and 18) were virtually solitary, and very rarely interacted with females; they were seen to attempt mounting on only three occasions during the entire mating season. Exclusion of these males from the analysis revealed a significant correlation between male rank and female refusals/terminations of mounts for the remaining males; the higher a male's rank, the less his attempts to mate were likely to be curtailed by females ($N = 6$; $r_s = 0.958$; $P < 0.05$).

It should be kept in mind that these findings derive from observations of a relatively small mandrill group, numbering between 45 and 68 individuals, as compared to the 700–800 animals that comprise a typical free-ranging supergroup during the annual mating season (Abernethy *et al.*, 2002). Webs of sexual interactions are doubtless very

Table 8.1. Mating behaviour and social rank in semi-free ranging male mandrills, during two mating seasons

Male no.	14(F)	13(NF)	7(F)	15(F)	2B(A)	5B(A)	9(NF)	18(NF)	
Rank no.	1	2	3	4	5	6	7.5	7.5	SR
Mount attempts	146	14	38	3	16	12	2	1	0.828*
% Initiated by male	94	100	100	100	100	89	50	100	0.375
% Attempts refused	14	21	26	21	44	58	0	0	0.482
Ejaculations (total)	97	8	19	11	6	3	1	0	0.922**
% Mounts with Ejaculation	67	57	50	35	37	25	50	0	0.744*
Ejaculations: Females At max swelling	90	4	14	6	5	2	1	0	0.851*
Ejaculations/hr #	0.57	0.05	0.13	0.03	0.03	0.01	0	0	0.970**

F, fatted male; NF, non-fatted male; A, adolescent; SR, Spearman rank correlation coefficient, social rank vs mating behaviour; # , 1990 mating season only; *, $P \leq 0.05$; **, $P < 0.01$. (Author's data.)

much more complex, and multiple partner matings are more frequent in a mandrill supergroup, which may include dozens of adult and sub-mature males, and well over 200 females. A tentative model of the social organization and mating system of wild mandrills is presented at the end of this chapter.

In Chapter 3, I noted that fatted male mandrills have very large testes and other genital traits indicative of the effects of sperm competition, upon the evolution of the male reproductive system. In this context, the very high rate of ejaculations (0.57/h) sustained by the dominant fatted male in the CIRMF Group 1 is of interest. Such a high ejaculatory frequency is consistent with occurrences of multi-partner matings and sperm competition in a variety of primate species, including macaques, baboons and chimpanzees (Dixson, 1995, 2012). On 20 occasions during the 1990 mating season, the alpha male in Group 1 was observed to ejaculate 2–3 times during daily observations that lasted for 90–120 minutes. Intervals between consecutive ejaculations were sometimes very brief (1–6 min). This male was especially likely to remount a female if he had seen another male attempting to gain access to her. Such behaviour, which involves marked shortening of the post-ejaculatory refractory period, also occurs in some other primate species that have multi-partner mating systems.

The menstrual cycle and behaviour

Female anthropoids do not exhibit oestrus, as there is no restriction of mating to the peri-ovulatory period of the ovarian cycle in monkeys and apes. Nonetheless, there are rhythmic fluctuations in sexual interactions during the menstrual cycles of Old World anthropoids, and the mandrill is no exception to this rule. In all Old World monkey and ape species that have been studied to examine this problem, copulations are more frequent during the follicular phase than during the luteal phase of the cycle; peri-ovulatory peaks may occur and, in a few cases, secondary peaks have been measured during the days preceding menstruation. These fluctuations are situation dependent, however, so that observational conditions, social rank, partner preferences and a variety of other factors may influence the patterning of sexual interactions (Dixson, 2012).

In the case of the mandrill, there is a profound impact of male rank on relationships between the female's cycle and patterns of male pre-copulatory and copulatory behaviour. Rhythmic changes in behaviour are very much more pronounced when females interact with high-ranking partners, whereas lower-ranking adult and sub-adult males copulate infrequently and opportunistically, rarely gaining access to females during the late follicular phase when ovulation is most likely to occur. These points are illustrated in Figures 8.4 and 8.5, with reference to the adult males of Group 1 and their sexual behaviour during the menstrual cycles of the group's 14 females.

Only the alpha male in the group (no.14) showed a marked rhythm of pre-copulatory behaviour, following females, examining their swellings and making grinning displays with increasing frequency during the follicular phase of each cycle (Figure 8.4). Likewise, he mounted and ejaculated more often, so that these behavioural patterns peaked at the height of female swelling, and then declined as swellings detumesced. Note that sexual activity decreased on the day of sexual skin breakdown (BD), but was not completely absent; the alpha male continued to mate with two of the 14 females during the breakdown day.

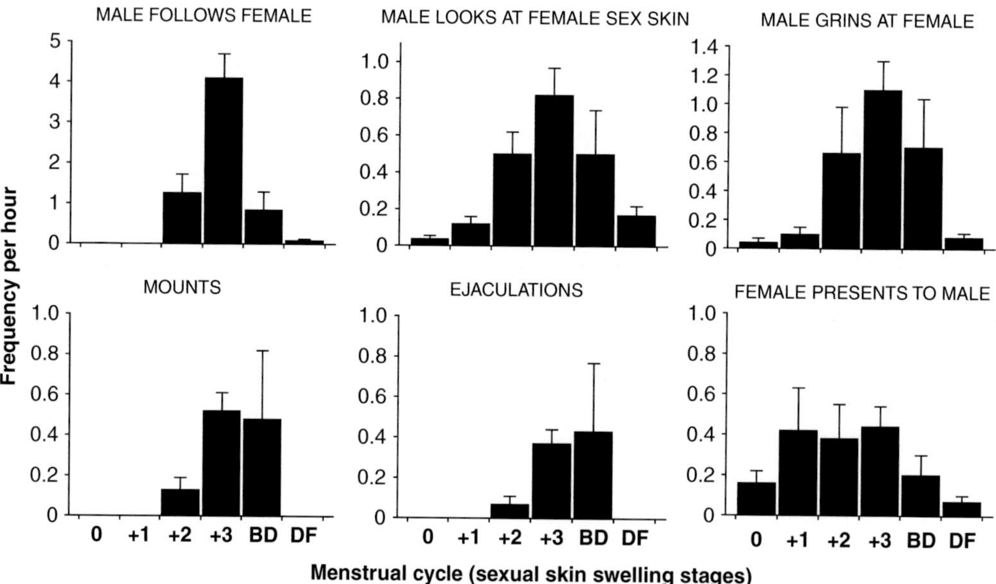

Figure 8.4. Sexual behaviour during the menstrual cycle. Means (+SEM) per hour for behavioural interactions between an alpha male mandrill, and 14 adult females at various stages of their menstrual cycles (0, sexual skin flat; +1, a small swelling; 2, medium swelling; 3, maximum swelling; BD, day of sexual skin breakdown; DF, swelling deflates (luteal phase or early post-conception). (Author's data.)

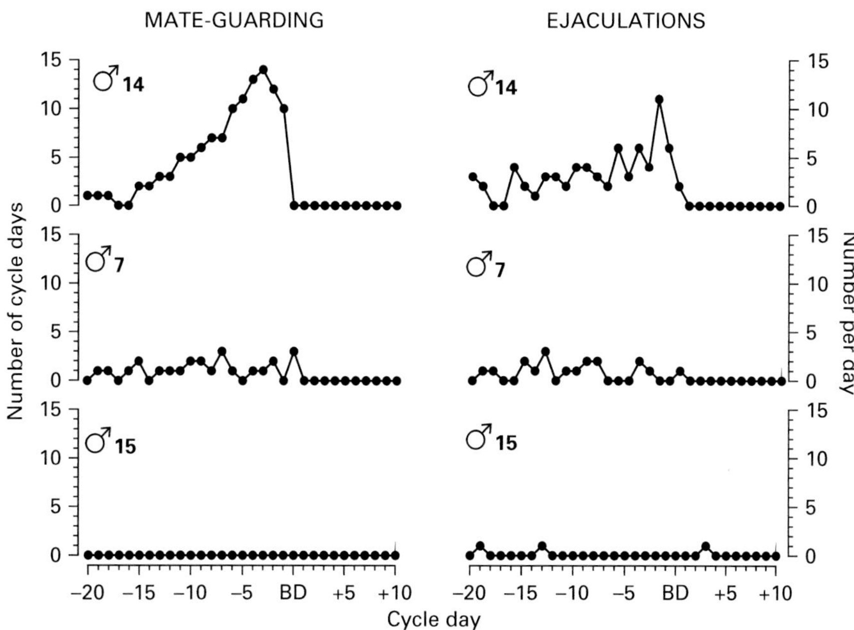

Figure 8.5. An alpha male mandrill (no.14) shows increased mate-guarding and ejaculations with females during their fertile periods, just prior to sexual skin breakdown (BD). By contrast, the second- and third-ranking males (nos. 7 and 15) exhibit very low frequencies of sexual activity with the group's 14 adult females. (Author's data.)

Data on daily patterns of mate-guarding and ejaculatory mounts for all three adult males are shown in Figure 8.5. Here it can be seen that only the alpha male showed a progressive increase in his mate-guarding behaviour and ejaculations during the 15–20 days leading up to the BD day. Mate-guarding was most intense during days –6 to –1, which, together with the BD day, constitute the period when ovulation is most likely to occur. The exact timing of ovulation has not been determined for the mandrill, but neuroendocrine and laparoscopic studies of baboons and macaques indicate that it occurs during this window of time (e.g. Wildt *et al.*, 1977).

The behavioural data presented in Figure 8.5 show that the alpha male did indeed invest more effort in guarding females during their putative fertile periods. The second-ranking male (no.7) was much less successful at mate-guarding during the fertile period. He exhibited a much more muted version of cyclical changes in pre-copulatory and copulatory activity, while the third-ranking fatted male (no. 15) showed no such changes during females' cycles.

Details of the sexual skin swelling cycles of four female mandrills are shown in Figure 8.6, together with the timing of matings and mate-guarding episodes by males. These diagrams provide vignettes of the types of interactions that characterized the mating season in this mandrill group. The marked mating success of the alpha male (no.14) is evident, but note that he only effectively guarded one female at a time. As each female's cycle reached its breakdown day, the alpha male tended to transfer, and guard a new partner. When more

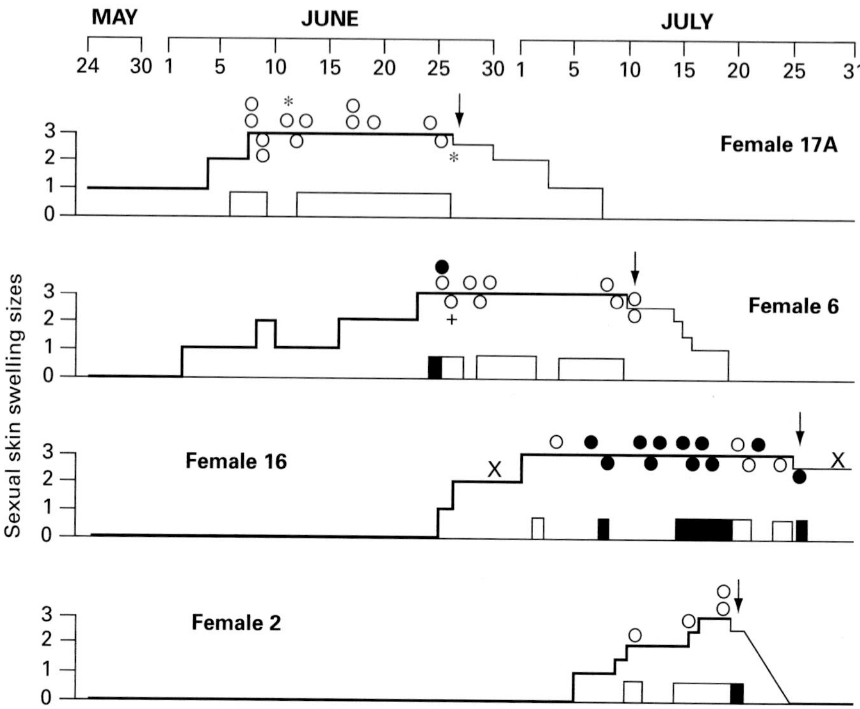

Figure 8.6. Details of sexual interactions during sexual skin swelling cycles in four female mandrills of Group 1, during the annual mating season, in Gabon. Swelling sizes are rated on a four-point scale (0–3), as shown in Figure 8.1, and discussed in the text. Sexual skin breakdown day is indicated by a vertical arrow. Mate-guarding episodes by males are shown as horizontal bars (□, male no. 14; ■, male no.7). Ejaculatory mounts are indicated by symbols for each adult male (○, no.14; ●, no. 7; X, no.15; ✻, no.13) and subadult male (+, no. 2B). (From Dixson, 2009.)

than one female had a large sexual skin swelling (e.g. females nos. 6 and 16; females nos. 16 and 2), conflicts occurred and the second highest-ranking of the three fatted males (no. 7) sometimes achieved greater mating success. This was most apparent in the case of female no. 16. It is interesting to note that, of the four conception cycles depicted in Figure 8.6, male no. 7 sired two of the resulting offspring (with females nos. 16 and 6), and male no. 14 accounted for the remaining two (with females nos. 17A and 2). Several other males mated only sporadically, and were rarely successful in doing so during the fertile period. Note also the long follicular phases of females nos. 17A, 16 and 6, the three lowest-ranking females in the group, as compared to the much shorter cycle of the alpha female (no. 2). Relationships between female rank, cycle length and fertility are discussed later in this chapter.

The dynamics of mate-guarding

Figure 8.7 presents data on the contribution made by each sex to moving apart ('walking away' >3 m) or following one another, during the series of mate-guarding events that

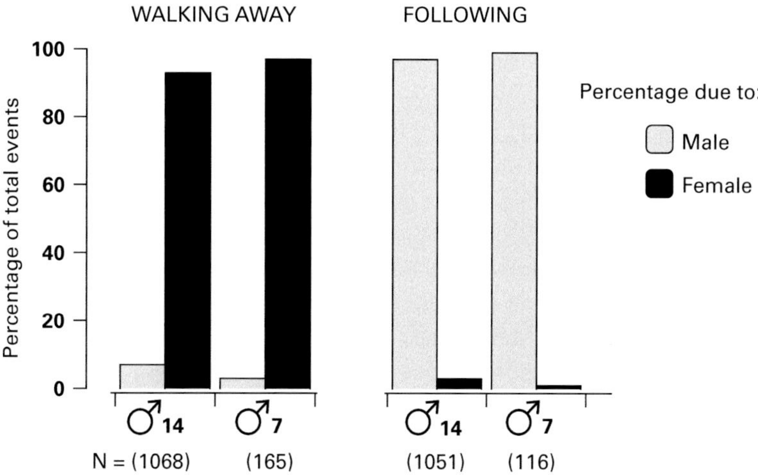

Figure 8.7. Maintenance of proximity between the sexes during mate-guarding. Data show percentages of leaving (walking away) or following, due to either the male or female partners during mate-guarding episodes. (Author's data.)

spanned a single mating season. In overall terms, females accounted for 95% of total 'walking away' scores, while males made 98% of all the following scores. Although the two highest-ranking fatted males initiated similar numbers of mate-guarding attempts, the alpha male (no.14) was markedly more successful at guarding individual females for prolonged periods (N = 13; range 1–21 days; mean 10.5 days) than was the group's beta male (N = 11; range 1–13 days; mean 2.8 days). These differences in mate-guarding durations are statistically significant (Wilcoxon test: $P < 0.01$).

The initiative for establishment and maintenance of proximity between a male and female mandrill during mate-guarding episodes thus rests primarily with the male partner. Females clearly influence these spatial relationships, as male mandrills do not herd or control their movements in any obvious way. Instead, the female moves freely; she often walks away from the male, and he usually follows her (see Plates 21–23). During the follicular phase of the menstrual cycle, females tend to be especially 'restless' and more mobile than at other times. This tendency for females to become more mobile during the mid-cycle occurs in some other primate species as well (e.g. in chacma baboons: Saayman, 1970). It is not associated with increases in foraging behaviour; indeed, female chacma baboons spend less time feeding during the follicular phase of the cycle, and treatment of ovariectomized females with oestradiol suppresses food intake (Bielert and Busse, 1983).The greater activity of the female probably serves, instead, to advertise her presence to males. This appears to be the case in the mandrill, and the pace of the female's movements largely determines the male's following responses. Following may be relatively slow, with close proximity being maintained between the pair, or the male may follow at a distance of 30–40 m, but always keeping the female in view. A third type of following is extremely rapid, with the female running and the male in hot pursuit; this type of following by a male often occurs immediately before a mount attempt. Unfortunately, frequencies of these variations of following behaviour were not measured separately, so that it is not possible to say how they might vary throughout the follicular phase of the female's cycle or in relation to her social rank.

It seems unlikely that grooming interactions could play any significant role in maintaining proximity between the sexes during mate-guarding. Thus, the alpha male was only observed to groom a female on one occasion during the 126 days he spent mate-guarding during the 1990 mating season. Females groomed this male on just eight occasions. Male mandrills often made grinning displays while following and mate-guarding females, as well as emitting frequent, deep, two-phase grunting vocalizations. Both sexes engaged in subtle forms of eye contact, and monitored each other's movements. As well as presenting sexually, females sometimes engaged in approach-retreat sequences ('parading' behaviour, as was described in Chapter 5). These apparently served to show off their swellings, and stimulated males to follow and attempt to mount. Increases in some pre-copulatory displays were measured during the period of maximum sexual skin swelling, as has been discussed, and these interactions were especially pronounced on mate-guarding days.

Prolonged mate-guarding interactions, lasting for more than two weeks in some cases, did not always evoke positive reactions from females. Indeed, they sometimes avoided males that attempted to guard them, as well as refusing their attempts to copulate. One interesting type of avoidance involved climbing into the trees, where it was difficult for the much larger adult males to follow. This 'treeing' tactic by females that were guarded by males nos. 14 and 7 was observed on a total of 69 occasions. Most females resorted to this tactic only occasionally, but in three cases (females nos. 17B: N = 12; 17A: N = 20; 16: N = 15) treeing occurred on most days. These females accounted for 68% of all observations of this behaviour, and they occupied positions 11, 12, and 13 in the female rank order. As low-ranking females, they were subject to considerable harassment from other group members during periods of mate-guarding; these factors may have accounted for their greater tendency to seek temporary refuge in the trees. Males tended not to pursue females that behaved in this way; in 67% of cases males remained on the ground and continued to monitor the female's position from a distance; they resumed following only after the female had returned to ground level.

One circumstance in which males may be more likely to follow females into the trees, in order to continue mate-guarding them, is at dusk. All the animals tended to move into their sleeping sites during the late afternoon, by about 17.00 h, which was roughly an hour before sunset. During this time, adult males continued to follow the female that they had been mate-guarding during the day. Behavioural observations of mandrills at their sleeping sites were very difficult to make, but in several cases (female nos. 6, 12D and 17) I was able to verify that the alpha male slept very close to his female partner. Continued guarding at the sleeping site is likely to be the rule, rather than the exception. The mandrill group became active soon after sunrise, and matings sometimes occurred before the monkeys had descended to the ground. Maintaining proximity to a female during the night may thus be an important extension of mate-guarding, given that other males might attempt to mate with her as soon as the animals awakened in the morning. A detailed study of the behaviour of mandrills at dusk and dawn, at their sleeping sites,

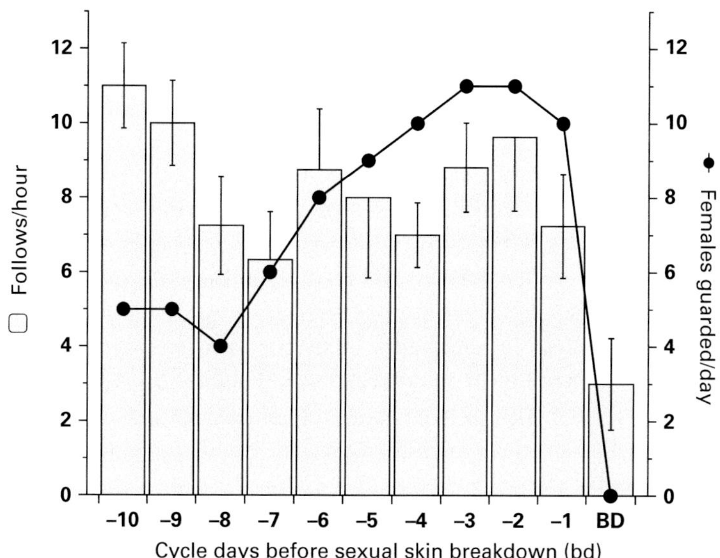

Figure 8.8. Mate-guarding and following behaviour by an alpha male mandrill (no. 14) on each of the last 10 days of the follicular phase (and sexual skin breakdown day). Data have been summed for all adult females in the group. (Author's data.)

would be most worthwhile. It would be difficult to undertake, but not impossible to achieve, by observing semi-free ranging groups.

The frequency with which the alpha male followed any particular female during mate-guarding remained fairly stable, ranging from 7–11 times per hour on days –10 to –1 before sexual skin breakdown (Figure 8.8). However, as noted earlier, he was more likely to guard a female during the putative fertile period of her cycle, so that most of the group's females were guarded during the five to six days immediately preceding sexual skin breakdown. As soon as the sexual skin swelling began to detumesce, however, the male's guarding and following usually ceased (Figure 8.8).

As discussed previously, the precise timing of ovulation in the mandrill is unknown, but studies of baboons have established that a significant number of ovulations (approximately 27%) occur on the day of sexual skin breakdown (BD). Most females ovulate during the five days that precede BD day, however. If this should also prove to be the case in the mandrill, then it raises the question of why males do not continue to guard females on the BD day. The findings I have described here, as regards the propensity of dominant males to mate-guard when females' swellings are at their maximum, are not idiosyncratic. They have been amply confirmed by subsequent studies, involving larger numbers of subjects (Setchell *et al.*, 2005b). The explanation probably lies in understanding how sexual selection has favoured the development of acute visual attention, in male mandrills, to the minute changes in sexual skin morphology that occur during the final days of maximum swelling. The swelling displays tiny increases in its turgescence, and its surface becomes shiny, while the pale area of skin surrounding the vulva is especially prominent during this time (Plate 12). If most ovulations occur under these conditions, then any male that is attuned to these subtle changes of the female swelling, and finds them to be additionally attractive, will be more likely to mate-guard, and hence more likely to sire offspring. However, changes during the breakdown (BD) day abruptly reverse these attractive cues, so that the swelling appears less shiny to the human eye as it loses turgidity. The rapid decrease in visual attractiveness of the swelling is probably responsible for the sudden cessation of mate-guarding. No decreases in female sexual receptivity or proceptivity were recorded during sexual skin breakdown. Some change in a chemical (olfactory) cue might be involved, but I think this much less likely to bring about such rapid changes in behaviour than alterations in the visual properties of the sexual skin. Interestingly, four females (nos. 17, 17A, 17B and 6) showed temporary loss of full turgidity of the sexual skin during the maximum swelling phase, but well before the fertile period. In the three cases where mate-guarding was in progress, the alpha male abruptly stopped following and guarding the females, until their swellings had recovered. He then resumed guarding in the usual way, until the day of sexual skin breakdown (Figure 8.9).

Although male mandrills that cease to mate-guard when sexual skin breakdown begins might risk 'missing' ovulations that occur on BD days, this is presumably offset by their ability to switch attention to more attractive females that have entered, or are close to reaching, the most fertile period of the cycle. Extensive overlap occurred between females in the timing of maximum swelling and likely fertile periods (Figure 8.10). As an example, the specific case of sexual interactions involving female no. 17 and her daughter (no.17B) is shown in Figure 8.11. Here the alpha male succeeded in guarding both females sequentially during their likely fertile periods. Lower-ranking males mounted only sporadically, and earlier in the follicular phase.

Considerable switching between female partners thus occurred among male mandrills; lower-ranking males took the opportunity to mate with females that were left unguarded, as well as harassing the alpha male. This naturally leads to a further consideration of the alternative mating tactics displayed by mandrills, and the outcome of such tactics in terms of reproductive success.

Alternative mating tactics and male reproductive success

Lower-ranking males often followed females that were being mate-guarded, and attempted to mount when the dominant male's attention was distracted. This occurred, for example, when infants squealed, and the alpha male broke off guarding in order to investigate. Once inside the forest, it was often not possible for him to maintain visual contact with the female he was currently guarding. Not infrequently, he also broke off guarding to threaten and chase away other males.

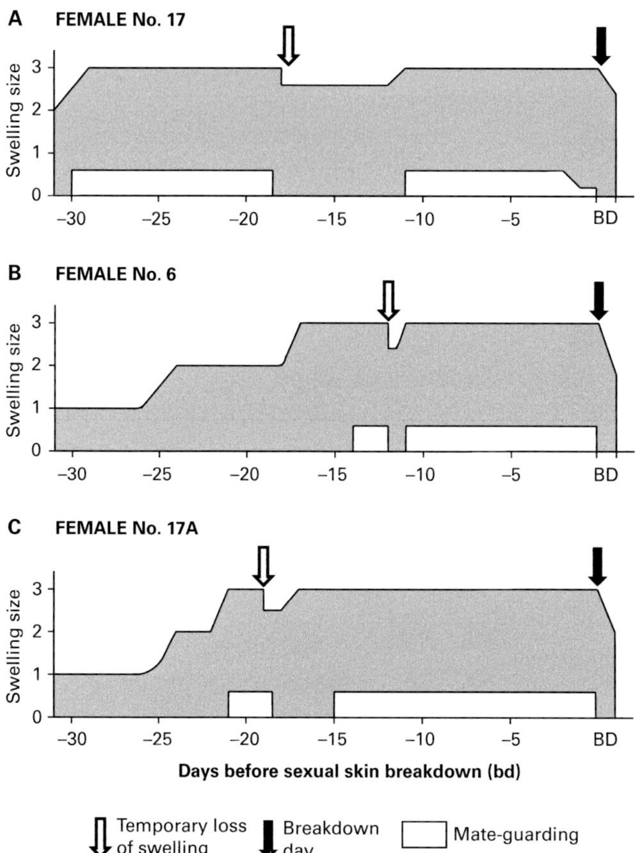

Figure 8.9. Examples of how 'interrupted' follicular phases and slight loss of swelling during the cycles of female mandrills may result in temporary cessation of mate-guarding by males. (Author's data.)

Episodes of non-contact aggression, involving facial threats, lunges and chases between the alpha male and other males occurred, on average, once every twenty minutes on mate-guarding days (range, 0.6–7.3 episodes/h; mean, 3/h, for all mate-guarding periods involving 13 females). Some of these aggressive episodes lasted for five minutes or more.

Twelve males were threatened and chased by the alpha male on mate-guarding days, three adults, three adolescent males (aged 5–7 y) and six juveniles (aged 2.5–4.25 y: Figure 8.12). Juvenile males persistently followed females and investigated their sexual swellings; 27% of the dominant male's aggressive responses were directed at these young individuals. However, the majority of his threats and chases involved just two resident five-year-old males in the group (nos. 2B and 5B); together these adolescent males accounted for 45% of the 312 aggressive episodes involving the alpha male on mate-guarding days. Avoidance of such

Figure 8.10. Timing of maximal sexual skin swelling during the annual mating season of 14 adult female mandrills, living as members of semi-free ranging Group 1, in Gabon. Considerable overlap occurs between females, so that males in the group often have to partition their sexual activity between a number of potential partners. (From Dixson, 1998a).

aggressive competition during the mating season may indeed be one factor influencing the emigration of pubescent male mandrills from their natal groups. Male mandrills began to peripheralize and emigrate from their groups when they reached six or seven years of age (Setchell and Dixson, 2002). One of the three adolescent males (no. 6A, aged 7 y) in Group 1 had already begun this process; he was frequently seen alone, and he interacted only sporadically with the dominant male (5% of aggressive episodes). Emigration and immigration by pubescent and sub-adult male mandrills will be discussed in more detail in the final section of this chapter.

The time and effort that a fatted male mandrill must invest in mate-guarding a female necessarily diminishes his ability to control access to other females for mating purposes. A secondary tactic, sometimes deployed by the alpha male during mate-guarding was to maintain visual contact with the guarded female, while attempting to copulate with a second partner. This was sometimes possible if the guarded female remained in an open, grassy area of the enclosure or when she climbed into the trees. Males that were unable to mate-guard competed for

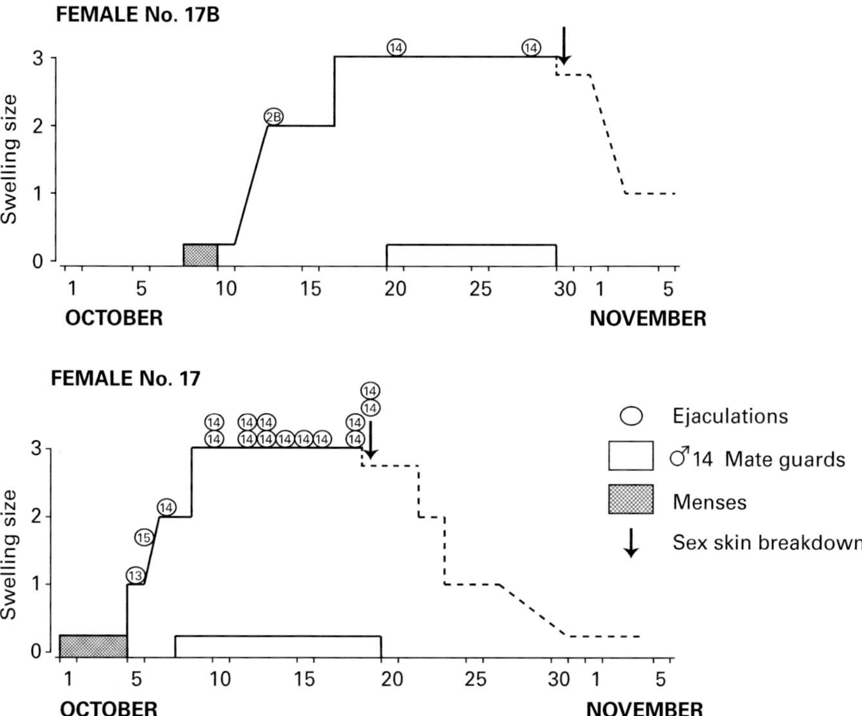

Figure 8.11. Partitioning of mate-guarding between two female mandrills (nos. 17 and 17B) by an alpha male mandrill (no.14) allows him priority of access to both partners during their likely fertile periods. Other males (nos. 2B, 13 and 15) mate opportunistically, earlier in the follicular phase. (Author's data.)

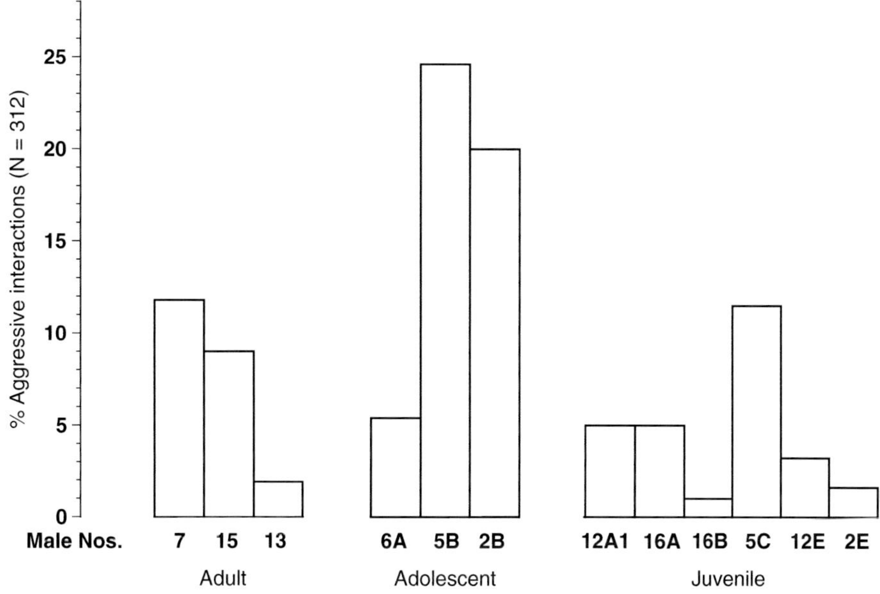

Figure 8.12. Aggressive interactions between the alpha male and other males during mate-guarding episodes. Note the high percentage of aggression involving the group's two adolescent males (nos. 2B and 5B). (Author's data.)

access to unguarded females, or engaged in covert matings. The great majority of these copulatory attempts occurred outside the fertile periods of the females concerned.

Female mandrills in this group were not necessarily the passive objects of the males' competitive mating tactics. Females influenced the nature of inter-male competition in subtle ways, and they were sometimes selectively proceptive to certain males, while avoiding others or refusing their mount attempts. A female that was followed and mate-guarded by a high-ranking male continued to move freely through the environment; she sometimes approached other males, and caused inter-male conflicts in the process (an example of such behaviour is shown in Plates 21–23). Avoidance or acceptance of a male's attempts to mate-guard also varied, depending upon the identity of the male concerned. Table 8.2 shows the very different reactions exhibited by two females (nos. 16 and 17B) towards the group's top-ranking fatted males. Female no. 16 was followed frequently by the alpha male no.14, but she consistently avoided him (sometimes taking to the trees to escape his proximity), and refused most of his mount attempts. His guarding and mating success was much lower than that of male no.7. Despite being subordinate to male no. 14, male no. 7 was much more successful at establishing a relationship with this female; she accepted his proximity, and mated more frequently with him. By contrast, female no. 17B showed a marked sexual preference for the alpha male, was guarded exclusively by him, and frequently invited him to mate. Although male no. 7 had persistently attempted to mate-guard this female (e.g. during the period when male no. 14 was occupied in guarding her mother (Figure 8.11), she avoided him and rarely tolerated his proximity (Table 8.2).

Examples such as these encourage the view that, despite the great size and social dominance displayed by males, female mandrills sometimes exhibit a limited degree of mate choice. It is important to note, however, that such examples in the groups at the CIRMF were infrequent and that, for the most part, male dominance rank determined priority of access to females. Males, not females, initiated and maintained proximity during mate-guarding episodes, and 90%–100% of mounts were likewise initiated by males.

The reproductive success of male mandrills, in terms of numbers of infants sired during the 1990 mating season and born in 1991, is shown in Figure 8.13A, which also summarizes the sexual interactions that were observed with individual females during their fertile periods. Data for conception cycles are shown in the upper part of this figure; non-conception cycles are represented in the lower part. The data refer, in each case, to the last six days of maximum sexual skin swelling and the day of sexual skin breakdown (BD day). The alpha fatted male (no. 14) succeeded in guarding 11 of the 12 females that are known to have conceived. The beta fatted male (no. 7) guarded three females, and only on days when priority of access was not monopolized by the alpha male. Although male no.7 guarded a total of nine females, most of this behaviour occurred during days that fell outside of the fertile period, and are thus not included in Figure 8.13.

Table 8.2. Effects of female choice upon the mate-guarding and mating success of two high-ranking fatted male mandrills

Example A. Female no. 16 exhibits a preference for the second-ranking male, no.7							
	Male's behaviour				Female's behaviour		
Male no.	Follow	Guard	Mount	Ejaculate	Avoid	Refuse	Present
14	3.84	5	0.66	0.22	1.64	58.3	0.16
7	2.85	7	0.77	0.44	0.99	28.6	0.33
Example B. Female no. 17B exhibits a preference for the alpha male, no. 14							
14	2.96	10	0.54	0.11	0	6.7	1.73
7	1.26	3	0.04	0	1.3	100	0.04

Data: Follows, mounts, ejaculations, avoids, presents = frequencies per hour; mate-guarding = number of days; female refusals of male's mounts and attempts =%. (Author's data.)

Between them, these two males sired all the offspring born subsequently, in 1991 (male no. 14, N = 9; male no. 7, N = 3). In five cases (female nos. 2C, 2D, 12, 17A and 16), more than one male had copulated with the female during her presumptive fertile period (this included opportunistic matings by non-fatted male no. 13 and adolescent male no. 5B). Three points may be made on the basis of these results. Firstly, no matter how careful behavioural sampling may be, it represents only an incomplete record of sexual interactions in such a mandrill group. Note, for example, although only male no. 14 was observed guarding and mating with female nos. 5 and 6, yet male no. 7 sired their offspring. Conversely, in the case of female no. 12C, male no. 14 was the father of her infant, yet he was not seen mating with her. Secondly, it follows from these caveats that multiple partner matings by female mandrills during the fertile period are probably much more frequent than indicated by Figure 8.13. Thus, the potential for sperm competition to occur is likely to be pronounced in this species. Finally, it is important to note that male no. 7 was the original alpha male in Group 1 (Feistner, 1989). He was deposed by male no. 14 prior to the mating season in 1989, after severe fighting had occurred between these two fatted males. As can be seen in Figure 8.14, over the course of the next two years, male no.14 gradually replaced no. 7 as the group's most reproductively successful male. Despite his loss of rank, however, male no. 7 continued to father some offspring (67% of those born in 1990, and 25% in 1991). A previously dominant male still retains the same genotype as during his period of tenure, and may still be preferred as a mating partner by at least some females. The continued mating success of male no. 7 may have, in part, been due to this fact, and provides another possible example of 'female choice'. While not denying that female choice occurs in mandrills, however, I do not find it to be exceptional or strongly expressed by comparison with many other primate species. The male mandrill's colourful secondary sexual adornments are exceptional, however. This naturally

A Conception cycles

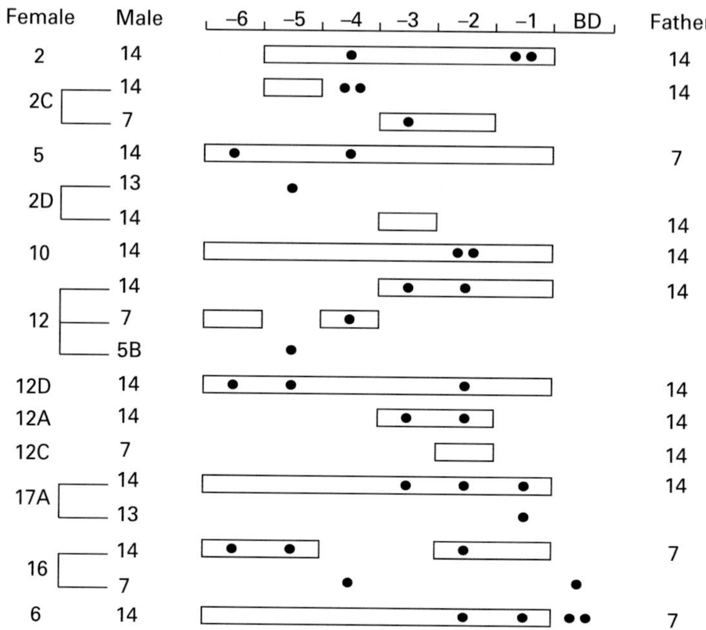

B Non-conception cycles

DAYS BEFORE SEX SKIN BREAKDOWN (BD)

	Female	Male	–6, –5, –4, –3, –2, –1, BD
Cycle 1	10	14	
Cycle 2	10	13	
	12D	14	
Cycle 1	17	14	
Cycle 2	17	14	
Cycle 1	17B	14	
Cycle 2	17B	14	

Figure 8.13. Mating behaviour during the presumptive fertile phases of the following. **A:** Conception cycles. **B:** Non-conception cycles, and reproductive success in male mandrills. Individual females are identified in the extreme left-hand column and those males that mated with them are listed in the adjacent column. Sexual behaviour was scored during the seven days leading up to and including the day of sexual skin breakdown (BD). Male mate-guarding of females is indicated by the open boxes, and occurrences of ejaculatory mounts by closed circles.The sire of each resulting offspring (as determined by DNA fingerprinting) is given in the right hand column for conception cycles. (Redrawn from Dixson *et al.*, 1993, with inclusion of the author's previously unpublished data.)

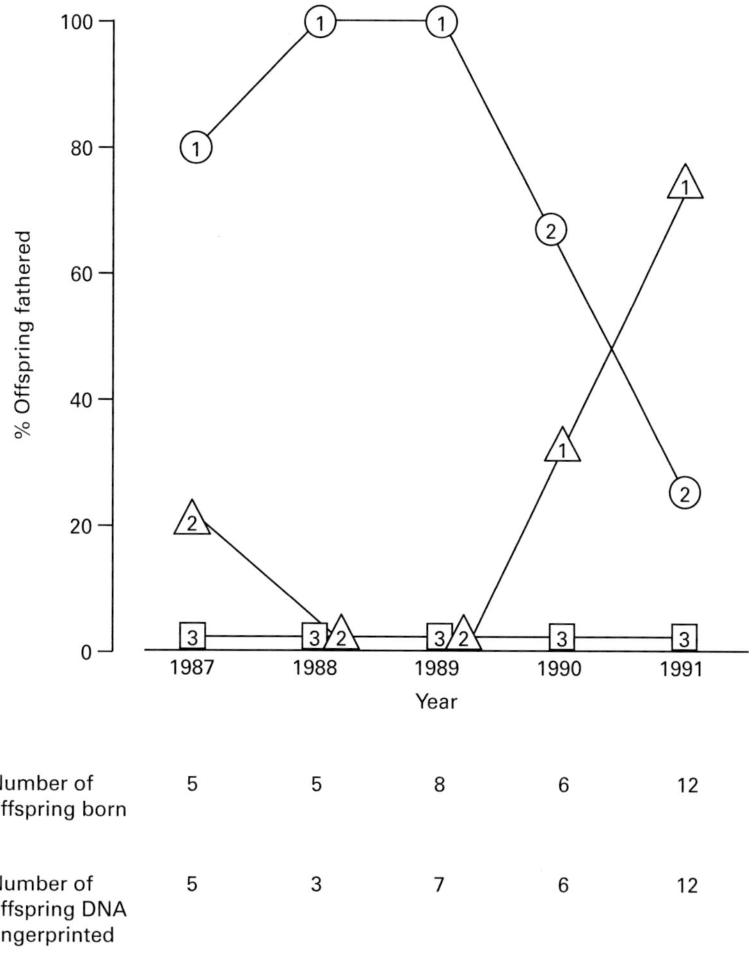

Figure 8.14. Association between male rank and paternity for three fatted adult male mandrills (nos. 7, 14 and 15) showing effects of rank changes upon percentages of infants sired over a five-year period. Each male is represented by a separate symbol (male no. 7, ○; male no. 14, △; male no. 15, □), and his rank (1, 2 or 3) is written inside the symbol. Note that male no. 15 remains at rank no. 3, and sires no offspring during the five-year period. Male no. 7 commences at rank no. 1, and then falls to the no. 2 position; his fall in rank is associated with a gradual decrease in reproductive success over the next two years. (Redrawn from Dixson *et al.*, 1993.)

leads me to question the assumption that sexual selection via female choice can account for the evolution of the male's adornments.

Such a pronounced reproductive skew, in favour of just two males in this group over the course of five years, may seem surprising, but it is necessary to keep in mind that Group 1 contained only 57 mandrills at this time, and that these monkeys inhabited a relatively small area of forest (six hectares). In the wild, we have seen that supergroups may contain as many as 700 or 800 mandrills, having a home range of more than 90 km^2! Under these conditions there is undoubtedly much greater

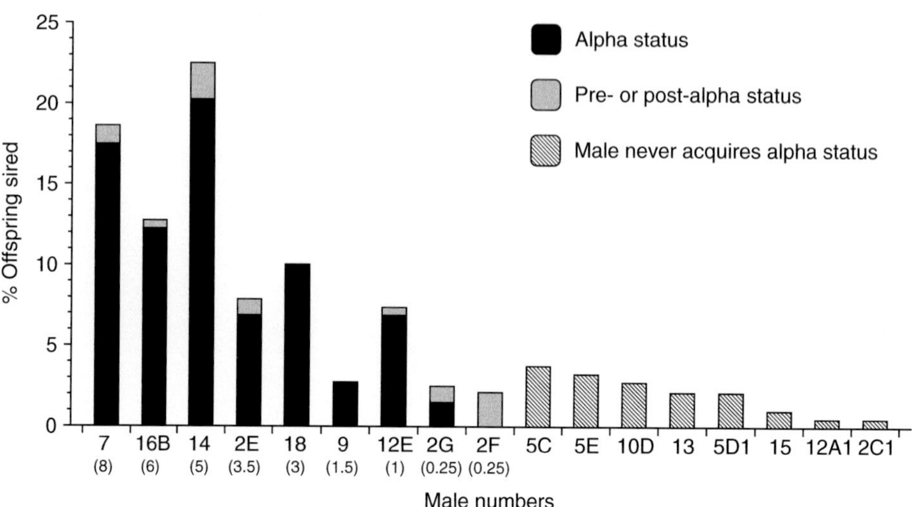

Figure 8.15. Long-term measurements of paternity and reproductive success of male mandrills in the CIRMF colony. Data for 17 adult males are shown here. Eight alpha males account for the majority of infants sired, but lower-ranking (including non-fatted) males also account for some paternities, as is discussed in the text. Numbers in parentheses indicate the length of tenure (years) achieved by each alpha male. (Redrawn from Charpentier *et al.*, 2005).

latitude for the expression of alternative mating tactics (by both sexes), so that more males, including some non-fatted and adolescent individuals, might be successful in siring offspring. Further fieldwork on mandrill supergroups is required to explore these questions.

Continued growth and development of the mandrill colony at the CIRMF has made it possible to conduct paternity studies on much larger numbers of animals over the years (Charpentier *et al.*, 2005). These studies have shown that, although eight dominant (i.e. alpha) males in Groups 1 and 2 have sired the majority of offspring, nine other males, including non-fatted individuals of lower rank, have also fathered significant numbers of infants. Results of Charpentier *et al.*'s genetic analyses are shown in Figure 8.15. Some males that were originally non-fatted adults in Group 1 (e.g. males nos. 9 and 18) later transitioned to become fatted males, with large testes, higher levels of testosterone and colourful secondary sexual adornments. The effects of gain and loss of social rank upon the secondary sexual traits and behaviour of male mandrills will be discussed later in this chapter.

Female rank and reproductive success

Data on selected reproductive parameters are shown in Table 8.3, and includes ranges and mean values for all females, irrespective of their social ranks. However, in the mandrill, social rank has considerable effects upon a female's reproductive success, and upon the growth and subsequent reproductive careers of her offspring. The matrilineal

Table 8.3. Reproductive parameters of female mandrills, irrespective of their social ranks (Effects of female rank are discussed in the text)

Measurement	N	Range	Mean±SEM
Age at first swelling (years)	10	2.75–4.5	3.6±0.6
Follicular phase duration (days):			
At puberty: first cycle	25	20–96	41.1±3.7
In multiparous females	82	12–40	25.2±0.8
During interrupted cycles	17	28–57	43±2.4
Cycling phase duration (days) *	—	13–500	102±14.9
Gestation period (days)	61	171–180	175±0.7
Age at first birth (years)	19	3.29–6.14	4.63±0.18
Postpartum amenorrhoea (days)	92	74–538	242±12
Inter-birth interval (days):			
If infant survives	92	—	473±17
If infant dies	11	—	346±44

*, the time elapsed between the commencement of cycles and conception. (Data sources: Setchell and Wickings, 2004a; Setchell *et al*, 2002; Wickings and Dixson, 1992a.)

system in mandrills largely determines the place that a female will occupy in the overall rank order of her group (Figure 8.16A).

Physiological advantages begin to accrue to the offspring of high-ranking (upper quartile) mothers from infancy onwards; such infants are significantly heavier than the offspring of low-ranking (lower quartile) females during pre-weaning and post-weaning developmental periods (Figure 8.16B and 8.16C). At the onset of puberty, daughters of high-ranking mothers tend to develop their first sexual skin swellings earlier than lower-ranking daughters do (median age, 3.5 years, as compared to 3.9 years in low-ranking females: Figure 8.17A). They also give birth to their first offspring 1.34 years earlier, on average (at a mean age of 3.89 years in high-rankers versus 5.23 years in low-rankers: Figure 8.17B).

Gestation length is not significantly influenced by rank in mandrills (mean, 175 days: see Table 8.3), so that an earlier age at first birth primarily reflects an earlier age at first conception in high-ranking females. On the basis of studies of the CIRMF mandrill colony, we concluded that status, mass and age appeared to interact in complex ways with a female's ability to conceive her first infant (Setchell *et al.*, 2002). Only 40% of females gave birth when they were aged between 3 and 4 years, but by 6.14 years, all females had produced at least one infant.

Once female mandrills reach adulthood, rank relationships continue to have an impact on their reproductive success. Differences in body weight are not so important in adults in this context; among females that are more than 10 years old, there is no correlation between body weight and social rank (Setchell *et al.*, 2002). Intervals between consecutive births are significantly longer in lower-ranking mothers, compared to those of high- and mid-ranking females, as can be seen in Figure 8.18. This figure includes data for surviving offspring only. Should her infant die, then a female's interbirth interval (IBI) is reduced, from a mean of 473 days to 346 days (Table 8.3). This major reduction

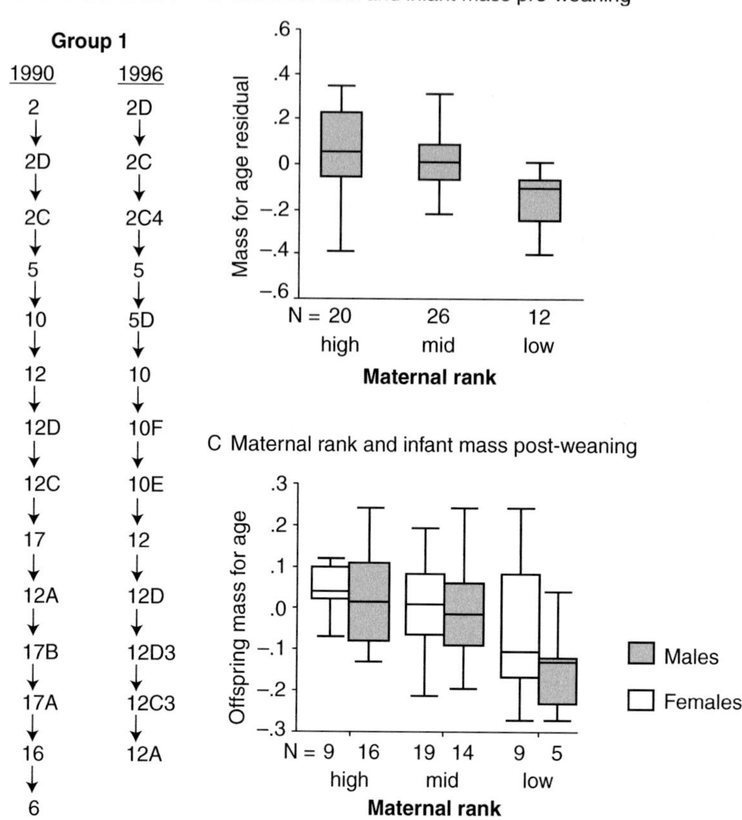

Figure 8.16. **A:** Rank orders of female mandrills in Group 1 at two stages during its development. Females with capital letters after their ID numbers are the daughters of founder females (e.g. 2C and 2D are the daughters of female no. 2; 10E and 10F are the daughters of female no. 10). In mandrill matrilines, daughters tend to inherit their mothers' ranks. **B:** Maternal rank affects infant body weight during the pre-weaning period, and also **(C)** during post-weaning development. (Redrawn from Dixson *et al.*, 2003 and Setchell *et al.*, 2001.)

in the IBI occurs because, once a suckling infant dies and its mother ceases to lactate, suppression of the hypothalamic–pituitary–ovarian axis is gradually relaxed. The female's ovarian cycle is thus more likely to resume, and she is more likely to become pregnant again during the subsequent mating season.

Given that high-ranking females have shorter interbirth intervals than mid- and low-ranking ones, what physiological mechanisms might be involved in mediating these differences? The IBI represents the sum of three sequential phases, as shown diagrammatically in Figure 8.19:

1. Lactation and the duration of postpartum amenorrhoea.
2. The cycling phase: the time from resumption of the ovarian cycle until conception.
3. The gestation period.

Plate 1. Mandrills (*Mandrillus sphinx*). An adult female with a large sexual skin swelling grooms an adult male.

Plate 2. Male and female drills (*Mandrillus leucophaeus*). Like the mandrill, the drill is highly sexually dimorphic, but the adult male's face is black, and red colouration is confined to an area of skin bordering the lower lip. Note this male's 'mane' of hair on the shoulders and back. (Photograph by courtesy of Dr Kathy Wood and the Tengwood Organization).

Plate 3. An alpha male mandrill; the world's largest monkey and the most colourfully adorned of all the mammals.

Plate 4. The rump of a 'fatted' adult male mandrill, to show its colourful sexual skin and the large size of its testes.

Plate 5. A non-fatted adult male mandrill. These males tend to be longer and leaner in overall appearance. They have less colourful adornments and smaller testes than dominant males.

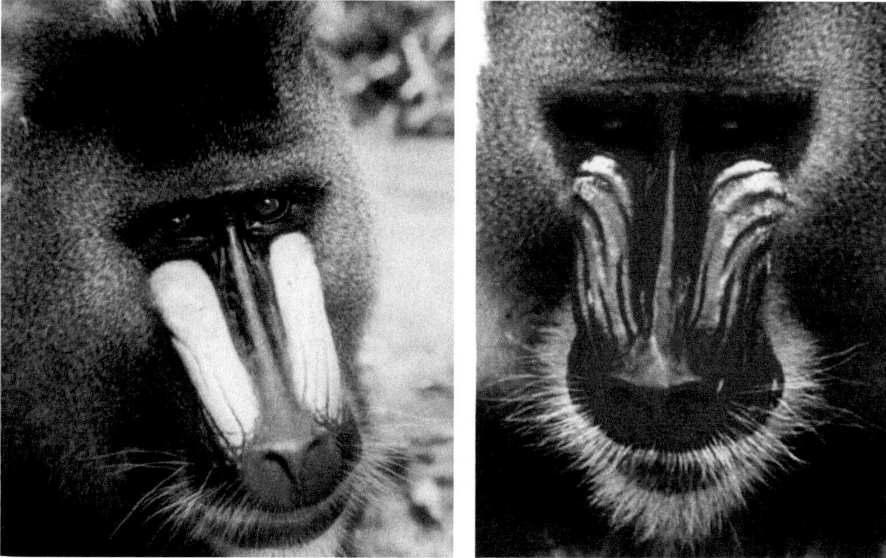

Plate 6. Changes in social rank are associated with changes in red sexual skin colouration in male mandrills. On the left: a subordinate male prior to the removal of all the higher-ranking males in the group. On the right: the same male, some weeks after removal of more dominant males. (From Dixson, 2012, after Setchell and Dixson, 2001b).

Plate 7. The former alpha male in a semi-free ranging mandrill group. This male's red facial colouration diminished in the months that followed his loss of rank.

Plate 8. A sub-adult male drill, to show the development of the face mask, and also the distinctive 'grin-face', which occurs in both *Mandrillus* species. (Photograph by courtesy of Dr Kathy Wood and the Tengwood Organization.)

Plate 9. An adult male mandrill yawns; its huge upper (maxillary) canines are longer than those of a leopard.

Plate 10. An adult female mandrill and her offspring. The face of this female is dark, and quite similar to that of a female drill.

Plate 11. Members of matrilines stay together. Here a mother suckles her infant, while grooming her adult daughter. Note that this female has much brighter facial colouration than the example shown in Plate 10.

Plate 12. The sexual skin of an adult female mandrill at maximum swelling. Note the paler area of skin surrounding the vulva.

Plate 13. The female's sexual skin as it appears during menstruation.

Plate 14. The sexual skin of a pregnant mandrill.

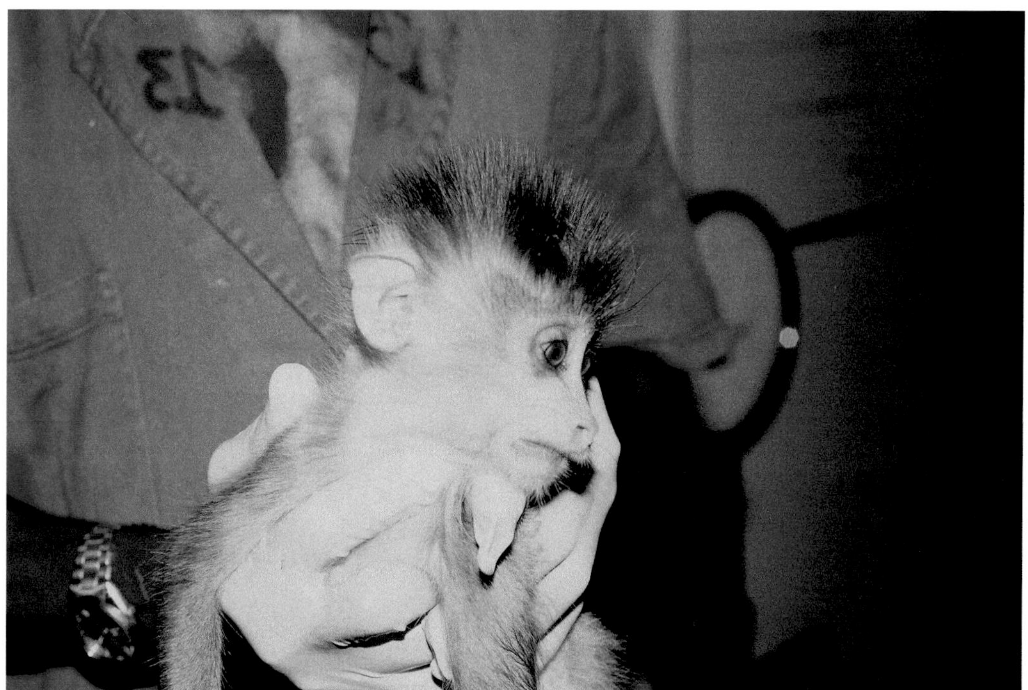

Plate 15. A three-week-old infant mandrill, showing the generally lighter pelage and black hair on the scalp.

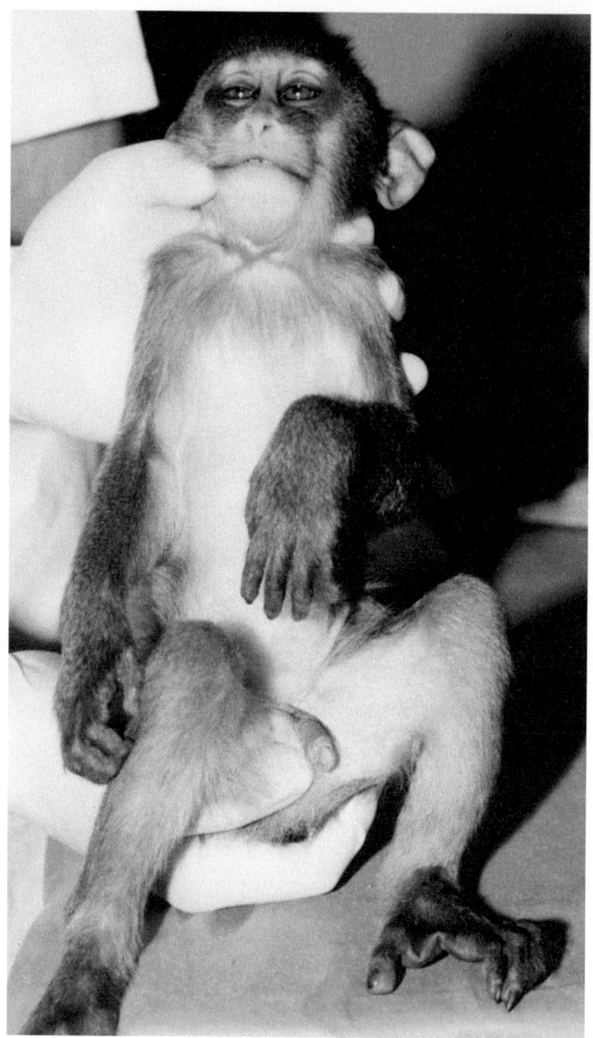

Plate 16. By approximately three months of age, the infant mandrill's pelage has developed its typical juvenile colouration, and the face is beginning to darken, especially in the paranasal area and around the eyes.

Plate 17. A pubescent male mandrill. Although capable of impregnating females, pubescent males in social groups rarely succeed in doing so, owing to high levels of inter-male competition for access to females. This male was caged alone with a female, however, and he sired her offspring when he was 5.5 years of age.

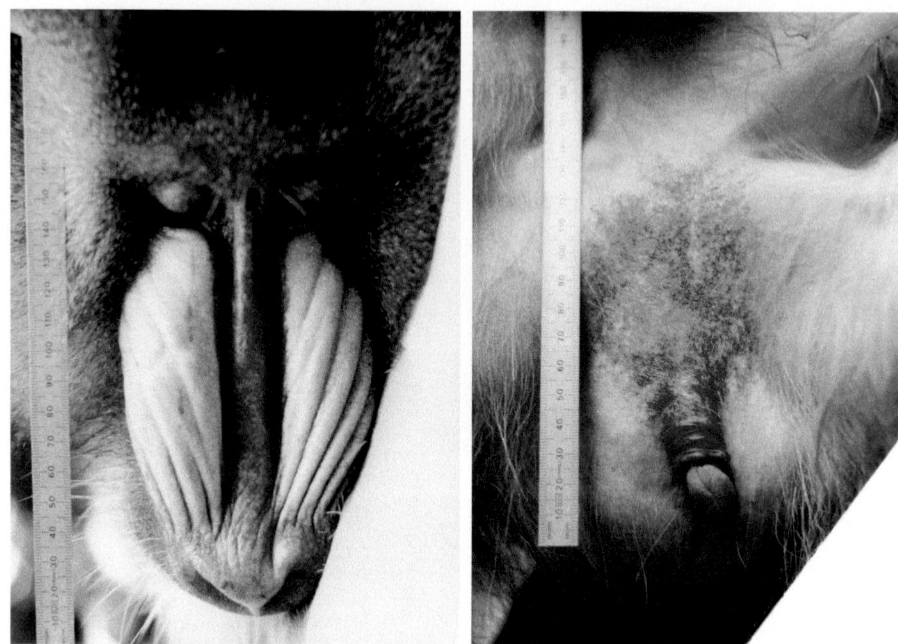

Plate 18. Any resemblance between the male mandrill's brightly coloured muzzle and his genitalia is most likely an example of convergent evolution, rather than being because of socio-sexual mimicry.

Plate 19. Copulatory posture of the mandrill. In this species, ejaculation usually occurs during a single mount.

Plate 20. Socio-sexual behaviour. A female mandrill presents to a higher-ranking female, and the latter inspects her sexual skin swelling.

Plate 21. Mate-guarding. A female with a large sexual skin swelling approaches an adolescent male. Meanwhile, the group's alpha male (on the left in the foreground) continues to follow her, as does a lower-ranking adult male (on the right).

Plate 22. Mate-guarding. The alpha male closes in, and the adolescent male turns and prepares to flee.

Plate 23. Mate-guarding. The adolescent male has fled. The alpha male continues to guard the female, while the lower-ranking adult male stays to one side and observes.

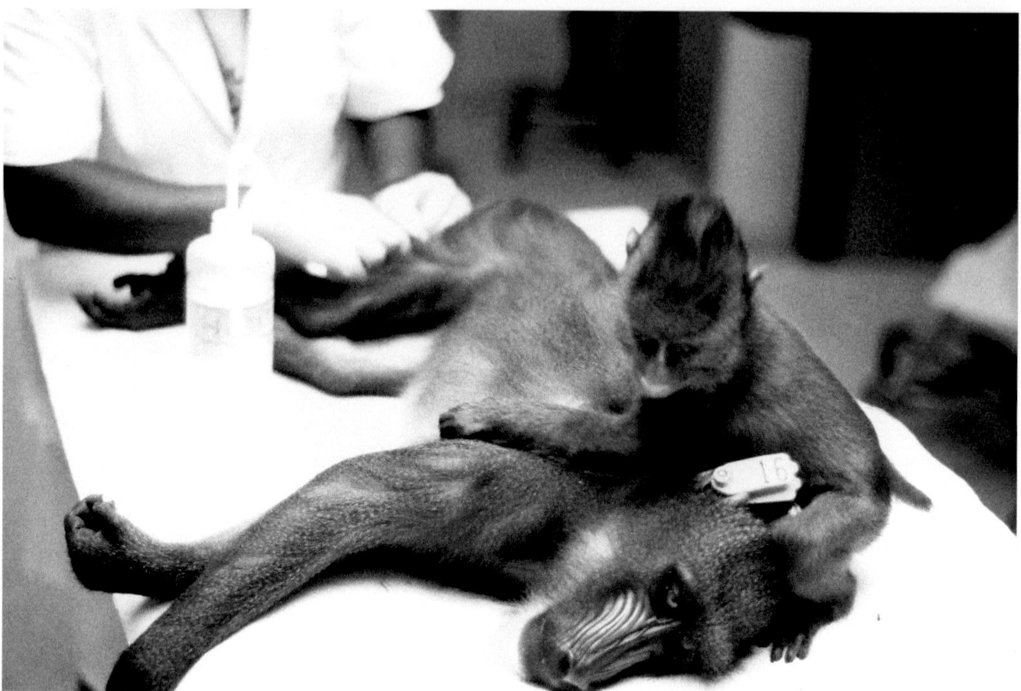

Plate 24. Veterinary staff at the International Medical Research Centre (CIRMF) in Gabon collect blood samples from a female mandrill and her offspring, as part of a study to determine paternity and to measure reproductive success in semi-free ranging mandrills.

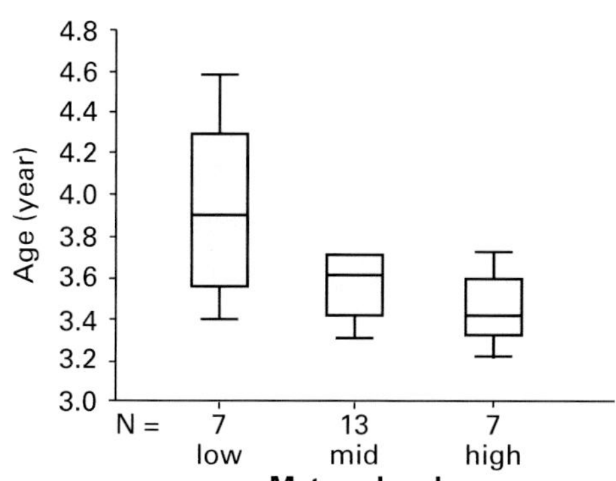

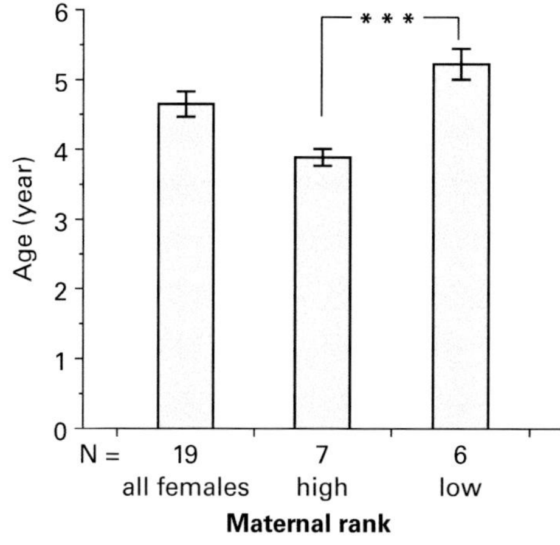

Figure 8.17. High-ranking female mandrills begin to reproduce earlier than lower-ranking females. **A:** Age at first sexual skin swelling and female rank. **B:** Age at first birth and female rank. ***, $P < 0.001$. (**A**: Redrawn from Setchell and Wickings, 2004a; **B:** Based upon data in Setchell *et al.*, 2002.)

Phase no. 3 is easily addressed, as we have seen that the mandrill's gestation period is of consistent duration and that its length is not influenced by female rank. Turning to phase no. 1, the period of postpartum amenorrhoea (PPA), by contrast, is highly variable in mandrills (range: 74–538 days, see Table 8.3). If a female fails to resume cycling during the mating season after she gives birth, then another year will elapse before she can become pregnant. In their analysis of factors affecting PPA duration, Setchell and Wickings (2004a) noted that PPA is significantly longer in primiparous females than it is after subsequent pregnancies. However, they could discern no significant correlations between PPA duration and female rank.

This leaves phase no. 2 (the cycling phase: the time taken between resumption of the female's ovarian cycle until conception) as the most likely determinant of shorter

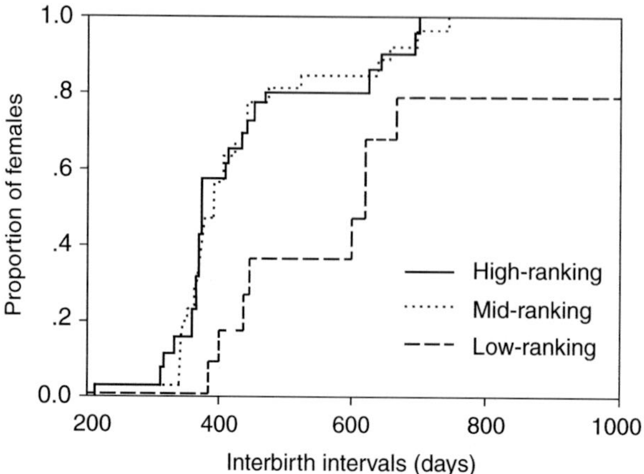

Figure 8.18. Low-ranking female mandrills have longer interbirth intervals than females of high or medium rank. (Redrawn from Setchell *et al.*, 2002).

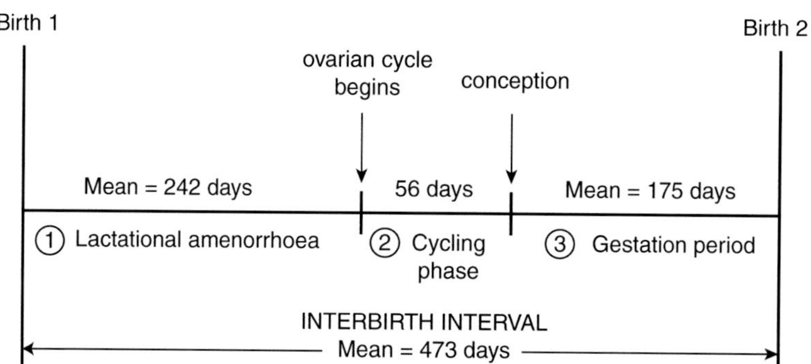

Figure 8.19. Diagram to show the three components that together constitute the interbirth interval (IBI) of the female mandrill. The duration of the IBI varies considerably between females, and possible reasons for this are discussed in the text.

interbirth intervals in high-ranking females. As can be seen in Figure 8.19, we might expect an average duration of 56 days for the cycling phase in female mandrills. In reality, their cycling phases average 102±14.9 days, and range in duration from 13 to 500 days (Setchell and Wickings, 2004a). Nulliparous females have significantly longer cycling phases than do parous females. Although no statistically significant effects of female rank upon cycling phase duration were found by Setchell and Wickings, they noted that 'the cycling phase was also markedly longer in low-ranking parous females than in high- or mid-ranking females'. It appears that lower-ranking females are less likely to conceive during the initial cycles that occur once the PPA phase ends. Setchell and Wickings observed that 'low-ranking females were disproportionately represented among females that took more than two cycles to conceive'. Indeed, some of these cycles were probably anovulatory; this is especially likely to be the case for cycles occurring at the end of the annual mating season (e.g. the non-conception cycles of female nos. 17 and 17B during October and November, as shown in Figure 8.11).

In the absence of any studies of hormonal changes during the mandrill's menstrual cycle, it is impossible to confirm how often females might cycle but fail to ovulate. However, the propensity for some females, and especially low-ranking individuals, to have multiple cycles prior to conception or to fail to conceive late in the mating season indicates that some cycles are anovulatory. It is also intriguing that the follicular phase of the cycle can be very prolonged and irregular in mandrills. During puberty, for example, the time taken from initial sexual skin swelling until the day of sexual skin breakdown ranges from 20 to 96 days, with a mean of 41 days, as compared to 25 days in multiparous females. In some cases, the swelling shows temporary periods of slight deflation during the follicular phase (see Figure 8.9 for examples of this phenomenon). These interrupted follicular phases are also longer than normal (43 days on average: see Table 8.3). Compared to the follicular phase of the menstrual cycle, the luteal phase is much more consistent in duration (averaging 15 days). If a female conceives, then her swelling detumesces in only seven days (Setchell and Wickings, 2004a).

Is the length of the follicular phase of the menstrual cycle influenced by female social rank in the mandrill? Social environment is known to affect the length of the follicular (swelling) phase of the menstrual cycle in captive baboons (Rowell, 1970). Thus, in a captive group studied by Thelma Rowell, lengthening of the follicular phase occurred as a result of being attacked, and when females were removed from the group. In the mandrill, one might posit, for example, that low-ranking females exhibit extended follicular phases owing to 'stressful' agonistic relationships with other females, or nutritional constraints resulting from feeding competition. 'Clumping' of food resources, such as occurs in the provisioned mandrill groups at CIRMF, might be expected to increase rank-related competition for high-quality food items. There is no doubt that members of high-ranking matrilines had priority of access to the small area where extra fruits and vegetables were provided each day. However, no simple correlation between female rank and follicular phase duration has yet emerged, either from my own studies or from subsequent work involving much larger numbers of females (Figure 8.20). It is interesting that follicular phases regularly exceed 40 days in female mandrills; nine of the 29 (31%) cycles represented in Figure 8.20A fall into this category. After a lengthy period of postpartum amenorrhoea, when females are still carrying their infants for substantial periods during the long dry season, it may be that the hypothalamic–pituitary–ovarian axis struggles to initiate and maintain follicular development. First cycles are longer than subsequent ones, therefore, and although 60% of conceptions occur on the first cycle (87% of females have conceived after two cycles), low-ranking females are more likely to require additional cycles and they may fail to conceive.

Higher-ranking female mandrills thus achieve greater reproductive success. Their daughters tend to be heavier, as infants, than those of low-ranking mothers. Age at onset of puberty and at first conception is significantly younger, and the interbirth intervals are shorter in high-ranking females, for the reasons discussed earlier. Thus, although the

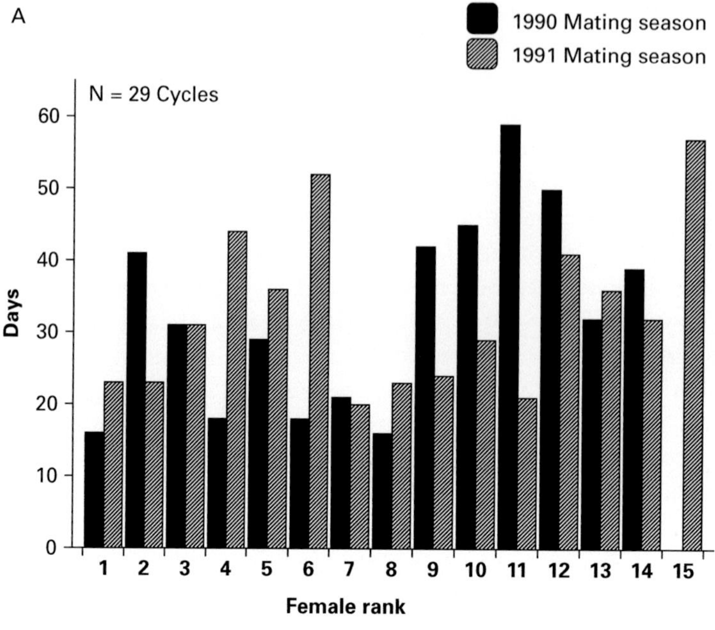

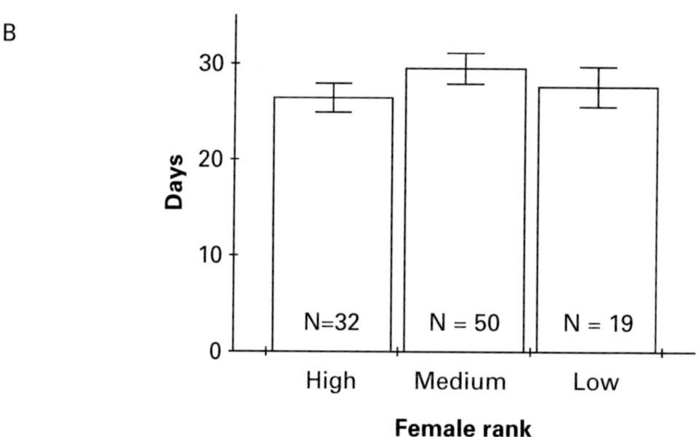

Figure 8.20. Duration of the follicular phase of the menstrual cycle in relation to social rank in female mandrills. **A:** Author's data for females in Group 1, during two consecutive years. **B:** Analysis of a larger data set confirms that there is no statistically significant relationship between female rank and length of the follicular phase. Numbers of cycles analysed are at the foot of each bar. (Based upon data in Setchell and Wickings, 2004a).

average annual reproductive output of 46 sexually mature female mandrills in the CIRMF colony was calculated to be 0.8 infants per year (maximum = one infant per year), high-ranking females had, on average, 0.26 more offspring per year than did low-ranking females (Setchell *et al.*, 2005a). Infant mortality was very low in this colony, and was not significantly affected by female rank (Setchell *et al.*, 2002). What might occur in wild mandrill groups remains unknown, but it is likely that infant mortality rates are a great deal higher in animals living under natural conditions.

Reproductive careers across the lifespan

Female mandrills begin to reproduce much earlier than males do, and they may continue to have infants until they are more than 20 years old. By contrast, very few males in the CIRMF colony have survived beyond 20 years of age. Figure 8.21 shows how reproductive output changes across the lifespan in the two sexes. First births occur, on average, at 4.63 years (Table 8.3) and, once females reach adulthood, their reproductive output remains fairly stable (a maximum of one infant per year, as discussed previously).

Across their known reproductive lifespan, females may produce as many as 17 offspring (mean±SEM: 4.7±0.6 for the 51 animals represented in Figure 8.21). Older founder females (i.e. those females present from the formation of Group 1 in 1983) achieved greater reproductive success (mean 13.0±1.1 offspring). Some founders were still alive and producing infants when Setchell *et al.* (2005a) conducted their analysis. Of course, longevity may have contributed to the greater cumulative reproductive output of these females. However, it is likely that other factors were also involved. The founder animals had been captured as infants and juveniles, and were thus subject to a good deal of stress prior to their release into the rainforested enclosure. However, from then onwards, they lived in a small group with a less competitive social environment. They received excellent veterinary care, as well as nutritional benefits not available to wild mandrills. It is likely that their higher reproductive output might have been because of their unusual history. Indeed, founders of both sexes exhibited higher growth rates and were significantly heavier as adults than mandrills that were born into the colony (Figure 8.22). I shall discuss next how environmental factors may have affected the reproductive development of the founder males in Group 1.

It will be apparent from an examination of Figure 8.21 that, across the lifespan, male mandrills have very different reproductive careers to those of females. They begin to

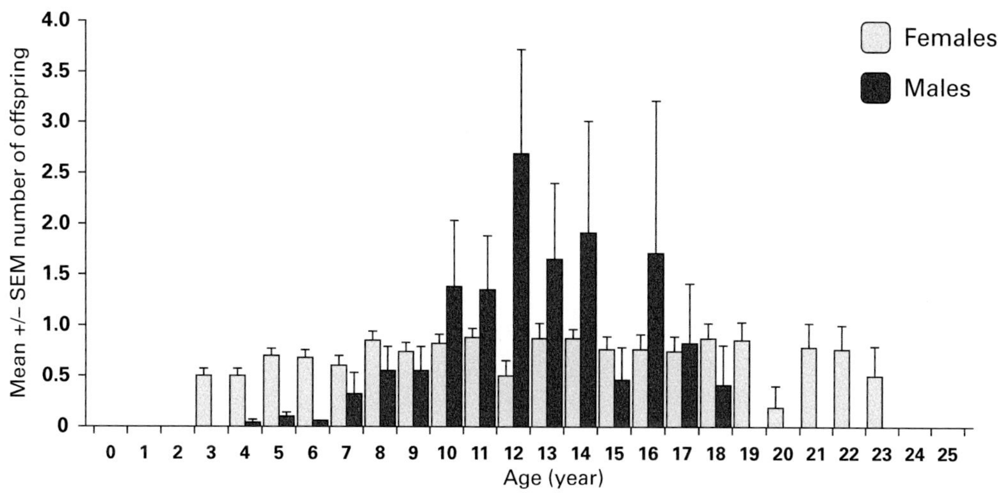

Figure 8.21. Reproductive output across the lifespan in male and female mandrills at the CIRMF, in Gabon. (Redrawn from Setchell *et al.*, 2005a).

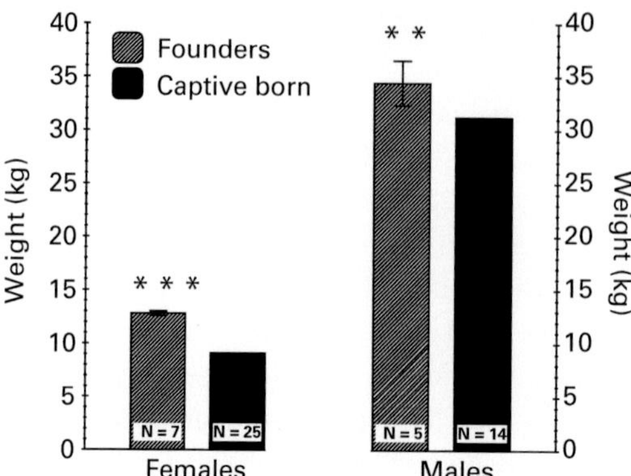

Figure 8.22. The original members (founders) of the mandrill colony at the CIRMF developed to be significantly heavier (as adults) than males and females that were born in the colony. **, P = 0.027; ***, P<0.001. Possible reasons for these differences are discussed in the text. (Based on data in Setchell *et al.*, 2001, and Wickings and Dixson, 1992b.)

reproduce later than females, and reproductive success is highly skewed in favour of adult males that succeed in attaining alpha status. Across all age groups, 53 male mandrills sired from 0 to 41 offspring (mean±SEM; 3.5±1.1 offspring). Of the 22 males that reached adulthood, only nine succeeded in becoming alpha males. These top-ranking males sired 85% of all offspring (163 of the 193 genetically resolved paternities: Setchell *et al.*, 2005a). Top rank was acquired by these males at ages ranging between 9 and 14 years. It can be seen in Figure 8.21 that male reproductive success was greatest for males aged between 10 and 16 years. Indeed, any male that survived until 18 years of age was already nearing the end of his reproductive career. Setchell *et al.* noted that alpha males held top-ranking positions for as little as one month up to six years (mean 34±9 months). The two highest-ranking fatted males discussed earlier in this chapter (male nos. 7 and 14) each held alpha rank for five to six years. Subsequent loss of rank in both these males was associated with marked loss of body condition and muting of their secondary sexual adornments, as will be discussed later.

Returning to events at puberty, we now know a good deal about how male mandrills develop physically, and when they emigrate from their natal groups. Testicular volume begins to increase at approximately 5.5 years of age, and by six years, males are beginning to develop their secondary sexual adornments. Full development of secondary sexual traits and maximum body size are not attained until an average of nine years of age, however (Setchell and Dixson, 2002). The sequence of physical changes that unfolds during puberty and adolescence in male mandrills was discussed in Chapter 3 (see Figure 3.5). As regards sexual behaviour, we have seen that pubescent males in Group 1 attempted covert matings, but were rarely successful. Long-term genetic data show that male mandrills aged four to six years only occasionally sired offspring. An exception is provided by male no. 7, a founder male and the first alpha male in Group 1. This male matured exceptionally early, and he sired his first offspring earlier than any of the males that were subsequently born in the semi-free ranging groups.

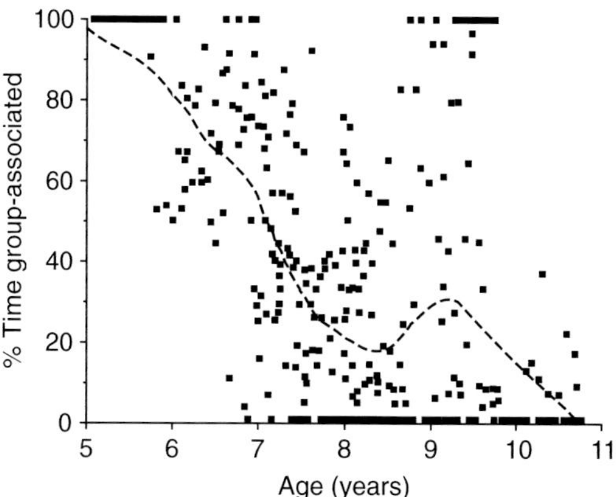

Figure 8.23. Age-related changes in group-association for males born into semi-free ranging groups at the CIRMF, in Gabon. Males typically emigrate at puberty, and rejoin groups for varying periods once they mature. (Redrawn from Setchell and Dixson, 2002.)

In the absence of competition from other males, it is likely that most pubescent male mandrills would be physiologically and behaviourally capable of reproducing by five or six years of age. An example is shown in Plate 17 of one such male that was housed with a female, separately from the main colony. He sired an offspring when he was only 5.5 years old. For males in the social groups at CIRMF, however, inter-male competition was intense and reproduction was delayed.

At six to seven years of age, most male mandrills began to move to the periphery of their groups and to emigrate. Emigration appeared to be an active process, initiated by the young males themselves, but it may also have been precipitated by their increasing conflicts with the group's mature males, especially during the annual mating season. During our initial studies of Group 1, I had naïvely assumed that low-ranking pubescent males might be more likely to emigrate, and subsequently to develop as non-fatted adults. However, further studies revealed that all young males emigrate, and that high-ranking adolescents are the first to do so. Daily records were kept of whether males were group-associated (moving and feeding as part of the group), peripheral (on the edge of the group, often 100 m from other members), or solitary (travelling and feeding alone: Setchell and Dixson, 2002; Wickings and Dixson, 1992b). Figure 8.23 shows the changes in group-association that occurred as males, from six years onwards, moved to the periphery and gradually transitioned to the solitary stage. By seven years of age, males were peripheral or solitary on 50% of days. As can be seen from the spread of data points in Figure 8.23, individual variability in this process of emigration was pronounced. At 8.5–9 years, some males began to re-enter the social group whereas, as far as we could tell, others remained alone. The final, downward curve in group association shown in Figure 8.23 is because of the presence of two older males that remained as steadfastly solitary individuals.

During their years of relative solitude, emigrant male mandrills in the rainforested enclosures at the CIRMF continued to interact sporadically with their natal groups. This was especially the case during annual mating seasons when, as described previously, some emigrant males engaged in opportunistic copulations. Non-fatted adult males associated with Group 1 (males nos. 13, 9 and 18) did not rejoin the group on a permanent basis, however. They exhibited muted red sexual skin colouration, and low testosterone levels, whereas group-associated adults (males nos. 14, 7 and 15) were brightly adorned and had large deposits of fat in the rump, tail and flanks (see Plates 3–5). Naturally, we are led to ask why male mandrills are prone to develop along such divergent physical and reproductive pathways.

Fatted and non-fatted males

> Fat and merry, lean and sad,
> Pale and pettish, red and bad.
>
> Proverbs of Alfred the Great

Given that relatively few male mandrills succeed in achieving high social rank, and that reproductive success is heavily skewed in favour of such individuals, it should not surprise us that selection has also favoured the evolution of alternative, and less physiologically costly, reproductive strategies in this species. We have seen that male mandrills do not form alliances with one another, as male chimpanzees or baboons sometimes do (Bercovitch, 1988; De Waal, 1982). The acquisition and maintenance of high rank is dependent upon individual success during inter-male competition; 'the law of battle' as Darwin (1871) called it. As an emigrant male approaches maturity, at 8–9 years of age, time is of the essence. If he is in good physical condition and can quickly signal this when entering a group during the annual mating season, he can guard individual females and secure preferential access to matings during the fertile period. However, his ability to sustain dominance over other males may last for only a few years, and his reproductive career is likely to be over well before he reaches 20 years of age. The divergence between males that are likely to become dominant, fatted adults, and those that have more muted secondary sexual adornments, smaller testes and lower testosterone levels is a gradual process that may sometimes begin well before puberty. Studies of the six founder males in Group 1 support this view. It should be recalled that these founders developed to be significantly heavier than captive born adult males (Figure 8.22). Yet, despite this advantage, 3 founders matured as group-associated fatted males, while 3 others developed as solitary, non-fatted males. Figure 8.24 charts the physical development of three age-matched pairs of these males; in each case one male became fatted and the other did not. Differences in body mass and testicular volume were already apparent by 4 years of age, with future fatted males (nos. 7, 14 and 15) developing faster than future non-fatted individuals (nos. 13, 9, and 18). Once testosterone levels began to rise at puberty, they were also consistently higher in those males destined to become fatted as adults.

In adulthood, these six males continued to show significant physical differences, as well as behavioural differences. Fatted males had larger testes and higher testosterone

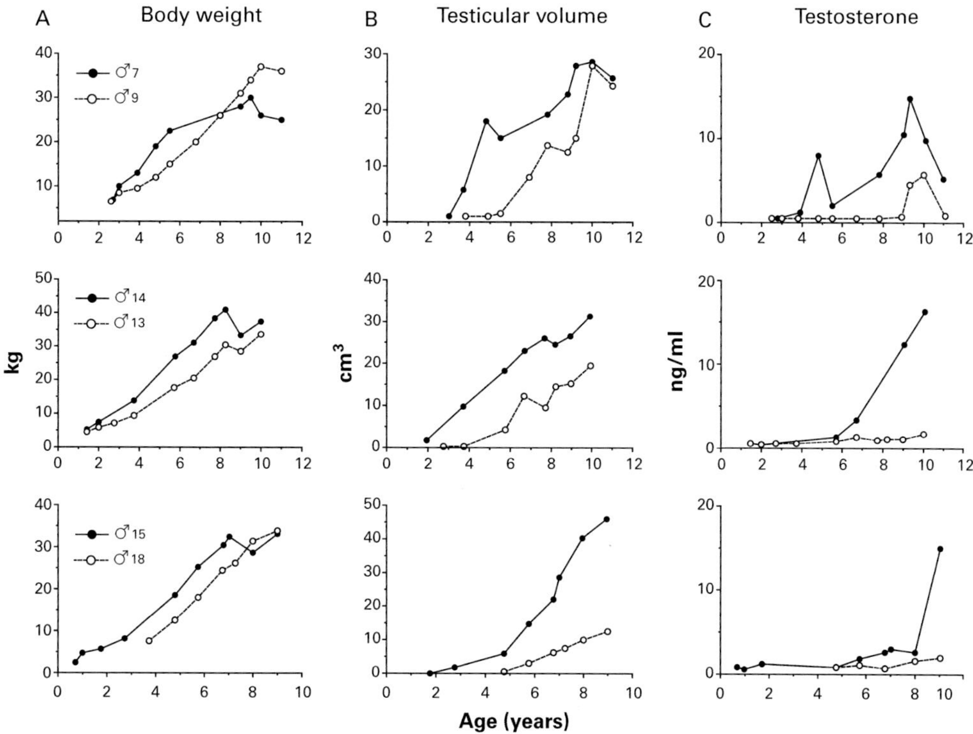

Figure 8.24. Developmental changes in (**A**) body weight, (**B**) testicular volume, and (**C**) plasma testosterone levels in three age-matched pairs of founder males in the CIRMF mandrill colony. The closed symbol of each pair denotes the male that developed as a fatted adult, and the open symbol denotes the male that became a non-fatted adult. (Redrawn from Wickings and Dixson, 1992b.)

levels; non-fatted males were slightly longer and leaner, with significantly greater crown-rump lengths. These two types of male did not differ in their body mass, however (Figure 8.25A). As well as having brighter colouration, especially where red sexual skin was concerned, some differences in skull morphology were noted, particularly in the highest ranking fatted male (no.7), as compared to non-fatted individuals. This dominant male had entered puberty first, and started to sire offspring unusually early. He also developed larger paranasal (maxillary) swellings. His bony swellings were broader dorsally, and there were deeper grooves between the longitudinal folds of the overlying blue sexual skin (male no.7 is compared to the non-fatted male no. 18 in Figure 8.25B). Early onset of puberty, and consistently higher testosterone levels in male no. 7 may have resulted in precocious development of his paranasal swellings.

In Chapter 3, the occurrence of unusually robust cranial morphologies in some captive-raised male mandrills and drills was discussed (see pages 25–27). Singleton (2012) has shown that these cranial features, which include greater dorsal expansion of the paranasal ridges, only occur in the male sex, and she suggests that they may be a result of endocrinological factors. In free-ranging mandrill groups, where

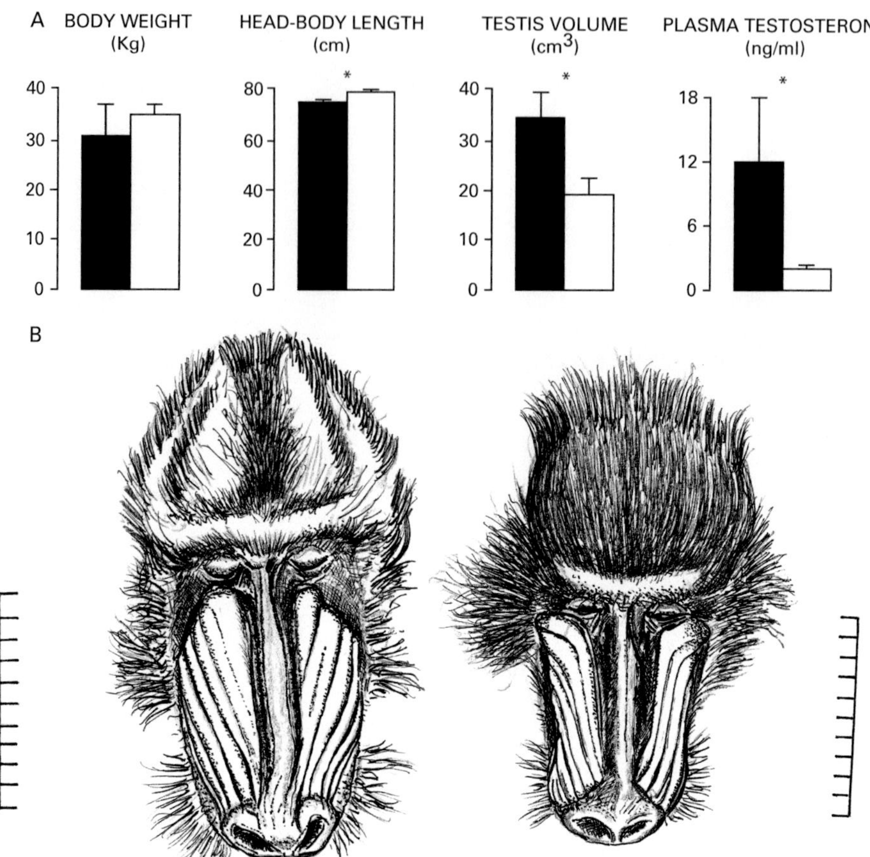

Figure 8.25. **A:** Comparisons of body weight, head-body length, volume of the left testis and plasma testosterone levels in (closed bars) three fatted adult male mandrill and (open bars) three non-fatted individuals. Data are means (±SEM); ✻, $P < 0.05$. (From Dixson, 2012). **B:** Differences in size and shape of the paranasal swellings of (left) a fatted, alpha male mandrill (no.7) and (right) a non-fatted, solitary male (no.18). Scales are 10 cm long. (Author's drawings.)

competition for food and for mating opportunities is intense, such extreme growth of the male's maxillary swellings probably does not occur, as reflected by its absence in skulls of wild mandrills that are currently held in museum collections.

Under natural conditions, intense inter-male competition results in some individuals opting for a slow rate of secondary sexual development. The longer and leaner appearance of non-fatted males may be accentuated by an extended period of skeletal growth and delayed fusion of the epiphyses. The reproductive strategy pursued by non-fatted male mandrills may have some disadvantages, as they are usually less competitive. However, a longer-term benefit may accrue to non-fatted males by lessening the risk of damaging aggressive conflicts with dominant males, by delaying investment in costly secondary sexual adornments, and facilitating opportunistic matings.

I am not suggesting that non-fatted males necessarily remain so for their lifespans, or that fatted males always remain brightly adorned. On the contrary, long term studies of the CIRMF mandrill groups have shown that the secondary sexual adornments of male mandrills present as a spectrum of developmental types, and often undergo changes in adulthood, in association with gain or loss of social rank (Setchell and Dixson, 2001b). Gain of alpha rank results in increased testicular growth and greater concentrations of circulating testosterone, reddening of the sexual skin on the face and genitalia, and heightened secretion of the sternal cutaneous gland. By contrast, deposed alpha males exhibit decreases in testicular volume, reductions in the extent of their red (but not blue) sexual skin colouration, and decreased sternal glandular activity (Figure 8.26). Deposed alpha males also lose overall body condition, and their pelage appears drab. These more gradual changes were observed, for example in male no. 7 and subsequently in male no.14, following loss of alpha rank in both cases.

A striking demonstration of the degree of physiological suppression exerted by the presence of more dominant males upon lower-ranking individuals is shown in Plate 6. In this case, removal of all the more dominant males from Group 1 resulted in the single remaining non-fatted individual (no. 18) showing supra-normal flushing of the red sexual skin, such that even the blue paranasal areas of his face became suffused with blood. This male's testosterone levels also rose markedly, his testes increased in size, and the sternal gland became fully active, as he rapidly transitioned to alpha status. Long term studies of the CIRMF groups have also shown that fattedness sometimes correlates positively with the time that males spend foraging outside the group. Indeed, solitary male no.18 had showed signs of increased fat deposition (but not colour change) long before more dominant males were removed from his group. Reduction of feeding competition may be crucial in this context. Perhaps, in the wild, males that emigrate at puberty or in adulthood once the mating season ends, are able to forage more effectively and thus to put on weight and improve their physical condition.

Measurements of rump fattedness and other morphological traits, frequencies of group association, and plasma testosterone levels for adult males in Groups 1 and 2 are shown in Table 8.4. Note that by this stage of the colony's development, the alpha males in both groups (Group 1: male no. 18; Group 2: no.9) had previously spent long periods as non-fatted, solitary individuals. Acquisition of high rank by both these males was associated with large increases in plasma testosterone, and with marked behavioural and morphological changes. Likewise, previously dominant males (such as male no. 14 in Group 2) declined physically over time. Some males that were previously group associated later switched to a solitary way of life (e.g. male no.15 in Table 8.4). Thus it is likely that, in free-ranging mandrill groups, adult males may also vary across a spectrum of traits, rather than being divided rigidly into just two 'fatted' and 'non-fatted' morphotypes.

What is the mandrill's mating system?

Studies of semi-free ranging mandrill groups have allowed us to glimpse what might occur in the wild, as regards the social organization, sexual behaviour and the mating

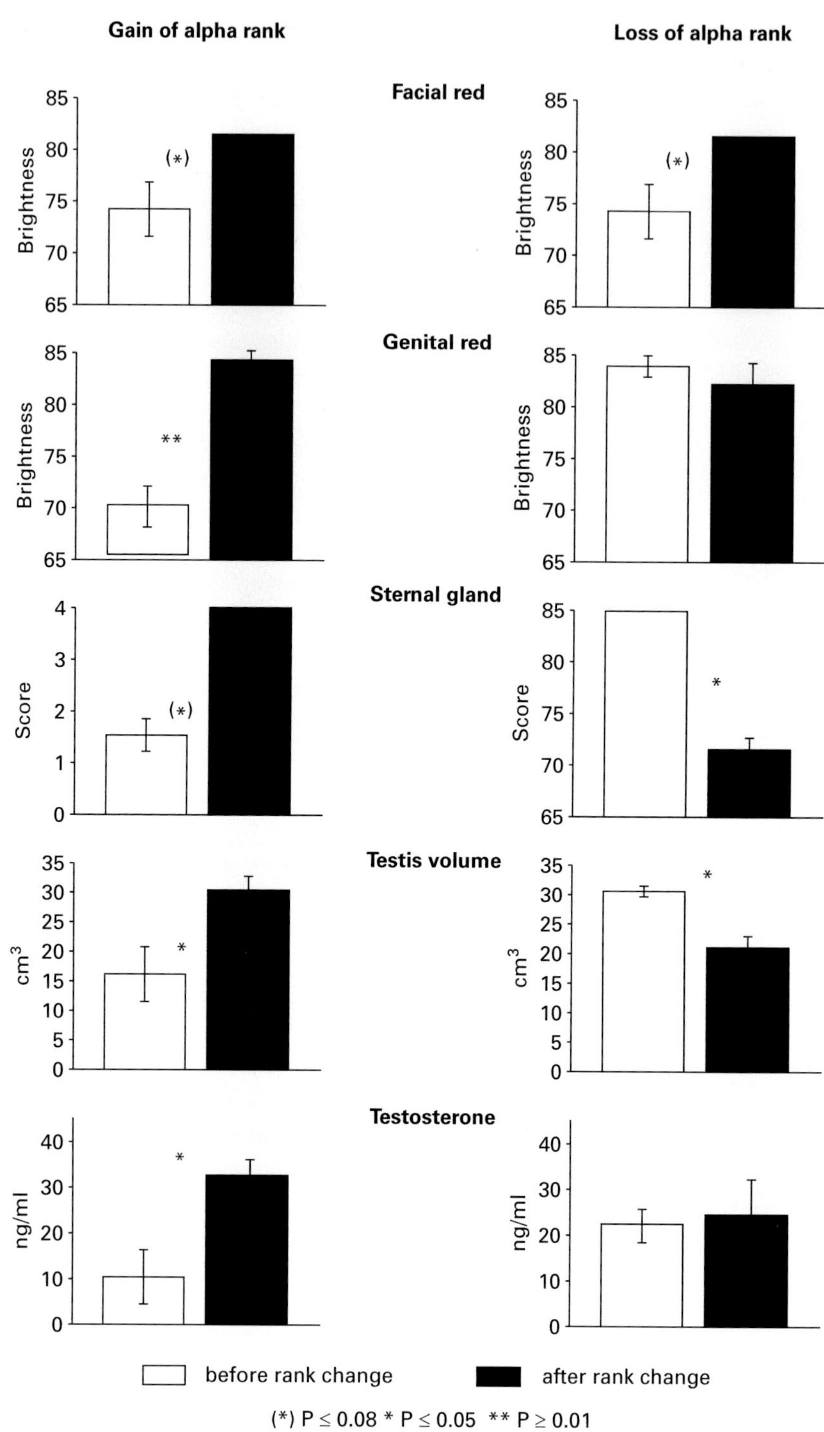

Figure 8.26. Effect of gain or loss of alpha status upon the expression of secondary sexual traits, testicular volumes and circulating testosterone in adult male mandrills. Data refer to four males that gained rank, and four males that lost alpha status. (*), $P \leq 0.08$; *, $P \leq 0.05$; **, $P \leq 0.01$. (From Dixson, 2012.)

Table 8.4. Measurements of morphological traits, circulating testosterone and group-association in ten adult male mandrills during later years in the development of Groups 1 and 2 at the CIRMF in Gabon

		GROUP 1				
	Male no.	18	2E	12A1	5C	12E
Testosterone (ng/ml)		28.9	7.9	4.2	18.1	7.9
TV (cm^3)		25.3	9.3	9.5	19.0	8.9
Body mass (kg)		30.0	33.0	32.0	34.0	28.0
Body length (cm)		70.0	75.0	66.0	72.0	69.0
Rump area (cm^2)		570	680	740	550	420
In group (% days)		100	18.0	8.0	11.0	40.0
Solitary (%days)		0	57.0	68.0	57.0	19.0
		GROUP 2				
	Male no.	9	13	16B	14	15
Testosterone (ng/ml)		29.7	4.6	10.9	22.5	1.7
TV (cm^3)		33.3	18.8	4.4	19.6	47.4
Body mass (kg)		31.0	29.0	30.0	27.0	35.0
Body length (cm)		76.0	76.0	72.0	70.0	72.0
Rump area (cm^2)		620	710	460	510	1010
In group (% days)		99	41	39	86	0
Solitary (%days)		0	34.0	30.0	6.0	90.0

TV, mean testis volume. (Data are from Setchell and Dixson, 2001a.)

system of these extraordinary animals. The ecology and behaviour of free-ranging mandrills was discussed in Chapter 4. There I reviewed data on group size and composition, derived from the fieldwork of Abernethy *et al.* (2002), on the mandrill supergroups in the Lopé National Park. Filmed records of group progessions had made it possible for Abernethy *et al.* to obtain much more accurate counts of the various age/sex classes in mandrill groups than had hitherto been possible.

These huge groups (mean size = 620) contained an average of just 19 sub-mature males (estimated to be between six and nine years old) and seven fully adult males. Indeed, the majority of males spent most of the year foraging alone and entered supergroups during the mating season months, from June through to November (see Figure 4.2). During these months, females developed their sexual skin swellings, and Abernethy *et al.* found that the number of adult and sub-mature males present in the groups on any given day was positively correlated with the number of females with 'visible sexual tumescence' (Figure 8.27). Thus, the timing of the mating period during the long dry season in the Lopé National Park was similar to that observed for the semi-free ranging mandrill groups at the CIRMF. Moreover, observations of wild groups reinforce the view that the onset of sexual activity is likely to be triggered primarily by changes in female reproductive condition. Once females begin to cycle and to develop their sexual skin swellings, then increasing numbers of males re-enter the supergroups.

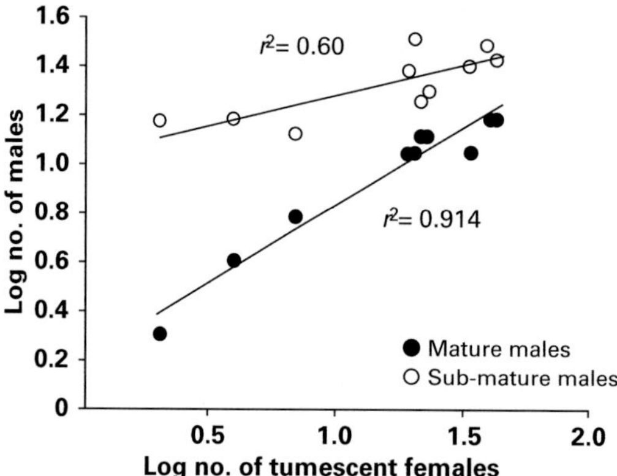

Figure 8.27. Positive correlations between numbers of mature and sub-mature male mandrills that are present in mandrill supergroups during the annual mating season, in Gabon, and the numbers of females that have developed sexual skin swellings. The Pearson moment correlation coefficients are significant, for adult ($P < 0.001$) and sub-mature males ($P < 0.01$). (Redrawn from Abernethy *et al.*, 2002.)

Abernethy *et al.* (2002) were unable to observe the sexual behaviour of free-ranging mandrills. They noted, however, that 'there is no evidence in wild mandrill hordes of baboon-like one male units' (see also Hongo, 2014). The same is true of semi-free ranging mandrill groups. Dominant males in mandrill groups housed in rainforested enclosures at the CIRMF, mate-guarded individual females at the height of sexual skin swelling, and moved on to guard a new partner once sexual skin breakdown had occurred (Dixson *et al.*, 1993). In the wild, adult males that enter supergroups during the annual mating season presumably compete for access to females that have large swellings, and especially those females that are in the peri-ovulatory phase of the menstrual cycle. Competition between males is likely to be intense; those individuals that have benefitted most from their long periods of solitary foraging, and which have fatted rumps and brightly coloured sexual skin, are most likely to succeed in mate-guarding a succession of females. Mate-guarding is an energetically costly but highly effective mating strategy involving prolonged following of each female, long bouts of two-phase grunting to advertise the position of the guarding male to others, and frequent agonistic encounters with rival males.

Males in the six- to nine-year 'sub-mature' category, as defined by Abernethy *et al.*, are more likely to engage in opportunistic mating tactics. We have seen that in the CIRMF mandrill colony, six-year-old males begin to peripheralize and to emigrate from their groups. By the time they reach nine years of age, some of these emigrant males have begun to revisit groups (Figure 8.23). In the wild, mortality is likely to be high among these inexperienced pubescent and young adult males. They may be more vulnerable to predation, and many probably fail to acquire sufficient food to transition to the fatted condition and to complete secondary sexual development. Yet, it is clear from Abernethy *et al.*'s data that significant numbers of males in the sub-mature category also enter groups during the annual mating season (Figure 8.27), and it is likely that some of them succeed in siring offspring. I say this because mandrill supergroups contain such large numbers of adult females (range, 94–288; mean 182 in Abernethy *et al.*'s study), that possibilities for alternative mating tactics, including opportunistic, covert and multiple partner matings, are likely to be much greater under such conditions. Multiple partner matings do occur in the semi-free ranging groups at the CIRMF, including matings by females during the fertile period (Figure 8.13). Adult males struggle to monopolize access to one female at a time. Under natural conditions, total numbers of adult males in supergroups are low, however, so that multiple partner matings by females are likely to be the rule, rather than the exception. This would explain the evolution of very large relative testes sizes as well as other reproductive traits in male mandrills as adaptations for sperm competition. These features again argue for the existence of a multimale–multifemale type of mating system in the mandrill, rather than for a complex multilevel social organization comprised of polygynous one-male units.

From December onwards, once the mating season ends, adult and sub-mature males leave the supergroups and forage alone. A few remain, however, and Abernethy *et al.* recorded an average of 1.05 ± 0.8 mature males in supergroups during non-mating months. We have seen that male mandrills appear to lose physical condition during the lengthy mating season. Subsequent periods of solitary foraging may allow them to recover physically, and to regain body mass before the next mating season begins. It should not be assumed that such males live an entirely isolated existence during the non-mating season months. They mark trees with their sternal glands, for example, and it remains unknown what role these scent-marks might play in social communication. Males might also remain in social contact with supergroups, and re-enter them from time to time. There is very little known about these matters. However, as was noted in Chapter 7 (Dr Kate Abernethy: personal communication), solitary males have occasionally been seen to enter supergroups during non-mating season months. Kate Abernethy has also told me that lone males do not occupy discrete ranges, rather they 'overlap with each other, the natal group, and neighbouring group's ranges. They are not territorial, and there is no evidence of competition outside the mating season.'

As yet, nothing is known about lifetime reproductive success in members of mandrill supergroups. On the basis of the studies conducted on semi-free ranging groups, we might expect males to enter their most productive years from the age of nine years onwards, to hold high rank for perhaps four to five years, and to die in their late teens or early twenties. Males that delay full physical development may opt for opportunistic mating tactics over a longer period of time. Given the huge numbers of females in supergroups, the limited visibility imposed by the rainforest environment, and the possibilities for deploying alternative mating tactics, it is likely that reproductive success is less skewed in favour of small numbers of dominant males in wild groups than is the case for captive groups.

It is clear, from long-term research conducted at the CIRMF, and from analyses of supergroups in the wild, that adult females and their immature offspring form the core of mandrill social group structure (Abernethy *et al.*, 2002; Bret *et al.*, 2013; Hongo, 2014). Mandrills have a matrilineal social organization, as do macaques, baboons and many other Old World monkeys. We have seen that a female mandrill's social rank is, in large measure, determined by that of her mother. An individual's position in her matriline has important consequences for feeding competition and reproductive potential. Even relatively well-fed females in the provisioned groups studied at the CIRMF exhibited significant effects of rank upon age at first conception, duration of interbirth intervals, growth rates of offspring and overall reproductive success. Many females, irrespective of rank, also experienced very long follicular phases during their menstrual cycles. Lower-ranking females were more likely to require multiple cycles in order to conceive, and they failed to reproduce during some years. Such effects are likely to be much greater in wild groups containing hundreds of permanent members, where competition for resources is intense and the animals must range over large distances in search of food. The 'splitting' of mandrill groups into smaller subgroups is thus almost certainly determined by ecological constraints, rather than occurring for reproductive reasons. When a group does split, we should expect this to occur along matrilineal lines, as members of each matriline tend to act in concert, moving through the forest and foraging together.

In Figure 8.28, I have attempted to incorporate the information discussed above into a diagrammatic 'model' of the mandrill's social organization and mating system. Note that once males have left their natal groups at puberty, there is no implication that they will necessarily re-enter the same group once they attain adulthood. It is likely that dispersal of solitary male mandrills results in some of them contacting and entering new groups. Indeed, this may be one important mechanism by which gene flow between mandrill supergroups occurs. These questions, like so many others, can only be answered in future by conducting detailed behavioural and genetic studies of free-ranging mandrills.

Finally, in concluding this discussion of the mandrill's mating system, I should like to emphasize again that our current knowledge of its reproductive biology derives predominantly from studies of semi-free ranging groups. Mandrill supergroups in the wild are likely to show many important differences, especially where the quantitative aspects of behaviour and reproduction are concerned. However, I predict that, in qualitative terms, wild mandrills will be found to have the same kind of mating system, patterns of sexual behaviour and social organization as discussed here, with respect to the groups at the CIRMF. The fact that the CIRMF mandrills are provided with ample food and excellent veterinary care, however, must significantly benefit their health, longevity and reproduction. In 2002, a group of 36 mandrills was translocated from their six-hectare enclosure at the CIRMF to the Lékédi Park, which is situated 100 km from Franceville. There, the animals were released into a huge enclosed area, encompassing 1750 hectares of natural rainforest. Yet, during the first year after the group was released, 33% of the monkeys died, and it was necessary to re-establish provisioning in order to prevent further losses owing to malnutrition (Peignot *et al.*, 2008). Clearly, this small semi-free ranging group, consisting entirely of captive born monkeys, lacked the necessary foraging and ranging skills required for survival under completely natural conditions. It should also be kept in mind that the colony at the CIRMF represents a limited sample of this species' genetic diversity. Mandrills at the CIRMF are the descendants of just 14 founder members. Thus, the results of genetic studies (e.g. to

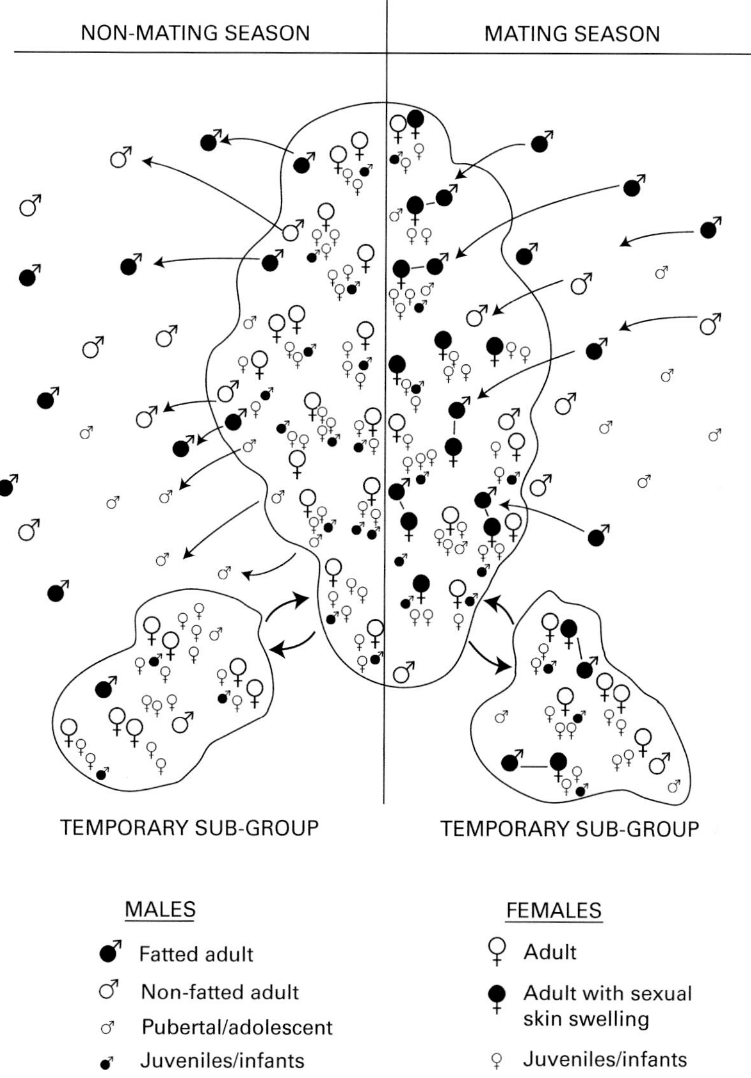

Figure 8.28. The social organization of a mandrill supergroup, including changes that occur during its annual mating season. During non-mating season months, many adult males emigrate, leaving females and offspring as the permanent core of the supergroup. This has a matrilineal social organization; younger individuals are shown closely associated with their mothers. Pubertal males also emigrate from the group and, although this is indicated as occurring in the non-mating season, it is likely that this also occurs during mating periods. The mating season involves increased immigration by adult and sub-mature males, in response to the presence of sexually attractive adult females with sexual skin swellings. Dominant males (including fully fatted adults) are posited to have a reproductive advantage, and to mate-guard individual females, whereas lower-ranking individuals (including non-fatted adults and sub-mature males) engage in opportunistic matings. The supergroup may split temporarily into subgroups and this occurs for ecological (not reproductive) reasons, as is discussed in the text.

examine the possible role of *MHC* genes in mate choice or post-copulatory selection: Setchell *et al.*, 2010a; Setchell *et al.*, 2011b; Setchell *et al.*, 2013) should also be interpreted with appropriate caution. Field research on natural populations will be required to determine whether MHC genotype affects mate choice in mandrills. Clearly, this represents an important and technically very challenging task for future research.

Part III

Evolution and sexual selection

In the last eight chapters, I have attempted to draw together as much information as possible, bearing upon the mandrill's natural history, including its distribution range, social organization, behaviour and reproductive biology. One goal of this exercise has been to provide some basis for discussing the evolution of this remarkable animal. Whenever possible, I have also included information about the drill, as this is the mandrill's closest phylogenetic relative; these two species being the only extant members of the genus *Mandrillus*. Although Charles Darwin (1871, 1876) thought that sexual selection must have played an important part in the evolution of the male mandrill's bright colouration, he had access to very little information about its behaviour or its reproductive biology. Knowledge concerning these matters during the nineteenth century was anecdotal. Indeed, although the mandrill's existence had been recognized by Conrad Gesner during the sixteenth century, it is only during the last 30–40 years that science has advanced beyond descriptions of its morphology and anatomy, to explore its behaviour, ecology, reproduction and evolutionary biology. Even now, much less is known about the drill than about the mandrill. However, enough has been learned to facilitate comparisons between the two species, and to discuss those traits that would most likely have been present in their common ancestor.

The following set of conclusions is intended to provide the reader with a brief synopsis of the evolutionary history of the genus *Mandrillus*, based upon the material presented in earlier chapters. Following this, the special topic of sexual selection, and its extreme expression in the mandrill, will be addressed in Chapter 10.

9 A brief evolutionary history of the genus *Mandrillus*

There is no fossil record of the evolution of the genus *Mandrillus*. This is not surprising, given that these monkeys originated in the tropical forests of western Central Africa, where very few primate fossils have ever been discovered. A viable alternative to direct fossil evidence may be secured by pursuing multiple, comparative lines of enquiry. For example, the application of molecular genetic techniques, and the use of 'molecular clocks', has made it possible to estimate when the genus *Mandrillus* arose and subsequently give rise to its two extant species. Likewise, zoogeographical researches have increased our understanding of how the mandrill and drill came to occupy their present distribution ranges. Cycles of climate change in the remote past have had important effects upon rainforest distribution in Africa, and these events, as well as the barriers imposed by major rivers, have affected speciation and shaped distribution ranges. Taxonomic studies place the genus *Mandrillus* firmly within the Tribe Papionini, along with the mangabeys, macaques, baboons and several other genera of the cercopithecine monkeys. It is highly instructive to consider the functional anatomy of the cercopithecines, and to compare and contrast various traits (e.g. skeletal, dental and genital) that occur in *Mandrillus* with homologous structures in other papionins and in less closely related genera. Likewise, it is now possible to compare the behavioural repertoires of Old World monkeys, and especially basic patterns that serve for visual and vocal display. This ethological approach has produced some valuable insights concerning the evolution of social communication. Now that the social organization, ecology and reproductive biology of the mandrill are better understood, comparative studies in these areas may also open the way to a better understanding of its evolutionary history.

Mandrills and drills were traditionally classified as 'forest baboons' and, as such, they were thought to be closely aligned with the savannah baboons of the genus *Papio*. However, it has become apparent that many of the baboon-like traits that occur in the genus *Mandrillus* are because of convergent evolution. As large-bodied, sexually dimorphic and primarily terrestrial monkeys, mandrills and drills may look superficially like baboons, but there are many anatomical and genetic differences between the two genera. Studies of mitochondrial DNA as well as skeletal anatomy have shown that *Mandrillus* is most closely related to the mangabeys, and especially to the semi-terrestrial forest mangabeys of the genus *Cercocebus*, within the Tribe Papionini. The mandrill and drill should be thought of, not as baboons, but as the very large and robust, terrestrial descendants of mangabey-like ancestors. The 'molecular clock evidence' reviewed in Chapter 2, places the split between *Mandrillus* and the mangabey

(*Cercocebus*) lineage at some time around four to five million years ago (Mya). Subsequently, the divergence between *M. sphinx* and *M. leucophaeus* took place at approximately 3.17 Mya.

Changes in climate and associated cycles of rainforest contraction and expansion have likely played important roles in these events. During the late Miocene and early Pliocene epochs, between 6.4 and 4.6 Mya, the climate in equatorial Africa gradually became much drier, with the result that areas covered by rainforest were reduced in extent. In some areas, however, local conditions favoured the survival of forests in mountainous areas, or close to major rivers. The forests of the Ogooué River system in Gabon, the Sanaga River, the Cameroonian mountains and the Niger delta are thought to have acted as refuges for rainforest species of many kinds. Isolation of ancestral mangabey populations in the various forested areas may have profoundly affected their speciation. When the climate became wetter during the later Pliocene (4.6–2.3 Mya), and the forests expanded once more, there had emerged a new, larger-bodied, mainly terrestrial, mangabey-like monkey. This ancestral *Mandrillus* stock later gave rise to the mandrill and drill.

Major rivers form geographic barriers to the distribution of many mammals, including the primates (Harcourt and Wood, 2012), and thus affect their speciation. It is thought that during further cycles of rainforest contraction and expansion, between approximately 2.43–1 Mya, the Sanaga River played a critical role in speciation of the genus *Mandrillus*. Ancestral mandrill populations became confined largely to rainforests to the south of the Sanaga river, whereas ancestral drill populations dispersed northwards into what is now Cameroon, southeast Nigeria and the Island of Bioko. It is possible, but unproven, that the two species might once have been sympatric in forests bordering the Sanaga River.

The mandrill's recent distribution range includes southern Cameroon, Gabon, the mainland of Equatorial Guinea, and parts of Congo Brazzaville north of the Congo River. Rivers have also restricted the spread of mandrill populations in Gabon, where the upper reaches of the Ivindo and Ogooué Rivers constitute barriers to the eastward dispersal of this species. All mandrills are currently classified as belonging to a single subspecies (*M. sphinx sphinx*). However, recent studies of *cytochrome-b* sequences have revealed that an ancient split must have occurred between mandrill populations living to the north and to the south of the Ogooué River. These populations have been separate for at least 800,000 years. Further research may result in them being assigned to separate subspecies.

Turning now to questions of morphology and anatomy, it was noted in Chapter 3 that the mandrill displays the greatest body weight sexual dimorphism of any extant primate species. Adult males weigh 30–35 kg and adult females 9–10 kg, on average. Male mandrills are slower to mature than females, and they continue to grow for much longer. Many of the distinctive morphological and anatomical traits that occur in the mandrill are also present in the drill. Where adult males of these two species are concerned, the following features are common to both species:

- Very large body mass; males are three times larger than females.
- Very large maxillary canine teeth and extreme sexual dimorphism in canine size.

- The presence of a large, sternal scent-marking gland.
- A beard, crest of hair on the scalp, a mane, and an epigastric fringe of hair.
- Bony, fusiform paranasal swellings (larger in mandrills).
- Very large heads, and striking secondary sexual pigmentation of the face (red and blue in the mandrill; predominantly black in the drill).
- An unusually short, stumpy tail.
- Striking development of red, blue and violet sexual skin on the rump and genitalia (especially so in the drill).
- Large relative testes sizes and large seminal vesicles.
- 'Fatting' of the rump and adjacent areas in dominant males (best documented for the mandrill).

The occurrence of so many distinctive traits in adult males of both *Mandrillus* species indicates that they were more than likely present in their common ancestor that existed some 3.17 million years ago. The most striking differences between mandrills and drills relate to the facial adornments of adult males. The black face mask of the male drill is a result of eumelanic pigmentation. This is likely to represent the ancestral condition, although the peculiar expansion of the facial disc of the drill is probably a derived specialization. The blue and red colouration of the mandrill's snout are also derived traits. The red nasal colouration is owing to vascular specializations and is testosterone dependent. The striking blue colouration of the male mandrill's paranasal areas is not hormone dependent, however. Recent research has shown that the blue colour results from coherent (i.e. non-random) scattering of light by highly organized, dense arrays of dermal collagen fibres. The same specialized arrangement of dermal collagen is found in some other mammals, including the vervet monkey (*Chlorocebus aethiops*), and in *Marmosa mexicana*, which is a New World marsupial species. It was noted here that mandrill populations to the North and South of the Ogooué River in Gabon have been separated for at least 800,000 years. Because adult males of both these populations have blue and red sexual skin on the face, we can be confident that these traits were already present by that time. They are likely to date back to well before then, however, and probably arose quite early during speciation of the mandrill and drill.

Examination of female sexual skin morphology also provides confirmation of the close evolutionary ties between both *Mandrillus* species and the semi-terrestrial mangabeys of the genus *Cercocebus*. The female sexual swelling in mandrills and drills involves the skin surrounding the clitoris, vulva and anal region. A pale elliptical area of skin surrounds the vulva when the swelling is at maximal size, in contrast to its overall pink or reddish colour. Females of the genus *Cercocebus* have a very similar sexual skin morphology to that of *Mandrillus*, whereas in the arboreal mangabeys (*Lophocebus*), a perianal swelling is lacking. Sexual skin morphology of the type seen in both the extant *Mandrillus* species thus has an ancient origin, dating back at least four to five million years, to the common ancestor of *Cercocebus* and *Mandrillus*.

Research on the osteology of the limbs and limb girdles of *Mandrillus* has also confirmed that a close phylogenetic relationship exists between *Cercocebus* and *Mandrillus*. Shared features of the scapula, humerus, radius and ulna of *Mandrillus* and *Cercocebus* reflect similarities of their musculature and locomotor capabilities. These adaptations have evolved in relation to a mode of life that involves climbing in the trees, as well as vigorous manual foraging on the forest floor. Comparative measures of the pelvis, femur and tibia also unite *Mandrillus* and *Cercocebus*, and distinguish them from *Papio* and *Lophocebus* within the Papionini. In addition, examination of the dentition has shown that the second (posterior) premolars of *Mandrillus* are notably larger and better adapted as crushing teeth than is the case in *Papio* and *Lophocebus*. Again, both the mandrill and drill share this condition with *Cercocebus*, and this is probably because they are all specialized 'forest floor gleaners' that include substantial amounts of hard shelled items in the diet.

These conclusions regarding questions of functional anatomy, foraging and feeding behaviour lead naturally to a consideration of the mandrill's ecology, and of those ecological factors that might underlie the evolution of group size. These topics were addressed in Chapter 4. Research, conducted by Kate Abernethy and her colleagues in Gabon's Lopé National Park, has shown that mandrills form the largest social groups yet recorded for any primate species. These supergroups vary in size between 338 and 845 individuals. Numbers of females remain constant throughout the year, whereas numbers of mature males vary seasonally. More males are present during the long dry season months (June–November) when females develop sexual skin swellings and matings occur. At other times of year, most of the males emigrate from the groups and spend long periods as solitary foragers.

Field studies conducted in southern Cameroon, as well as in Gabon, have established that mandrills are omnivores, utilizing more than 100 plant species as food, and including many types of invertebrates and small vertebrates in their diet. More than 84% of the diet consists of fruits, and especially seeds; the large molar and posterior premolar teeth of the mandrill are highly effective in crushing such hard items. Foraging for fallen fruits and seeds on the forest floor is a major strategy; as much as 70% of the time is spent foraging within five metres of the ground. Mandrills are indeed 'forest floor gleaners'. Trees in fruit have a wide but patchy distribution in the rainforest, and many of the mandrill's major foods are subject to seasonal changes in their availability.

The enormous size and overall biomass of supergroups probably induces a high level of feeding competition among their members. The temporary splitting of these large groups into smaller sub-units is, therefore, likely to be a consequence of the mandrill's feeding ecology and of seasonal changes in food availability. It used to be thought that smaller subgroupings of mandrills might represent polygynous one-male units that sometimes separate from the main group for reproductive purposes. However, it is now clear that the supergroup itself provides the arena for mating activities, which occur during the long dry season. When the mating season ends, however, many adult males emigrate from supergroups, in order to forage on their own and regain their physical condition.

Consistent with the patchy distribution of the mandrill's major foods and the huge biomass of its groups is the existence of very large home range areas utilized by supergroups. Radio-tracking studies conducted over a six-year period in the Lopé National Park by White *et al.* (2010) showed that one group of 700 mandrills had a maximum home range of 182 km^2. However, the area of suitable forest habitat within this range was limited to 89 km^2, and the group spent more than 50% of its time in less than 10% of its total range. This study also confirmed that the supergroup sometimes split into two to four subgroups, which reunited after varying periods.

Comparative information on the ecology and behaviour of the drill is quite limited. However, it is worthy of note that large groups, containing at least 400 drills, existed until recently in the montane forests of Bakossiland in Cameroon. Temporary fissioning of drill groups has been recorded in the Korup National Park, as well as in the (now extinct) Bakundu Forest population. At all these sites, lone males were also recorded. Studies of semi-free ranging drill groups in Nigeria have established that they are also seasonal breeders; like mandrills they mate during the drier months and give birth during the wet season. Thus, it seems likely that drills originally had the same type of social organization as mandrills, but their supergroups are now very rare, owing primarily to hunting and deforestation. Likewise, drills are omnivores and forest floor gleaners that rely primarily upon fruits and seeds for the major portion of their diet. It is likely that many of these shared features of ecology, social organization, and reproductive biology in the drill and the mandrill would also have been present in their common ancestor.

Little is known about mortality rates owing to predation in the mandrill or drill. Leopards are known to kill mandrills, while crowned hawk eagles and rock pythons are likely predators. Long periods of solitary foraging by pubescent and adult male mandrills are likely to increase their predation risk, especially where leopards are concerned. Thus, it is probable that natural selection (as well as sexual selection) might have played its part in the evolution of large body size, and canine size in male mandrills, as a means of deterring predators. It is certainly the case that female mandrills live for longer, on average, than do males. Even when predators are absent, as in the semi-free ranging groups at the International Medical Research Centre (CIRMF) in Gabon, male mandrills do not often survive beyond 20 years of age. Yet females continue to reproduce into the third decade of life, and the oldest post-reproductive female survived until 37.3 years of age.

The social communication and the sexual behaviour of mandrill supergroups in the wild remains very poorly understood. However, much has been learned by studying semi-free ranging mandrill groups in the large rainforested enclosures at the CIRMF. The visual displays and vocalizations of mandrills are, broadly speaking, homologous with those of other cercopithecine monkeys, and especially with other papionins such as the macaques, baboons and mangabeys. However, some distinctive patterns have arisen during the evolution of the genus *Mandrillus*, and these deserve special comment. For example, adult male mandrills often emit deep two-phase grunting (2PG) vocalizations during group progressions, and also when mate-guarding females. High-ranking males are more likely to vocalize in this way, and to advertise their positions to other group members. Females and younger animals of both sexes give high-pitched ‘trills’ or ‘crowing’ vocalizations as contact calls. Choruses of such calls are often heard when a mandrill group is on the move.

Both the male 2PG vocalization and group contact call find their homologues in the drill, and both *Mandrillus* species also exhibit a distinctive grinning facial expression, which is displayed in affiliative or submissive contexts. The mandrill's grin is a variant of the 'silent bared-teeth face' that has been described by Van Hooff (1967), as occurring in many Old World monkeys. However, the mandrill and drill are unusual because their lips are often kept closed at the centre of the mouth, and raised at its corners to produce a '∞' shape, which has a superficial resemblance to a snarl. Yet, the grinning expression is not aggressive; it is often exhibited during play by younger animals, and as an affiliative or submissive expression between adult males. Adult males also frequently grin at females during pre-copulatory behaviour. The display is sexually dimorphic, in the sense that males often combine the grin with stereotypical head-shaking movements and piloerection of the crest on the scalp and the nape of the neck. Females very rarely give such complex or prolonged grinning displays. Head-shaking, or 'head flagging' has also been reported to occur as a sexual invitation in males of *Cercocebus sanjei* and *Lophocebus albigena*. These displays are less stereotyped than those of male mandrills. However, the mandrill's display may derive from such simpler patterns that were present in its common ancestor with the mangabeys.

Mandrills and drills are the only Old World monkeys that are known to scent-mark using a sternal gland, so that this specialization presumably arose after the genus Mandrillus had separated from the mangabey lineage. The glandular complex develops during puberty and dominant adult males have the largest sternal glands. Sternal glandular secretions contain a mixture of volatile chemicals, the communicatory functions of which remain unknown. They may play some role in communication during periods when males emigrate from groups and live a solitary existence. There has been one report that olfactory communication in the mandrill includes flehmen in response to sternal glandular and vaginal odours. Flehmen occurs in some male mammals, as in ungulates, where it serves to transfer chemical cues, present in the female's urine, to the male's vomeronasal organ. However, I have never observed such behaviour in the mandrill; nor has flehmen been reported in other Old World monkey species, none of which is thought to possess a functional vomeronasal organ.

Sexual behaviour occurs during the long dry season in mandrills, when high-ranking adult males often mate-guard individual females during periods of maximal sexual skin tumescence. The mandrill's copulatory pattern consists of a single mount, with intromission and pelvic thrusting, culminating in ejaculation. Copulations are relatively brief (<1 min duration), and neither sex vocalizes during or after mating. No behavioural correlates of female orgasm have been observed in this species. From a phylogenetic perspective, it is interesting that mate-guarding by high-ranking males has been reported in a least one mangabey species (*Cercocebus sanjei*). During mate-guarding, male sanje mangabeys follow females and emit grunting vocalizations. These may represent more simple homologues of the following and 2PG displays that characterize mate-guarding by male mandrills.

In common with most Old World anthropoids, elements of the mandrill's sexual behaviour, such as its presentation and mounting postures, have become ritualized during the course of evolution, and incorporated within the broader sphere of its social communication. Thus, socio-sexual presentations and mounts occur between members of the same as well as the opposite sex, and they fulfil a variety of roles in social communication. A strong correlation exists between social rank, based on measures of agonistic behaviour, and rank as reflected by socio-sexual presentations. More subtle indicators of rank are provided by measurements of displacement and avoidance between group members. Rank orders are primarily linear, and social rank has important consequences for reproductive success in both sexes. This is especially the case among adult males, in which reproductive success is highly skewed in favour of dominant, brightly adorned individuals.

Information presented in Chapters 7 and 8, concerning reproduction in semi-free ranging mandrill groups, offers us valuable insights concerning what may occur in the natural state, as regards social organization, sexual behaviour, and the evolution of the mandrill's unusual mating system. It is apparent that females form the core of mandrill society. The mandrill has a matrilineal social organization; females are philopatric and probably spend their entire lives as members of supergroups. Young females derive their rank status from their mothers, and high rank confers significant advantages in terms of the growth rates of offspring and lifetime reproductive success. Females may continue to reproduce until they are more than 20 years of age. Young males, on the other hand, emigrate at puberty and forage alone in order to improve their body condition. As adults, from the age of nine years onwards, they begin to re-enter social groups periodically, during annual mating seasons, when females develop their sexual skin swellings. Competition between males is intense; the most dominant ('fatted') males achieve high reproductive success, and in some cases high rank may be sustained for five or six mating seasons. Priority of access to females is fiercely contested, especially during the most fertile period of the menstrual cycle. Even high-ranking males can only effectively mate-guard a single female at any one time. Subordinate males mate opportunistically, and some exhibit an alternative reproductive strategy (the 'non-fatted' condition), which involves arrested development of their secondary sexual adornments. Multiple partner matings occur in semi-free ranging mandrill groups, and they are probably much more frequent in the wild, where adult females in supergroups far outnumber adult males. 'Harems' or 'one-male units' do not occur, however, and mandrills do not have a polygynous mating system or a multilevel type of social organization. Their mating system is more accurately defined as being an unusual variant of the multimale–multifemale systems that occur in many other cercopithecine monkeys. These include the mangabeys, macaques and most baboons, which, like the mandrill, are members of the Tribe Papionini.

10 Sexual selection

> Having carefully weighed, to the best of my ability, the various arguments which have been advanced against the principle of sexual selection, I remain firmly convinced of its truth.
>
> Charles Darwin

Romanes (1893) tells us that this statement constituted Darwin's 'very last words to science', as they were written shortly before his death, in 1882. Well over a century later, the importance of sexual selection has been amply confirmed and much more is now known about its scope, although the exact processes through which it might operate are still debated. Darwin identified two principal types of sexual selection, intra-sexual competition for access to mates and inter-sexual selection for secondary sexual adornments and displays as enhancers of sexual attractiveness. Both these processes were thought to act mainly upon males, leading to the evolution of larger body size, greater weaponry and aggressiveness, as in the case of red deer stags, or to extravagant masculine adornments and courtship displays, as seen in the peacock. The evolution of extreme body size and canine size sexual dimorphism in the mandrill will be considered first, and then I shall discuss the evolution of the adult male's colourful secondary sexual traits.

Although Darwin was preoccupied with the effects of intra-sexual selection upon males, it is apparent that females might sometimes compete among themselves for mating opportunities, as well as more indirectly for the resources required to nurture their offspring. Inter-sexual selection has also favoured the development of sexually attractive adornments in females of certain species, as well as in males. Indeed, the sexual skin swellings that occur in females of many Old World anthropoids provide powerful evidence for the existence of this type of inter-sexual selection (Dixson, 1983a, 1998a, 2012). The effects of sexual selection upon the evolution of female sexual skin in mandrills, and in other species, will be discussed in this chapter.

Darwin considered that the primary genitalia of both sexes had been moulded by evolution in response to natural selection, rather than as a result of sexual selection. However, it is now known that sexual selection sometimes operates at the level of the gonads (via sperm competition: Parker, 1970) and some other parts of the reproductive system (via cryptic female choice: Eberhard, 1985, 1996), in addition to influencing the evolution of copulatory patterns. These types of sexual selection have had widespread consequences for the evolution of reproductive anatomy and physiology in both sexes. There is increasing evidence that sperm competition has played an important role during the evolution of the mandrill.

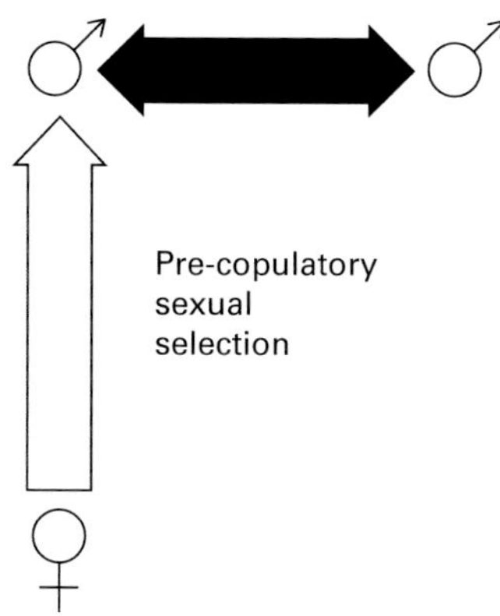

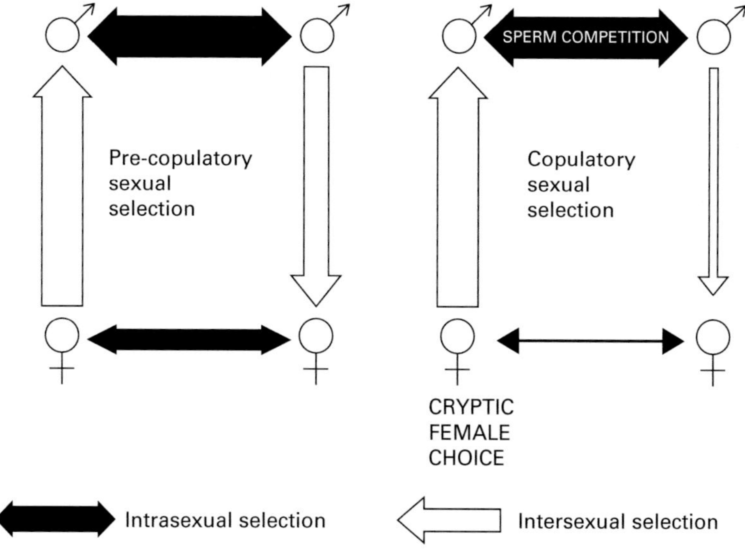

Figure 10.1. Traditional and modern views of the processes underlying sexual selection. The modern Darwinian view recognizes that females may also engage in intra-sexual competition, and sometimes produce extravagant adornments to attract males (e.g. female sexual skin in mandrills and other Old World anthropoids). Most important, however, has been the realization that sexual selection also operates at copulatory and post-copulatory levels, via sperm competition and cryptic female choice. (From Dixson, 2012).

This modern view of sexual selection, which incorporates copulatory and post-copulatory processes in addition to pre-copulatory competition and mate choice, is shown in Figure 10.1. Given the marked sexual dimorphism in body size and the secondary sexual traits of the mandrill, as well as occurrences of multiple partner matings and genital specializations, this species clearly has a great deal to tell us about the nature of sexual selection. A discussion of the various hypotheses that seek to account for the evolution of secondary sexual adornments across various vertebrate groupings is included towards the end of this chapter.

Male body size and weaponry

We have seen that, among its many unusual traits, the mandrill is notable for displaying the most extreme body weight sexual dimorphism of any living primate species. Male to female body weight ratios for primates are shown in Figure 10.2, which is taken from the work of Martin *et al.*, (1994). The mandrill appears as an outlier, having a male:female weight ratio of 2.5. Impressive as this sex difference is, it is an underestimate. By the time that semi-free ranging mandrills at the CIRMF reach eight to ten years of age, they exhibit male:female body weight ratios of 3.2–3.3. This is the result of the rapid increases in growth shown by male mandrills once they enter puberty. Some disparity in body weight between the sexes is apparent by the time that mandrills are three to four years of age (see Figure 3.4, page 21), but this gap widens rapidly from the age of six years onwards.

Male mandrills thus continue to grow for a longer time, and they attain their maximum body weights four to five years later than do females. Matching this bimaturism in somatic growth is a marked sex difference in the age of attainment of sexual maturity. We have seen that female mandrills have their first infants at an average age of 4.63 years; indeed, some high-ranking females give birth as early as 3.89 years of age. Although male mandrills may be fertile by five or six years of age, very few of them sire offspring then. The majority of infants are sired by fully grown, high-ranking males aged between 10 and 16 years.

Later attainment of puberty and sexual maturity in males of sexually dimorphic species is a widespread phenomenon, especially among the Old World monkeys and the great apes (Table 10.1). Delayed sexual maturation of male mandrills has evolved as a physiological 'trade-off' in favour of investment in prolonged growth and attainment of very large body mass. This may be attributable to effects of extreme intra-sexual selection, coupled with the unusual ecological conditions affecting the development of males in this species. From puberty onwards, male mandrills spend considerable amounts of time foraging alone, increasing in body mass and accumulating fat reserves in preparation for the intense inter-male competition that occurs when they enter super-groups during the mating season. Male emigration from natal groups occurs in many Old World monkeys, such as baboons, macaques and mangabeys. For example, in *Cercocebus torquatus atys*, some males spend long periods foraging alone, and enter social groups during the annual mating season (Benneton and Noë, 2004), as is the case in mandrills. In the long-tailed macaque (*Macaca fascicularis*), van Noordwijk and van Schaik (1988) observed that 'the only males growing to maximum body size were those who were semi-solitary for several months at the age of about nine years.' Indeed, in long-tailed macaques it appeared that it was 'impossible for a male to achieve rapid growth while a full-time member of a social group'. Adolescent and young adult male

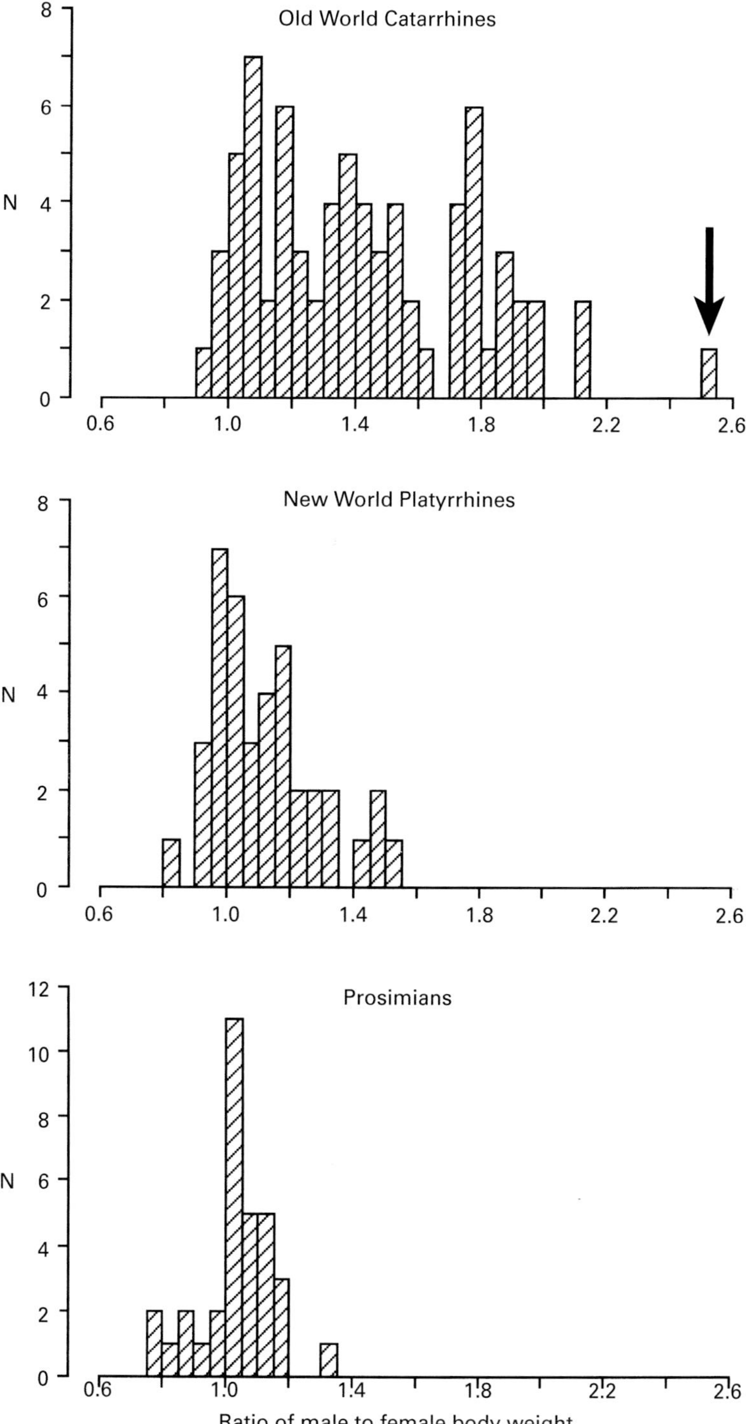

Figure 10.2. The extent of sexual dimorphism in body weight in primates, based on ratios for adult males and females in Old World anthropoids (N = 73 species), New World anthropoids (N = 39 species), and prosimians (N = 33 species). Data for the mandrill are indicated by an arrow. (From Dixson, 2012; redrawn from Martin *et al.*, 1994.)

Table 10.1. Sexual bimaturism in Old World monkeys, apes and humans

	Females (age in years)		Males (age in years)	
Species	at puberty	full grown	at puberty	full grown
Chlorocebus aethiops	2.8	–	5	–
Erythrocebus patas	–	3	4	5
Miopithecus talapoin	4	5	–	6
Lophocebus albigena	3.6	5	4.5	7
Cercocebus atys	3	5	5	6
Mandrillus sphinx	2.75–4.5 (3.6)	5	4–8 (6)	8–10
Papio cynocephalus	4.8	7	5.7	9
P. hamadryas	4.3	5.6	5.8	10
Macaca mulatta	2.5	6	3.5	8
M. sylvanus	3.5	5.5	4	7.5
Hylobates spp.	<6	8	<6	8
Pongo pygmaeus	7?	10–11	6–7	14+
Pan troglodytes	7.5–8.5	10–11	6–9	16.5
Gorilla gorilla	7.75	10	6–8	12–15
Homo sapiens: P	15–16	–	17	–
H. sapiens: M	12.8–13.2	16–18	13–14	20+

P, pre-industrial populations; M, modern populations. (Data from Bercovitch, 2000; Dixson, 2009, and sources cited therein.)

mandrills provide a much more extreme example of this strategy. They greatly prolong these periods of solitary foraging, and in adulthood most of them continue to exit groups periodically, in order to forage alone. When one considers the huge size of mandrill supergroups, consisting of hundreds of adult females and immature offspring acting as ‘forest floor gleaners’, as well as foraging in the trees, there must be considerable ecological pressure for males to adopt alternative feeding strategies. Solitary foraging for extended periods may allow males to avoid the intense feeding competition within the supergroup, and to recover their body condition. Some males cross an, as yet, ill-defined physiological threshold that allows them to lay down extensive fat reserves in the rump and flanks. Such males may be better equipped to withstand the energetic challenges of the lengthy mating season, spent guarding and defending a series of females, during their periods of maximum sexual skin swelling.

The Old World anthropoids include by far the largest members of the Order Primates. Given that body size sexual dimorphism tends to increase with greater body mass (Rensch’s rule), it is possible that the extreme differences in body weight between male and female mandrills might reflect this fact. However, comparative studies indicate that Rensch’s rule accounts for only a small proportion of the observed differences in body weight sexual dimorphism among primates (Martin *et al.*, 1994).

An additional factor requiring consideration is that the mandrill is largely a terrestrial species; fully grown males in particular spend most of the daylight hours foraging at ground level. A long history of terrestrial life among species such as baboons, macaques and the African apes might have contributed to the development of larger body size as a means of defence against predators. This proposition deserves serious consideration where the mandrill is concerned. Solitary males might be more vulnerable to predation by leopards, for example, known to be predators of the mandrill (Henschel *et al.*, 2005).

Leopards also prey upon baboons. However, studies have shown that baboons usually constitute only 5% of a leopard's diet and that adult males in particular are rarely taken because they are so large and aggressive (see the discussion by Jooste *et al.*, 2012). Plavcan and van Schaik (1992), who conducted a detailed analysis of canine tooth dimorphism in 76 species of anthropoid primates (including the drill, but not the mandrill), concluded that predation had mainly affected canine dimorphism in savannah dwelling species, such as baboons. Their analysis assumed that forest dwelling species 'rely less on canines and more on flight for anti-predator defence than species dwelling in open country'. However, this generalization seems less likely to apply where drills or mandrills are concerned. Leopards are primarily nocturnal hunters; fleeing would seem an unlikely form of defence for a solitary male drill or mandrill. The massive size of the male mandrill, coupled with possession of dagger-like canine teeth that are 4–5 cms long, probably acts as a powerful deterrent to potential predators. Indeed, the adult male mandrill's maxillary canines are larger than those of many male leopards (e.g. maxillary canine lengths averaged 3.3 cms in eight adult male leopards measured by Stander, 1997). The degree of canine sexual dimorphism displayed by the mandrill is at least as pronounced as that which occurs in savannah baboons.

Plavcan and van Schaik's (1992) comparative studies confirmed that inter-male competition (measured on a four-point scale) is very strongly associated with canine size sexual dimorphism among the anthropoid primates, as is body weight sexual dimorphism. They noted that canine and body weight dimorphisms are linked 'through a common selective basis' (e.g. effects of intra-sexual selection acting upon males) and that 'body weight seems to play only a minor, indirect role in the evolution of canine dimorphism'. Where the mandrill is concerned, we have seen that males often display their enormous canines, when 'yawning' (which should not be dismissed as a behavioural epiphenomenon), during agonistic and sexual encounters. Males also sharpen their canines during audible bouts of 'tooth-grinding'. Much of the use of the canines in relation to inter-male competition may thus involve non-contact displays. However, as a last resort, male mandrills do fight, especially during rank-related disputes. Such fights were infrequent in the mandrill groups housed at the CIRMF, and they were serious affairs, resulting in deep canine slash wounds, especially when an alpha male was deposed.

Among the primates, body weight and canine size sexual dimorphisms are generally greatest in polygynous forms, less so in multimale–multifemale societies, and minimal in monogamous species (Alexander *et al.*, 1979; Clutton-Brock, 1991; Clutton-Brock *et al.*, 1977). This generalization applies to the anthropoids, and especially to the Old World monkeys and apes, rather than to the prosimians (Martin *et al.*, 1994). Yet, there are exceptions to the general rule even among the anthropoids. Figure 10.3 compares the socionomic sex ratio (numbers of adult females divided by numbers of adult males in the social group) with the degree of body weight sexual dimorphism in 111 species,

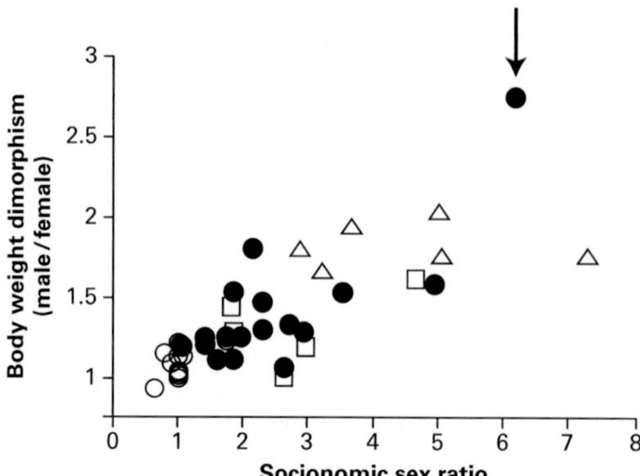

Figure 10.3. Relationships between socionomic sex ratios and body weight sexual dimorphism in 39 genera of monkeys and apes. Sexual dimorphism is greatest in those genera that have large socionomic sex ratios. The mandrill (indicated by an arrow) provides an extreme example of this phenomenon. Mating systems are indicated by symbols: ○, monogamy; △, polygyny; ●, multimale/multifemale; □, mixed: polygynous or multimale/multifemale groups may occur. (From Dixson, 2009.)

representing 39 anthropoid genera (Dixson, 2009). This confirms that there is a positive relationship between the degree of body weight sexual dimorphism and the socionomic sex ratio among the anthropoids. However, there is a wide scatter of data points on the graph, so that some multimale–multifemale forms are more sexually dimorphic and some have larger socionomic sex ratios than polygynous forms. The most obvious of these exceptional cases concerns the mandrill (indicated by an arrow in Figure 10.3), in which adult females outnumber adult and sub-mature males in supergroups by a ratio of 6:1, during the mating season, despite the existence of a multimale–multifemale mating system in this species (Abernethy *et al.*, 2002).

The mandrill thus exhibits a degree of sexual dimorphism in its body size (and its canine size), as well as a high socionomic sex ratio, such as one usually associates with a highly polygynous species. Yet all the behavioural evidence points to the contrary, as males do not compete to establish or guard 'harems' of females, either in semi-free ranging mandrill groups or in the wild. It might be objected that the evidence concerning wild supergroups of mandrills is still too limited to be certain about this issue, and that the behaviour of captive groups may be 'abnormal'. Yet the mandrill groups discussed in Chapters 7 and 8, living under relatively natural conditions, never formed one-male units. Instead, dominant males exhibited very specialized mate-guarding tactics involving individual females, while low-ranking or sub-mature males engaged in opportunistic copulations.

It is also relevant to consider that monkeys that form highly polygynous, one-male units continue to do so when held in Zoos or research centres, under conditions that are far less natural than those experienced by semi-free ranging mandrills at the CIRMF, in Gabon. This is the case, for example, in the hamadryas baboon. More than 50 years ago, Kummer and Kurt (1963) pointed out that the hamadryas baboon's 'one-male-group organization is not easily destroyed by even a radical change of environment', such as had occurred among the hamadryas baboons kept in the 'Monkey Hill' exhibit at the London Zoo (Zuckerman, 1932).

Part of the reason that male mandrills are extremely large and have enormous canines is probably a result of the unusual ecological constraints that have shaped the evolution of this species; for a male mandrill, survival as well as mating success depends upon his physical prowess, and individual behavioural skills. Individual behavioural traits and 'personality' are worth mentioning here, because some male mandrills at the CIRMF managed to acquire alpha rank despite having significant physical disabilities. Thus, male no.7 had lost the lower part of one leg and male no. 9 lacked one forearm, injuries they had sustained in infancy before they arrived at the CIRMF, yet both males rose to alpha status in their groups.

It has been emphasized in previous chapters that male mandrills do not form close social relationships with one another; they do not engage in alliances when establishing rank relationships, and when competing for access to females. When males enter supergroups during the annual mating season, they must establish themselves rapidly, as individuals, without benefit of the social networks that sustain females throughout their lives. Perhaps, at some point in the future, fieldworkers will succeed in recording sexual and aggressive interactions within mandrill supergroups during their annual mating season. They are likely to find that inter-male competition is extreme, perhaps even more so than in the complex multilevel societies and polygynous mating systems of hamadryas baboons and geladas.

Sexual segregation and body size sexual dimorphism

In many vertebrates, adult males and females in the population are spatially separated during non-mating periods. This is the case, for example, in large ungulates, such as mountain sheep (Ruckstuhl and Neuhaus, 2002), in some cetaceans (e.g. the sperm whale: Whitehead, 2003), and pinnipeds (e.g. elephant seals: Staniland, 2005). In many of the mammalian species that exhibit marked sexual segregation, males are much larger than females. It has been argued that sex differences in activity budgets and metabolic requirements linked to body size sexual dimorphism contribute to segregation of the sexes during non-mating periods. However, no single hypothesis has yet been shown to account for sexual segregation, either in mammals or in a wide range of other vertebrates (Ruckstuhl and Neuhaus, 2005).

Main and Du Toit (2005) have proposed a reproductive strategy hypothesis to account for occurrences of sexual segregation in some ungulate species. They argue that habitat usage by females and by males is driven by different selective forces, which relate to the different reproductive objectives of the two sexes. Habitat use by female ungulates is very much concerned with maximizing offspring survival. Habitat use by males, on the other hand, is dictated by 'decisions related to accumulation of energy reserves, in preparation for the rut and competition with other males for access to females'.

Main and Du Toit thus seek to account for the evolution of habitat segregation between the sexes in terms of its beneficial effects upon reproductive success. The marked differences in habitat usage between adult male and female mandrills are consistent with the reproductive strategy hypothesis. Indeed, it is probable that the mandrill represents an example, rare among the primates, of such pronounced sexual segregation. This possibility was mooted by Watts (2005), who noted that 'data to test whether segregation maximizes foraging efficiency and/or results from activity budget incompatibility are not available'. Further fieldwork on mandrills is required to examine the feeding ecology and ranging behaviour of lone males. Such studies are likely to show that male mandrills are better able to recover their body condition, and to engage in subsequent competition for mates, if they exit supergroups and forage alone for extended periods.

Male secondary sexual adornments

> The male mandrill has not only the hinder end of his body but his face gorgeously coloured and marked with oblique ridges, a yellow beard, and other ornaments.
>
> Darwin, 1876

Darwin singled out the male mandrill as being the most brightly coloured of all the mammals, and he speculated as to the role that sexual selection might have played in the evolution of its adornments. His view was that inter-sexual selection, and female mate choice, had been the crucial factor in this respect. Thus, he argued as follows:

> We may infer from what we see of the variation of animals under domestication, that the ornaments of the mandrill were gradually acquired by one individual varying a little in one way, and another individual in another way. The males which were the handsomest or the most attractive in any manner to the females would pair oftenest, and would leave rather more offspring than other males.

More than a century later, can we say with any certainty why male mandrills should be so 'gorgeously coloured?' Do their adornments indeed attract females, and do females actively choose to mate with such males? These are important questions that, in the absence of any studies of the mandrill's sexual behaviour, Darwin was unable to address. In Chapter 8, I discussed what is currently known about the mandrill's behaviour during its annual mating season, including some results of my own previously unpublished work on semi-free ranging groups in Gabon. My conclusion, based upon this evidence, was that patterns of female mate choice are not strongly expressed in the mandrill. I wish to assure the reader that I am not seeking to deny the existence of mate choice by female mandrills. Females may sometimes avoid certain males, refuse their attempts to mate-guard or to copulate, and are sometimes proceptive to those males that succeed in mate-guarding them (see Table 8.2, page 107). However, the most important factors determining sexual interactions in this species are male dominance rank and the non-behavioural qualities of female attractiveness. High-ranking, brightly coloured males initiate mate-guarding episodes. Then, the onus is very much upon the male partner to follow the female and maintain proximity between the pair. He also initiates almost all the copulations. Females, by contrast, do not invite males to guard them and only occasionally initiate copulations once guarding is in progress. Instead, it is a female's attractiveness, and specifically her sexual skin swelling morphology, that determines whether a male will guard and mate with her. If several females simultaneously develop swellings, they do not compete overtly with each other for the attentions of a dominant male, at least not in the semi-free ranging groups that I have observed. What might occur in free ranging supergroups of mandrills is not known. Multiple partner matings and sperm competition also occur in this species, rather than long-term sexual relationships. It is the waning of female attractiveness, from the day of sexual skin breakdown onwards, that causes males to cease initiating copulations, as they shift their attentions to more attractive and more fertile partners. A series of transitory mate-guarding relationships occurs throughout the mating season of a high-ranking and brightly adorned male. Less colourful, lower-ranking males attempt to mate opportunistically. Females are less likely to accept such attempts, however (see Table 8.1, and discussion on pages 95–102), so that, once again, there may be some degree of choice exerted by females.

These conclusions differ somewhat from those reported by Joanna Setchell (2005), concerning significant female mate choice for red facial colouration in males. Her study was carried out at the CIRMF in 1996, by which time Group 1 had been disrupted by removal of most of its adult males. The oldest remaining male (no. 18) exhibited supra-normal flushing of his red sexual skin, triggered by his rapid elevation to alpha status, as has already been discussed (see Plate 6). Only 19 mount attempts were observed during the 1996 study; 47% of these were made by male no. 18 and the rest involved four much younger and less brightly adorned males. All copulations were male-initiated, however, and the evidence concerning female choice for the male's red nasal colouration as distinct from male rank was, in my own view, problematic and inconclusive.

One of the many challenges that occur when attempting to study female mate choice in the mandrill, as in other mammals (Charlton, 2013) is that sexual interactions are much more complex than is the case for most other vertebrates. Thus, in contrast to the males of many bird, reptile or fish species, male mandrills do not engage in highly stereotyped, repetitive courtship displays. The behaviour of the male Temminck's tragopan provides a useful comparison here, because males of this pheasant species include among their secondary sexual adornments a lappet on the throat that, like the mandrill's muzzle, is strikingly coloured red and blue. The male Temminck's tragopan inflates the red and blue lappet as part of his courtship display (Islam and Crawford, 1998). Successful pairings and matings are dependent upon the female's positive responses to these displays; females actively choose males, based upon their behaviour and appearance. The grinning facial expressions and head-shaking displays of male mandrills, which occur prior to many copulations, pale into insignificance when compared to the displays of male tragopans, argus pheasants or peacocks. Moreover, male mandrills grin and head-shake in a great variety of social contexts, so that they often direct their displays at other males as well as at females.

In the rhesus monkey, it has been proposed that the adult male's red facial sexual skin may be sexually attractive to females. Thus, in laboratory tests, female rhesus monkeys spent longer examining images of male faces that were red (as is the case during the mating season) compared to pale non-mating season images (Waitt *et al.*, 2003). No consistent effects of male facial colour upon other aspects of females' behaviour (e.g. presenting and lip-smacking) were observed. In free ranging rhesus monkey groups on Cayo Santiago, females are reported to be more proceptive to males that have the darkest red facial colouration. However, such males are not necessarily the highest-ranking individuals in the group (Dubuc *et al.*, 2014a).

Most recently, Dubuc *et al.*, (2014b) have reported evidence that facial sexual skin colouration is partially heritable in rhesus monkeys. Adult males that are darkly coloured and are assessed (on the basis of their group residency durations) to be socially dominant, achieve greater fecundity. Dubuc *et al.* also noted that facial colouration is partly condition dependent in this species, with inter-individual variations in darkness being 'further maintained by balancing selection owing to an interaction between inter- and intra-sexual selection on males'.

It is unlikely, in a species as floridly adorned as the mandrill, that a single trait such as the male's red nasal colouration could, in isolation, determine his attractiveness to the opposite sex. It is much more likely, if physical traits do affect female mate choice, that gestalt perceptions of an amalgam of such traits might be involved in attractiveness judgements. As well as the male's red and blue snout and large paranasal swellings, we must consider his beard, the crest of hair on the scalp and nape of the neck, the mane and the epigastric fringe of hair, his colourful rump and genitalia, the large sternal gland, as well as his massive body size, degree of fattedness and the condition of his canines. The sternal gland is mentioned because it is now known to secrete a complex mixture of at least 97 volatile chemical constituents (Setchell *et al.*, 2010c). The gland is more active in dominant males, but its relevance to masculine sexual attractiveness, if any, is unknown. Nor do we know what role vocal cues might play in this context. Dominant males (but not subordinates) frequently emit deep two-phase grunts while they follow and mate-guard females. It seems likely that these vocalizations, as well as the striking appearance of dominant males, serve to advertise their presence to other males, and to reduce the likelihood that encounters with rival males might occur without some forewarning. These monkeys have evolved in environments where visual communication is limited by dense vegetation. Hence, their bright colouration, vocalizations and scent-marking behaviour may be highly adaptive under these conditions. In conclusion, I suggest that the male mandrill's adornments are more relevant to discussions of intra-sexual selection than of sexual selection via female mate choice.

Forty years ago, Pierre Jouventin (1975b) attempted to investigate the communicatory functions of the male mandrill's bright colours by making a life-size model, painted to resemble a living animal. However, responses shown by caged mandrills (of both sexes) to the model did not allow him to reach any firm conclusions. Just how difficult it may be to demonstrate, experimentally, whether the male mandrill's colourful adornments influence female mate choice may be illustrated with reference to a pilot study that I conducted. An adult female mandrill with a medium-sized sexual skin swelling was given the choice of approaching either a colourful fatted adult male or a non-fatted male (each housed in separate, adjacent, large cages). Four 30-minute tests were conducted; behavioural scoring consisted of recording the number of 30-second intervals during

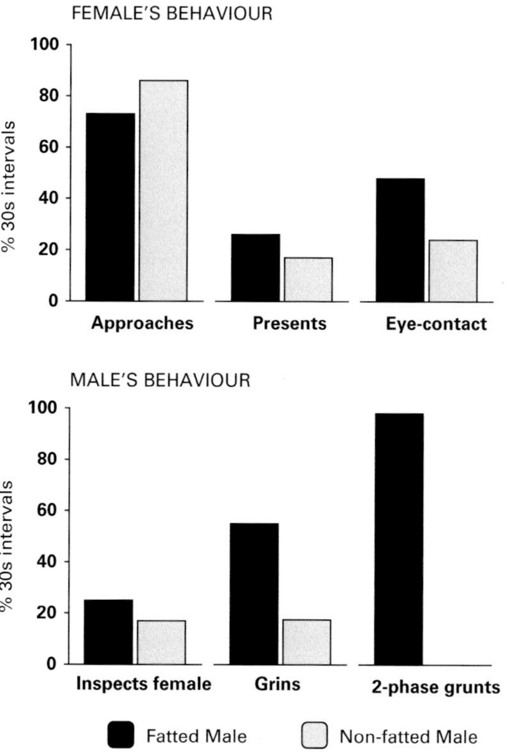

Figure 10.4. Results of a pilot study of behaviour, during tests in which a female mandrill could choose to approach either a fatted-male or a non-fatted male partner. Results are discussed in the text. (Author's data.)

which the female approached each male, presented to him, or engaged in eye contact. Males' genital inspections, grinning displays and two-phase grunting vocalizations were also scored. Results (Figure 10.4) showed that the female approached the non-fatted male slightly more often than the fatted individual (during 86% versus 73% of 30-s intervals). However, she presented more often to the fatted male (during 26% versus 17% of time intervals) and engaged in mutual eye contact twice as often (48% versus 24% of intervals).

The two males behaved very differently during these tests, however, and this confounded attempts to assess how morphological differences alone might have influenced female choice. Although both males inspected the female's sexual swelling, the fatted male grinned at her much more frequently, and only he made loud two-phase grunting vocalizations throughout every test (Figure 10.4). He behaved very much as a dominant male would during mate-guarding, and he even attempted to copulate with the female between the cage bars (on 23 occasions). The proximity and constant deep vocalizations of the fatted individual appeared to unsettle the non-fatted male in the adjacent cage; he gradually ceased to respond to the female's approaches and to withdraw from interactions (no genital inspections or grinning displays were scored during the final 30-min test).

Lacking the opportunity to employ a more sophisticated experimental approach (e.g. using operant techniques), or to access sufficient numbers of singly housed mandrills, I had to abandon these studies. Hopefully, researchers at the CIRMF might revisit this problem in the future.

Most of the secondary sexual traits discussed so far for the adult male mandrill, also occur in the adult male drill. It was noted in the last chapter that such concordance of secondary sexual traits between the two species indicates that they were likely to be present in adult males of their common ancestor. It seems most unlikely that so many unusual morphological features could have arisen by convergent evolution alone. The most obvious difference between male mandrills and drills, of course, concerns their facial appearance. The drill lacks the striking red and blue colouration of the mandrill's snout, although the skin bordering its lower lip is red. Instead, the drill's face is primarily black, and in dominant adult males, it is expanded laterally to create a striking disc-shaped appearance, fringed with white hair. The dark colouration of the face in the adult male drill, which is caused by epidermal deposits of eumelanin, is likely to represent the primitive condition for the genus *Mandrillus.* I say this because, among the Papionini, such dark facial pigmentation is the rule among the mangabeys, baboons, geladas and some macaques, as well as being present in many other genera of the Old World monkeys. Newborn infant mandrills typically have a flesh-coloured muzzle, which gradually darkens in hue during the first few months (see Plates 15 and 16); only later in life does the red and blue pattern emerge. Therefore, the common ancestor of the drill and mandrill, which existed around 3.17 million years ago (Telfer, 2006), would probably have also had a darkly pigmented face but lacked the lateral specializations and disc-like appearance, which developed later in the drill. The red and blue colouration of the male mandrill's nasal and paranasal regions also represent derived traits, not present in its common ancestor with the drill. The blue colouration is a result of non-random scattering of light by highly organized arrays of dermal collagen fibres (Prum and Torres, 2004). The same unusual arrangement of dermal collagen has arisen a number of times in mammals, by convergent evolution. Among the primates, for example, this type of convergence has occurred in some members of the Cercopithecini, such as the vervet monkey (*Chlorocebus aethiops*). In vervets, the adult male's scrotum is rendered blue by the same specialized arrangement of collagen fibres that occurs in the mandrill (Prum and Torres, 2004).

In his review of the socioecology of baboons, mandrills and mangabeys, Clifford Jolly (2007) pointed out that marked age and sex differences in the size and length of the snout mean that 'any papionin can immediately assess the age and sex of a conspecific in a face to face encounter'. Added to this, Emily Klopp (2012) has shown that, in male mandrills and drills, canine size and size of the bony paranasal swellings are likely to scale allometrically with body size. She considers that these features may therefore 'function to advertise a male's body size and fitness to other males competing for mates and potential discerning females'. The striking adornments on the muzzle and adjacent areas, red and blue in the male mandrill and black fringed with white in the drill, probably serve to enhance these advertisements. The effect is greatest in the mandrill, in which cutaneous colouration is concentrated in the nasal and paranasal areas. In the drill, black pigmentation has expanded to include the peculiar facial disc, as well as the snout.

Where the mandrill's facial colouration is concerned, there is now much better evidence for its importance as a badge of status in adult males, rather than it being a sexually attractive cue. Renoult *et al.* (2011) have reported that it is the degree of contrast between the blue and red colours of the male's snout that may be especially important in signalling dominance status. Although the mandrill's blue colouration is not androgen dependent, it is possible that increased androgen secretion during puberty might stimulate development of the dermal collagen fibres that underpin the colour changes (see Markiewicz *et al.*, 2007, concerning androgens and cutaneous collagen fibrillogenesis). We have seen that dominance interactions between males have powerful effects upon plasma testosterone and the degree of redness of the sexual skin. Changes in testosterone levels may thus occur in response to changes in dominance rank; males that ascend to alpha rank then show marked increases in their androgen-dependent secondary sexual traits, such as reddening of the sexual skin on the face and genitalia and secretory activity of the sternal gland (see Figure 8.26, page 124).

Similar effects occur in males of some other primate species. In the gelada, for example, a male that succeeds in ousting a resident male from his one-male unit, then exhibits marked flushing of the red sternal chest patch. The chest patch of the deposed male fades in colour (Dunbar, 1984). Fieldwork by Bergman *et al.* (2009) has shown that male geladas with larger harems have the brightest chest patches. Bachelor males and secondary ('follower') males have much more muted colouration. In the drill, Marty *et al.* (2009) found that dominant males, in four semi-free ranging groups, had significantly brighter red sexual skin than subordinates. This applied to the brightness of red skin in the supra-pubic area, which, in turn, correlated with measures of the brightness of red sexual skin on the lower lip. However, Marty *et al.* noted that 'these effects were largely related to male rank, and there were no relationships between male colouration and measures of association or sexual behavior independent of rank relationships'.

Another example of a colourful badge of status, which is best developed in dominant males, is provided by the blue scrotal colouration of the vervet monkey. Vervets display considerable variability in scrotal colouration, and Gerald (2001, 2003) has shown experimentally that, during paired encounters, males possessing darker blue colouration tend to be more aggressive and dominant.

There is an understandable temptation to assume that, if a striking cutaneous secondary sexual ornament develops at puberty in males of a primate species, inter-sexual selection via female mate choice is likely to have promoted its evolution. The human beard provides an interesting example of this type of bias, as exemplified by Darwin's view that:

> As far as the extreme intricacy of the subject permits us to judge, it appears that our male ape-like progenitors acquired their beards as an ornament to charm or excite the opposite sex, and transmitted them only to their male offspring.

However, modern research concerning this question indicates that the human beard is more likely to function as a badge of status, and that it evolved in that context, rather than as a sexually attractive ornament (Dixson and Vasey 2012). Thus, while women (in New Zealand and Samoa) assess pictures of men with full beards to be somewhat older and socially more dominant than images of the same men when clean shaven, they consistently rate the bearded images as being less sexually attractive.

If visually striking secondary sexual adornments have evolved primarily as badges of status, to intimidate rival males, then one might expect to find a more pronounced expression of such traits in mating systems where levels of inter-male competition are greatest. For example, marked competition among males to assemble and defend 'harems' of females occurs in species such as the gorilla, proboscis monkey, hamadryas baboon and the gelada. Quantitative studies of masculine secondary sexual adornments have rarely been attempted because they are much more difficult to measure than are the sex differences in body weight and canine size discussed earlier in this chapter. Ideally, for traits involving the pelage, such as capes of hair, beards or crests, it would be advantageous to obtain exact measures of hair length and colour for all of those species in which sexual dimorphism occurs. Likewise, quantitative information on the brightness and anatomical distribution of red and blue sexual skin would be valuable. When 'fleshy' features are of interest, such as the pendulous nose of the male proboscis monkey, or the bulbous flaps at the mouth corners of the male golden snub-nosed monkey, comparative measurements are especially challenging. If quantification and statistical comparisons are to be made, however, then some strategy is required to measure visual traits across a wide range of primate species.

One solution to this problem is to use a rating scale to measure the degree of male-biased sexual dimorphism that occurs for each visual trait (Dixson *et al.*, 2005). We used a six-point rating scale, ranging from zero (no difference occurs between adult males and females, e.g. both sexes have a crest of hair of the same size and colour) to five (maximum dimorphism, with males possessing a prominent visual trait that is absent, or virtually so, in females). Rating scales of this type have proven useful in comparative studies of genitalic morphology and to assess sex differences in the size of the larynx in primates (Dixson, 2009, 2012). It was possible to rate sexually dimorphic visual traits for a total of 124 species, representing 38 genera of the anthropoids (Dixson *et al.*, 2005); some results of this study are presented in Figure 10.5. These data show that polygyny is associated with significantly higher ratings for masculine visual traits than for those genera and species that have either multimale–multifemale or monogamous mating systems. For example, a polygynous species such as the hamadryas baboon achieves a score of 13.0 on this rating scale, whereas many multimale–multifemale species (such as macaques) have very low or zero ratings. There is also a robust correlation between the degree of sexual dimorphism in body weight and ratings for male-biased sexually dimorphic visual traits; this correlation is especially pronounced for polygynous genera ($r^2 = 0.78$, $P < 0.001$), as is shown in the lower part of Figure 10.5.

In our original study, the South American squirrel monkeys, belonging to the genus *Saimiri*, were rated as showing minimal sexual dimorphism in visual traits. However, in this instance our ratings were incorrect, for, although male and female squirrel monkeys look very similar to one another throughout most of the year, during the annual mating season some adult males undergo striking changes in their size and appearance, owing to

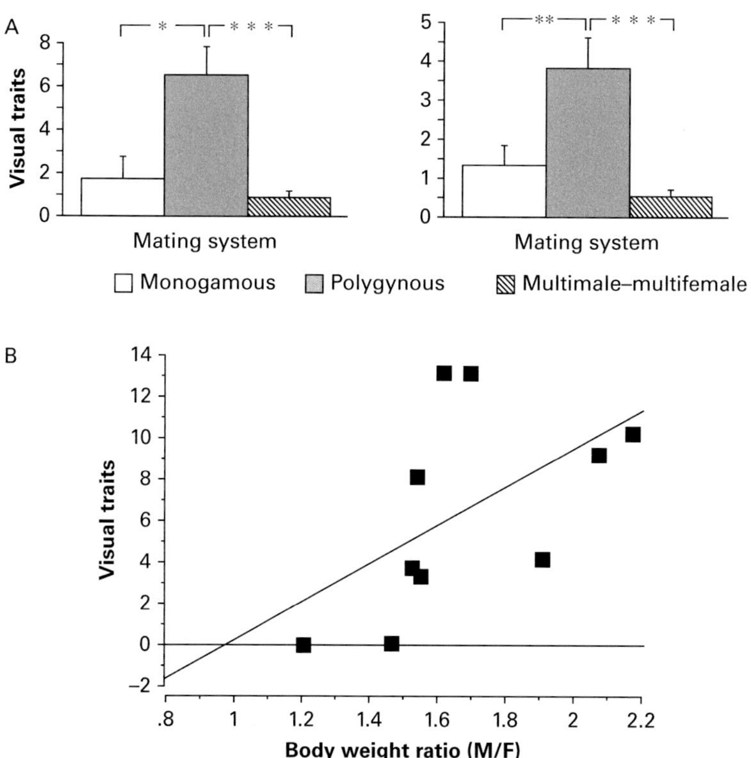

Figure 10.5. A: Ratings of the degree of male-biased sexual dimorphism in visual traits in monkeys, apes and humans. *, $P < 0.05$; **, $P < 0.01$; ***, $P < 0.001$, for analyses at the genus level and (on the right) at species level after statistical correction for possible phylogenetic biases in the data set. B: Correlation between ratings for male-biased sexual dimorphism in visual traits and degree of sexual dimorphism in body weight in polygynous anthropoid genera. Slope of the regression line: $y = 9.738 \times - 9.811$, $r^2 = 0.78$, $P < 0.001$. (Redrawn from Dixson *et al.*, 2005.)

a process called 'fatting'. This involves marked enlargement of the upper torso, arms and shoulders. Studies of squirrel monkey groups in captivity (Mendoza, 1978), and in the wild (Boinski, 1987, 1998) indicate that females find the highest-ranking fatted males to be most attractive sexually. Yet there is also evidence that inter-male competition is intense during the annual mating period and that maximally fatted males are most successful in attaining high rank, and have greater access to females (Stone, 2014).

The 'fatted' phenomenon in male squirrel monkeys should not be assumed to be the physiological equivalent of fat deposition in male mandrills. I say this because it has been suggested that subcutaneous water deposition, rather than an increase in adipose tissue, accounts for the bulky appearance of fatted male squirrel monkeys (Boinski, 1998; review: Jack, 2007). I am unable to confirm this, however, as fatted male squirrel monkeys that I have dissected showed marked enlargement of the muscles in their arms and shoulders. It is also worth noting that we have no objective measures of fat reserves in male or female mandrills. At the very least, it should be possible in future studies of the mandrill to use skinfold measures as an index of subcutaneous fat deposits. Fred

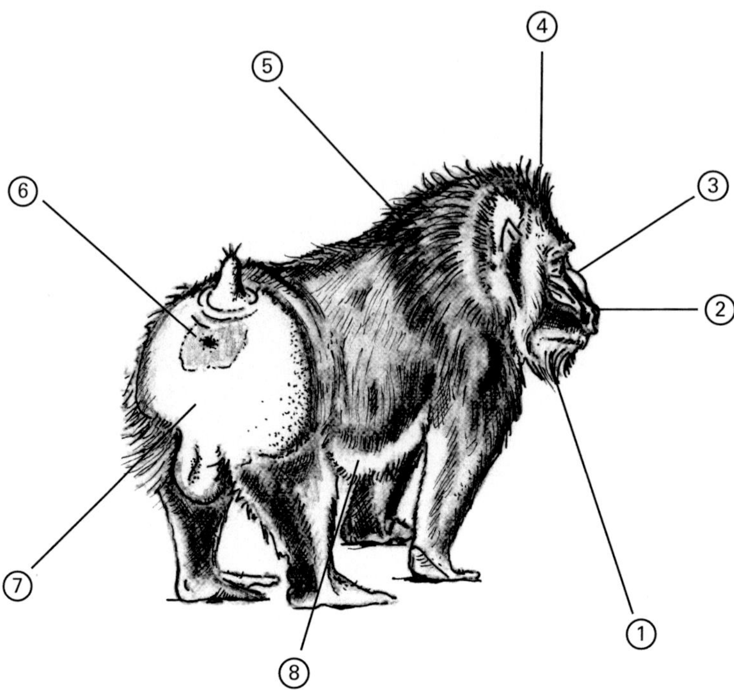

Figure 10.6. The degree of male-biased sexual dimorphism in visual traits is higher in the mandrill than in any other primate species. Its overall rating (32) is the sum of scores for eight traits: **1.** Yellow beard (3.5). **2.** Red and blue facial sexual skin (5.0). **3.** Bony paranasal swellings (5.0). **4.** Crest of hair on the scalp and nuchal region (3.5). **5.** Mane of longer hair on the upper trunk and shoulders (4.0). **6.** Red/blue/mauve/lilac sexual skin on the rump and genitalia (4.5). **7.** Fatted rump in some dominant males (3.0). **8.** Epigastric fringe of long white hair (3.5). (Author's drawing; data are from Dixson *et al.*, 2005).

Bercovitch (1992), for example, has successfully used this approach to monitor seasonal changes in the fat reserves of male rhesus monkeys.

Data for mandrills are not included in Figure 10.5, as they are exceptional and require a separate discussion. The male mandrill's battery of visual adornments, encompassing the head, body, rump and genitalia yields an extraordinarily high rating of 32 (comprised of eight traits, as detailed in Figure 10.6). This score is more than double that given to some other brightly adorned Old World monkeys such as the hamadryas baboon or gelada. The male drill, with its less colourful face, has an overall rating of 24.5 for visual adornments.These exceptional ratings in adult male mandrills and drills probably represent outcomes of extreme sexual selection for these animals to signal their rank and reproductive status rapidly and unambiguously when they enter supergroups. I suggest that their striking appearance, like their great size and large canines, is indeed owing primarily to intra-sexual selection. Although male mandrills do not compete to form one-male units, yet competition to guard a series of individual females, throughout the annual mating season, is every bit as intense as inter-male competition to hold harems in polygynous species. To be a male mandrill is to be very much alone, even in a supergroup. Adult males operate as individuals. Even for a male that is in excellent physical condition, with striking badges of status, high rank and mating success may be sustainable for only a few years.

Aside from the mandrill and drill, the highest scores for male-biased sexual dimorphism in visual traits occur in the orang-utans of Borneo and Sumatra. These huge and visually striking animals have large cheek flanges, a pendulous (laryngeal) throat sac, a beard, and long body hair. Ratings for male-biased sexual dimorphism in visual traits are 22 in the case of the Bornean orang-utan and 18 in the Sumatran species. Orang-utans are unique among the anthropoids in having a relatively non-gregarious social organization and mating system. Adults occupy individual, arboreal home ranges and adult males are highly intolerant of each other. Serious fights may occur (Utami Atmoko *et al.*, 2009), although males normally avoid one another and engage in vocal displays (the 'long call') and 'snag-crashing' as spacing mechanisms (Galdikas, 1983).

It should be acknowledged that the adult male orang-utan's striking visual adornments might play some role in sexual attractiveness, as females (and especially younger females) sometimes make persistent attempts to consort with flanged males (Dunkel *et al.*, 2013; Galdikas, 1995; Schurmann, 1982). However, the potential role of male adornments in intra-sexual display is better documented. If flanged males meet, they often behave aggressively, swaying trees and branches, piloerecting and glaring at one another (Galdikas, 1995; Horr, 1972;). Perhaps male orang-utans, like adult male mandrills, have developed especially prominent badges of status because their relatively solitary lives oblige them to communicate status quickly and unequivocally on those occasions when they are brought into proximity, as when conflicts arise over access to food or females. It is most interesting, therefore, that some male orang-utans opt for an alternative reproductive strategy, delaying development of the flanges and other secondary sexual traits, and copulating opportunistically with females. This recalls the alternative 'non-fatted' tactic that occurs in some subordinate male mandrills. The proximate (behavioural and physiological) mechanisms that might cause, and maintain, arrested secondary sexual development in male mandrills and orang-utans will be discussed later in this chapter.

Female secondary sexual adornments

> In the discussion on sexual selection in my "Descent of Man", no case interested and perplexed me so much as the brightly coloured hinder ends and adjoining parts of certain monkeys. As these parts are more brightly coloured in one sex, and as they become more brilliant during the season of love, I concluded that the colours had been gained as a sexual attraction.
>
> Darwin, 1876

The female mandrill's sexual skin is a remarkable structure, given that it undergoes considerable variations in its colour, size and turgidity throughout the menstrual cycle, as well as during pregnancy and lactation (Plates 12–14). The overall shape of the swelling also varies depending upon a female's age. Thus, the dorsal (peri-anal) portion of the swelling tends to become larger and wider in multiparous females. Whether these age-related changes have any consequences for female attractiveness is unknown. Likewise, the functions of sexual skin during pregnancy, if any, remain obscure. Mandrill females do not produce marked swellings during pregnancy; instead, the sexual skin attains a wrinkled appearance, and a dark red hue.

The widespread occurrence of sexual skin in female Old World Monkeys, as well as in chimpanzees and bonobos, presents an evolutionary question of some interest. As discussed previously, brightly coloured adornments are usually held to be the product of sexual selection acting upon males; it is less common to find such brightly coloured secondary sexual traits in female vertebrates. Hence, the existence of oestrogen-dependent, pink swellings in monkeys such as mandrills, mangabeys, baboons, the red and olive colobus monkeys, and talapoins has provoked considerable debate among evolutionary biologists.

A wealth of circumstantial evidence indicates that female sexual skin swellings are highly sexually attractive to males. Copulations occur most frequently when the sexual skin is fully tumescent, which includes the peri-ovulatory phase of the menstrual cycle (e.g. in *Miopithecus talapoin*: Rowell and Dixson, 1975; Scruton and Herbert, 1970; *Lophocebus albigena*: Rowell and Chalmers, 1970; Wallis, 1983; *Piliocolobus badius*: Struhsaker, 1975). Some species may mate throughout the cycle, but the majority of ejaculations occur during the follicular or peri-ovulatory stage (e.g. in *Macaca nemestrina:* Eaton and Resko, 1974a; *M. nigra*: Dixson, 1977; *M. silenus:* Kumar and Kurup, 1985; *Papio ursinus*: Bielert, 1986; Saayman, 1970; *Pan troglodytes*: Goodall, 1986; Young and Orbison, 1944; and *Pan paniscus*: Furuichi, 1987).

While such observations are consistent with the view that sexual skin swellings are visually attractive to males, it is necessary to acknowledge that other factors, such as odour cues and changes in females' behaviour, might also affect frequencies of copulation. The best evidence for a purely visual effect of sexual skin swellings upon sexual arousal in males, derives from experimental studies of chacma baboons (*Papio ursinus*), carried out during the 1980s by Craig Bielert and his colleagues at Witwatersrand University in South Africa. These experiments demonstrated that male chacma baboons show increases in seminal emissions, and elevations of circulating testosterone, if they are exposed to the sight of an ovariectomized female fitted with an artificial sexual skin swelling (Bielert and Van der Walt, 1982; Bielert *et al.*, 1989; Girolami and Bielert, 1987). Females do not exhibit any measurable changes in their behaviour (e.g. proceptivity) under these conditions, nor can changes in olfactory cues be implicated in the effects produced by an artificial swelling. The sight of the swelling is thus sexually arousing although, as the sexes were not allowed to contact one another, it is not known whether the model swellings used during these experiments would be sufficient to stimulate mounting behaviour by males.

Experiments to determine whether the colour of such model swellings affects sexual arousal in male baboons produced mixed results; naturally coloured models were attractive, but novel (black) models also stimulated male arousal (Bielert *et al.*, 1989). In the rhesus monkey, laboratory experiments by Deaner *et al.* (2005) showed that males will work for lower (fruit juice) rewards in order to view images showing the female perineum. Waitt *et al.*, (2006) have shown that males pay more attention to the photographs of the female sexual skin when its red colouration is maximal. How these findings might relate to their perceptions of female sexual attractiveness in real life situations is not clear, however. It was demonstrated many years ago that applying an oestrogen-containing cream directly to the sexual skin of ovariectomized rhesus monkeys causes it to redden, but that no increase in the male's ejaculations (during pair tests) results from this procedure (Herbert, 1966, 1970). Intra-vaginal application of oestrogen does stimulate the male partner's copulatory behaviour, however. I shall discuss the possible effects of vaginal cues upon female attractiveness next. As matters stand, there have been very few attempts to measure how female sexual skin might stimulate male sexual arousal in Old World anthropoids, and only Bielert's studies of chacma baboons have established unequivocally that swellings act as sexually arousing visual cues.

If one examines the female mandrill's sexual skin at full swelling (Plate 12), it is apparent that it presents an amalgam of potential visual, tactile and olfactory signals. The swelling is not uniformly pink/reddish in colour; there is a sharply defined, lozenge-shaped area of paler skin surrounding the vulva. The importance of this pale vulval area is unknown, but we do know that it becomes better demarcated during full swelling, especially so during the female's putative fertile period, and that males pay close attention to it. The colour contrast between the vulva and surrounding red sexual skin may have some special significance for female attractiveness. The swelling is also shiny and turgid at the height of tumescence. It is likely that subtle changes in its morphology during this phase provide males with additional cues concerning the approach of ovulation. Indeed, we have seen that the alpha male mandrill copulates most frequently, and guards females most attentively during the final days of swelling. His behaviour peaks just before sexual skin breakdown, and then an abrupt decline in copulations and mate-guarding occurs (see Figures 8.4 and 8.5). On a comparative note, observations of chimpanzees in the Taï National Park led Tobias Deschner to conclude that fine-grained changes in the attractiveness of female sexual skin occur during maximal swelling, and that these significantly affect the male's sexual behaviour (Deschner *et al.*, 2004). The same phenomenon occurs in mandrills.

Might olfactory cues emanating from the female genitalia influence sexual attractiveness in mandrills, or in other Old World Anthropoids? There is a substantial body of older experimental work, published in the 1960s and 1970s, which implicated an oestrogen-dependent vaginal pheromone ('copulin') in the control of sexual attractiveness in rhesus monkeys (Curtis *et al.*, 1971; Michael and Keverne, 1968; 1970; Michael *et al.*, 1975). According to these authors, increases in sexual attractiveness at mid-cycle, during the female's fertile period, depend in part upon the occurrence of (bacterially mediated) changes in the aliphatic acid content of her cervical secretions. However, a series of experiments conducted at the Wisconsin Primate Center, by David Goldfoot and his colleagues, failed to confirm these findings (Goldfoot *et al.*, 1976; 1978). Thus, rendering male rhesus monkeys anosmic did not affect the rhythmic changes in sexual interactions that occur during the menstrual cycle (Goldfoot *et al.*, 1978). Moreover, attempts to replicate some of Michael and co-workers' experiments indicated that while some males found oestrogen-dependent changes in female vaginal odour attractive, most of the 19 males tested by Goldfoot *et al.*, (1976) did not. There were no consistent effects of vaginal lavages, obtained from oestrogen-treated females, upon sexual attractiveness when applied to the rumps of ovariectomized females. The presence of ejaculate in the vaginae of females was shown to cause elevations of aliphatic acid levels and, in unmated females, aliphatic acids increased during the luteal phase of the cycle, several days after ovulation, rather than during the fertile period. It appears that, under laboratory conditions, some male rhesus monkeys may become conditioned to using olfactory cues of female attractiveness; the repeated use of small numbers of male subjects by Michael *et al.* (N = 6) in their experiments may have led to such conditioning, with resultant skewing of the behavioural effects (Goldfoot, 1981).

These studies of vaginal 'copulins' and sexual attractiveness in rhesus monkeys remain the most detailed reported for any Old World Monkey species. The conclusion that emerges from a critical evaluation of this work is that olfactory cues play, at best, only a minor role in female attractiveness in the rhesus monkey. Indeed, studies of sexual behaviour in captive rhesus monkey groups indicate that males do not exhibit increases in their olfactory inspections of the female sexual skin or genitalia during the fertile period. Rather, it is changes in female proceptivity, such as initiating proximity, presenting, 'head-bobbing', and so forth, which drive increases in males' mounts at this time (Wallen *et al.*, 1984). Yet, there is no reason to assume that rhesus monkeys are necessarily representative of all Old World monkeys, as regards cyclical changes in female attractiveness. Thus, we have seen that some male mandrills pay particular attention to the female's vulval field, sniffing it and looking closely at it prior to mounting. Such behaviour is especially marked during the phase of maximal sexual skin swelling in the mandrill. Are mandrills more attuned to olfactory cues than rhesus macaques? The vast majority of Old World monkeys do not have scent glands, or exhibit scent-marking behaviour, but again the mandrill is exceptional in this respect. Males have large sternal scent glands, and often mark trees using them; the gland is much smaller in females and they mark less frequently than do the males. The point is that olfactory communication is likely to be better developed in mandrills than in rhesus monkeys. Thus, we should not dismiss out of hand the possibility that vaginal odour might play at least some role in the sexual behaviour of the mandrill.

One possibility not discussed so far is that male mandrills may sniff the female's vulva in order to acquire information about her recent matings. It will be recalled that the presence of ejaculate in the vagina alters aliphatic acid levels, and hence affects olfactory cues, in rhesus monkeys (Goldfoot *et al.*, 1976). We have seen that female mandrills engage in multiple partner matings, despite the attempts of dominant males to exclude their competitors by mate-guarding individual females. Might the propensity for male mandrills to sniff the vulval area of female swellings regularly be indicative of attempts to determine whether they have recently mated with other males?

Oestrogen certainly has a vaginally mediated effect upon copulatory behaviour in rhesus monkeys. Thus, intra-vaginal placement of an oestrogen-containing cream in ovariectomized females stimulates their male partners to ejaculate more frequently during laboratory pair tests (Herbert, 1966, 1970). Conversely, intra-vaginal progesterone, administered to oestrogen-treated females at physiologically relevant dosages, decreases the frequency with which males ejaculate (Baum *et al.*, 1976). What causes these changes in female attractiveness? Although it was widely assumed, at the time, that such effects were a result of hormone-dependent olfactory cues, there is likely to be a more prosaic explanation. Ejaculation depends upon tactile sensory feedback, received as the male intromits and makes a series of pelvic thrusting movements. Oestrogen enhances vaginal lubrication, and it also stimulates changes in the physicochemical properties of the female's cervical mucus, which facilitates sperm transport during her fertile period. Male rhesus monkeys may ejaculate more readily if ovariectomized females receive intra-vaginal oestrogen, because this replicates the changes that normally occur in the cervical mucus and vaginal lubrication during the fertile period. Progesterone, by contrast, antagonizes the effect of oestrogen upon vaginal lubrication.

It is well known that oestrogen improves vaginal lubrication in women, and that vaginal dryness and discomfort may adversely affect sexual behaviour in menopausal women (Bancroft, 1989, 2009). It is not unreasonable to suggest that oestrogen-dependent tactile cues might also influence sexual interactions in monkeys or apes. Indeed, when Chambers and Phoenix (1987) were studying the sexual behaviour of ovariectomized rhesus monkeys, they treated females with a vaginal lubricant jelly 'because vaginal conditions in long-term ovariectomized females might make intromission aversive to the female and difficult for the male'. They acknowledged the importance of tactile vaginal cues, therefore, and it is noteworthy that some of the ovariectomized females studied by Chambers and Phoenix continued to copulate, and males continued to ejaculate more frequently with them, than during pairings with some oestrogen-treated females. In the case of the mandrill, it is relevant to note that subtle changes in the tactile qualities of the vagina may occur throughout the follicular phase of the menstrual cycle and during the fertile period when the sexual skin is largest. If this is so, then males may also gain subtle tactile cues that lead to increased mate-guarding and ejaculatory frequencies when the likelihood of conception is greatest.

Turning now from a discussion of the proximate mechanisms that control female sexual skin and attractiveness, it is necessary to consider the ultimate, evolutionary basis of these unusual structures. Firstly, it is important to acknowledge that female mandrills are in competition with one another to maximise their sexual attractiveness to males during the annual mating season. Just as male mandrills need to signal their high status and peak physical condition unequivocally, as they enter groups during the mating season, so females must signal their reproductive status effectively, if they are to attract potential partners and encourage inter-male competition.

Competition between females, for social rank and for resources (especially food) affects their physical condition and reproductive physiology, including external signals of their underlying fertility. Thus, in addition to being sexually attractive to males, swellings might have been moulded, in part, by the underlying resource competition that occurs between females. We have seen that female mandrills often have long follicular phases, and that their swellings are slow to enlarge and to become maximally attractive to males. It is likely, but unproven, that nutrition and energy balance may affect follicular phase durations. Thus, temporal factors may also influence female attractiveness; a female that fails to develop a full swelling over an extended period (lasting several weeks in some cases) may become less attractive owing to this delay. This, of course, relates to an underlying failure of maturation of a dominant follicle in the ovary, and a consequent delay in ovulation. Here we see in operation a pronounced example of Nunn's (1999) graded signal hypothesis as a possible explanation for the evolution of sexual skin swellings. This type of signal operates to convey graded information about female fertility during the current cycle, or perhaps during consecutive cycles, culminating in fertilization. There is no implication here that female swellings provide long-term signals concerning female reproductive fitness (numbers of offspring produced, and so forth), such as was suggested by Domb and Pagel (2001) for yellow baboons (*Papio cynocephalus)*. These authors reported a positive correlation between swelling size (or to be precise, the vertical height of the swelling) and long-term measures of female reproductive success. However, the correlation was tenuous at best. Height of the swelling also correlated with female body size, so that the results might simply have reflected the fact that larger females had greater reproductive success. This and other shortcomings of the Domb and Pagel study (Zinner *et al.*, 2002) limit its explanatory value. Indeed, subsequent work on baboons, mandrills and chimpanzees has failed to confirm any consistent relationships between female swelling size and reproductive history (Deschner *et al.*, 2004; Emery and Whitten, 2003; Higham *et al.*, 2008; Setchell and Wickings, 2004b).

It is worth noting that excessive swelling of the female sexual skin may, in some cases, be indicative of reproductive pathology. Rowell (1967b) observed that 'baboons, kept for long periods in captivity, under-going repeated cycles without breeding, seem to develop relatively enormous swellings'. The same phenomenon was observed in a nulliparous female mandrill, which had been kept as a pet and was given to the CIRMF when she was 10 years old. During the follicular phase of her menstrual cycle, this female developed a swelling of more than twice the normal volume.

A final point for discussion, as regards sexual selection and the evolution of female swellings, concerns their possible role as agents of cryptic female choice. Eberhard (1985, 1996) considered that in animals where females engage in multi-partner matings, the male's phallus may sometimes function as an internal courtship device, facilitating mechanisms by which the female tract receives, transports, or stores sperm. Males with more advantageous features of phallic morphology, copulatory patterns or chemical cues (transmitted via the semen) might thus gain a reproductive advantage through copulatory (or post-copulatory) cryptic female choice. The distinction between mechanisms that might arise via sperm competition, and those that are influenced by cryptic female choice, is not always clear-cut, and I think it likely that both processes may become entwined during co-evolution of the genitalia in males and females. Nonetheless, the crucial point made by Eberhard is that females are not passive in these situations and that female anatomy and physiology plays a vital role in gamete survival, and thus in determining which gametes are likely to fertilize ova.

Co-evolution of the male and female genitalia occurs in mandrills, as in other species; were this not so, there would be no correlation between vaginal and penile dimensions among females and males of the same species. In reality, penile and vaginal lengths are very similar, at least for the small sample of primate species measured to date (which includes the mandrill: Dixson, 2009, 2012). A close fit between complementary genital structures is essential if copulation is to be successful. Males are more likely to be reproductively successful if they are able to deposit sperm close to the female's os cervix, as by doing so they improve the chances that their gametes will survive and be transported to higher regions within the female reproductive tract. However, a potential problem for males in those species where females have large sexual skin swellings concerns the greater length of the vagina during the fertile period. The swelling adds considerably to the operating length of the vagina. This problem is even greater in those cases where a multimale–multifemale mating system occurs, because copulatory and post-copulatory sexual selection, via sperm competition and cryptic female choice, is magnified under these conditions. In chimpanzees, for example, there is a 50% increase in vaginal length at full swelling (Dixson and Mundy, 1994). Cryptic female choice may have driven co-evolution of genital dimensions; the chimpanzee studies showed that males with longer penes are better able to deposit semen close to the os cervix during periods of maximum swelling. The same may be true in mandrills, mangabeys, baboons, the red colobus and other species where females have especially large sexual skin swellings and engage in multiple partner matings.

The morphology of the female mandrill's sexual skin swelling closely resembles that of the more terrestrial mangabeys (genus *Cercocebus*). As we have seen, *Mandrillus* and *Cercocebus* are now considered to be closely related phylogenetically (Disotell 2000; Disotell *et al.*, 1992;). Male mandrills, like mangabeys, also have exceptionally long penes. Vaginal measurements are limited but, in the few cases for which I have been able to obtain such measures, it appears that the fully swollen sexual skin increases the operating depth of the vagina substantially. Thus, in addition to its role as a sexually attractive, graded signal of female reproductive status, the swelling may also play some ancillary role in cryptic female choice, favouring those males that are best able to place ejaculates close to the os cervix. This secondary role of sexual skin swellings is rarely considered in discussions concerning the evolution of female sexual skin in Old World monkeys and chimpanzees. However, examination of the sizes of swellings in a number of species shows that the posterior projection of the lower (pubic) region of the sex skin is sufficient to cause marked elongation of the vagina when swelling is maximal (some examples of this phenomenon are shown in Figure 10.7).

Fluctuating asymmetry and developmental instability

Fluctuating asymmetry (FA) involves small variations of bilaterally symmetrical morphological traits (including secondary sexual traits), which arise owing to environmental challenges (e.g. nutritional factors, parasites and diseases), that affect growth throughout development. Thus, although mandrills are genetically programmed to develop forelimbs that are of equal length or ears that are the same size and shape, tiny differences in such measures emerge during growth of the two sides of the body and head. Evolutionary biologists have explored this subject because measurements of such morphological asymmetries, which are relatively easy to make, may be useful as indicators of the effects of environmental stressors upon the growth of individuals or

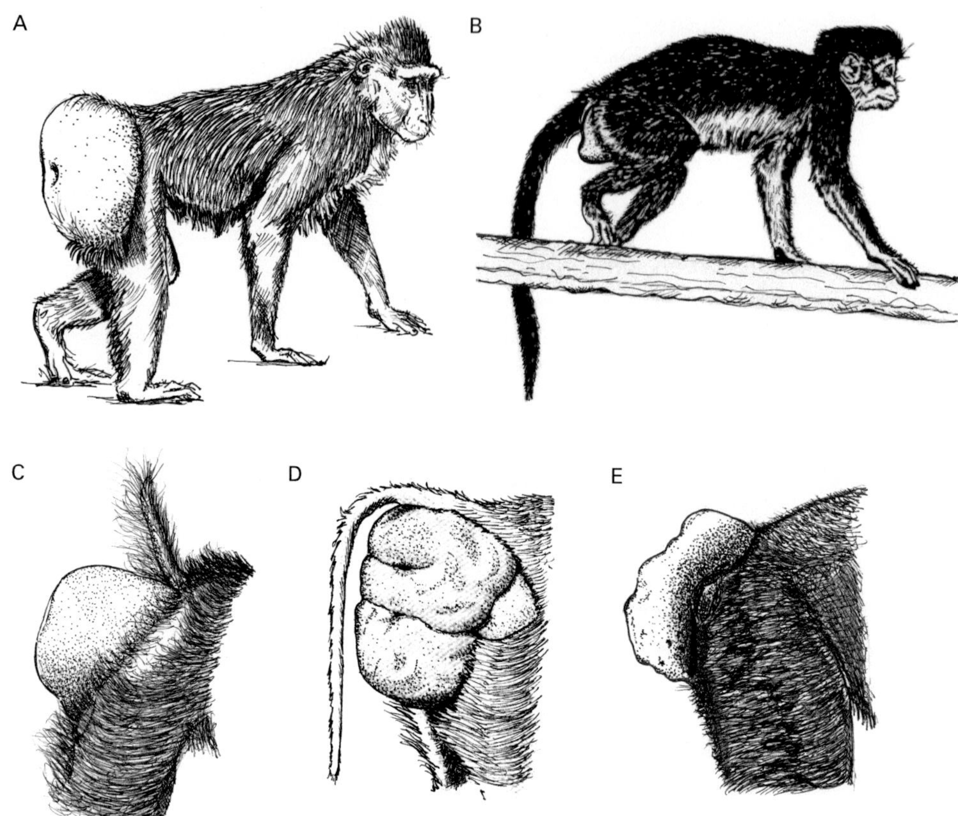

Figure 10.7. Lateral views of sexual skin swellings in five species of Old World anthropoids. **A:** *Macaca nigra*. **B:** *Miopithecus talapoin*. **C:** *Mandrillus sphinx*. **D:** *Papio hamadryas*. **E:** *Pan troglodytes*. (Author's drawings.)

populations of a species. For example, it has been reported that male (but not female) swallows develop greater FA of the forked tail in response to high mite infestations (Møller, 1992) and also as a result of environmental exposure to nuclear contamination, as occurred some years ago at Chernobyl in the former USSR (Møller, 1993).

That measures of FA really can make useful biomarkers of developmental effects owing to environmental stressors is supported by a recent meta-analysis of 53 studies conducted on various insects (Beasley *et al.*, 2013). However, much less is known about mammals in general, and very little work has been attempted on the non-human primates (see Dixson, 2012, for a review). Where humans are concerned, it has been reported that Polish women, who are 'more symmetrical' (in terms of measures of the fourth finger on each hand), exhibit significantly higher levels of salivary oestradiol during their menstrual cycles than women with higher FA of this trait (Jasieńska *et al.*, 2006). Some studies have shown that greater symmetry can enhance perceptions of attractiveness (e.g. facial symmetry in Hadza hunter-gatherers in Tanzania, as well as in people in the UK: Little *et al.*, 2007).

Can studies of fluctuating asymmetry of facial and bodily traits in general, or of secondary sexual traits in particular, tell us anything about how environmental stressors (including socially mediated stressors) might have an impact on reproductive success in mandrills and other non-human primates? Although there are currently no definitive answers to these questions, it is possible to suggest some worthwhile questions for future research. Firstly, in Chapter 8, it was noted that in the founding population of immature mandrills at the CIRMF, males destined to become fatted adults had attained greater body mass and larger testes sizes by four years of age, as compared to males that were destined to be non-fatted. It would be interesting to know whether such developmental differences in young males are associated with significant differences in FA, and whether such differences might become more pronounced during pubertal growth. Given that pubescent male mandrills begin to emigrate from their groups at six to seven years of age and to forage alone, nutritional stressors and other environmental factors might contribute to emerging differences in FA during this crucial developmental period. Are the most symmetrical males more likely to progress to the fully fatted condition and to develop brightly coloured secondary sexual adornments? A second question concerns the menstrual cycle of the female mandrill, and the pronounced individual differences that occur in the length of the follicular phase of the cycle. Given Jasieńska *et al.*'s (2006) findings on human females (described previously), might oestrogen levels, follicular phase duration and sexual skin morphology in mandrill females differ between highly symmetrical and less symmetrical individuals? Are lower-ranking females likely to exhibit greater asymmetry of their bodily or facial traits? Does FA differ in a predictable fashion between females that belong to high-ranking versus low-ranking matrilines?

Sperm competition

This aspect of sexual selection, involving competition between the gametes of rival males for access to a given set of ova, was unknown to Darwin. However, sperm competition is now widely acknowledged to have played a pivotal role in the evolution of reproduction in many groups of animals, including the primates (Birkhead and Møller, 1992, 1998; Dixson, 1998a, 2012; Parker, 1970; Simmons, 2001).

Anatomical and behavioural evidence indicates that sperm competition is likely to have been important during evolution of the genus *Mandrillus*. Some of this evidence was touched upon in Chapters 3 and 8. Thus, in common with many other anthropoids in which females engage in multiple partner matings during the fertile period, mandrills exhibit a suite of specializations of the male reproductive organs. These specializations include possession of very large testes in relation to adult body weight (especially so in dominant males), large sperm midpiece volumes, large seminal vesicles, greatly enlarged bulbocavernosus muscles (which are active during ejaculation), and rapid formation of a seminal coagulum once ejaculation has occurred. Dominant male mandrills also exhibit high frequencies of mounts and ejaculations during the annual mating season. Evidence implicating these traits in sperm competition in the Order Primates is

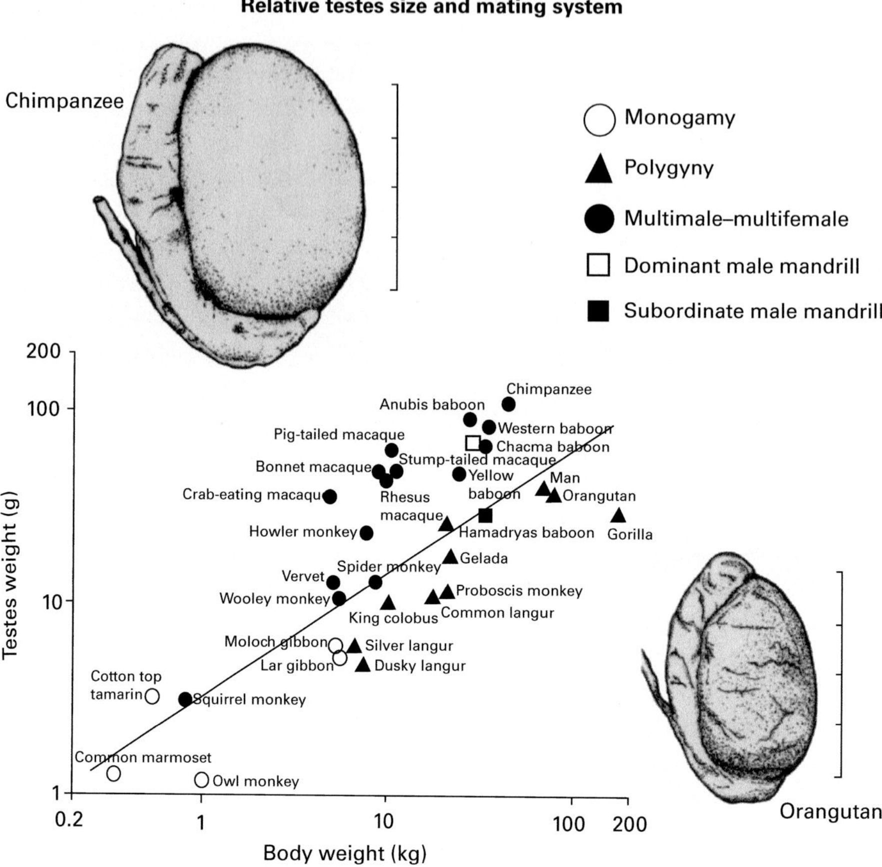

Figure 10.8. Relative testes weights and mating systems in anthropoid primates; a logarithmic plot of combined testes weight versus body weight for anthropoids having monogamous, polygynous or multimale–multifemale mating systems. Data (means) for adult male mandrills have been added to illustrate differences between dominant (fatted) males (N = 5) and subordinate (non-fatted) males (N = 5) in relation to other species. (Modified from Dixson, 2009; based on Short, 1985.)

reviewed briefly here; my goal being to place the information on mandrills within a broader, comparative perspective.

Short (1979) and Harcourt *et al.* (1981) were the first to apply Parker's (1970) ground breaking insights, concerning sperm competition in the insects, to studies of primate reproduction and evolution. Harcourt *et al.* examined relationships between relative testes sizes and mating systems in 33 species of monkeys and apes, as well as in humans. Their principal findings are shown in Figure 10.8. This double logarithmic plot of testes weights versus body weights shows that an allometric relationship exists between the two variables. Thus, testes size scales with body size in a predictable way. However, there is also considerable interspecific variability in the relationship, as indicated by the distances that the points representing particular species fall above or below the regression line. Adult males of those species that are situated above the regression line have larger testes than expected for their body weights. Almost all of these species have multimale–multifemale mating systems (indicated by the filled circles in Figure 10.8), as is the case in chimpanzees, various baboons and macaques, howler monkeys and vervets. I have added information on the body weights and testes sizes of adult mandrills to Figure 10.8. These data are mean values for five fatted males (three of which were the alpha males in their social groups) and five lower-ranking/non-fatted individuals (Dixson 1998a; Setchell and Dixson, 2001a; Wickings and Dixson, 1992b). Although body weights do not differ significantly between the two groups, the fatted males have much larger testes, more than twice the size of those measured in non-fatted adults. Indeed, their relative testes sizes are comparable to those measured in various baboons and macaques, which have multimale–multifemale mating systems.

Sexual selection has favoured the evolution of larger testes in males of those species that have the multimale–multifemale mating systems because females commonly mate with multiple partners during the peri-ovulatory phase of the cycle. Sperm from rival males compete for access to the ovum within the female's reproductive tract. Males that can maintain high sperm counts in the ejaculate may therefore gain a reproductive advantage. The occurrence of seasonal breeding does not influence relative testes size in multimale–multifemale mating systems; non-seasonal species, such as the chimpanzee, have testes just as large as those of seasonal breeders, such as rhesus macaques (Harcourt *et al.*, 1995). Greater testes size equates to larger volumes of sperm-producing tissue (seminiferous tubules) and with increased sperm production and greater ejaculate quality in primates, and in most other mammals (Møller, 1988, 1989, 1991). By contrast, in polygynous species such as the gorilla or monogamous forms such as gibbons, females mate primarily with a single male; hence, sperm competition pressures and relative testes sizes are low under these conditions (Figure 10.8).

Although high-ranking mandrills achieve great mating success, their mate-guarding tactics do not allow them to completely monopolize sexual access to females. Opportunistic and covert matings by other males are not unusual. Alpha males must also partition their mating efforts among a number of females that are simultaneously exhibiting maximal sexual skin swelling. Hence, sperm competition persists in what is still, fundamentally, the multimale–multifemale mating system of the mandrill. The same is probably true for drills, although we do not yet know enough about their behaviour to be certain. It is likely that their mating system is very similar to that of the mandrill. Thus, Marty *et al.* (2009) briefly note that they observed mate-guarding in semi-free ranging drill groups, and large relative testes sizes have also been reported to occur in this species (Dixson, 1987; Harcourt *et al.*, 1995).

Comparative studies of sperm morphology in mammals have demonstrated that the volume of the sperm midpiece is positively correlated with relative testes size. These findings derive from measurements of a large sample of non-primate mammals as well as measurements of primate sperm, including those of both the mandrill and drill (Anderson and Dixson, 2002; Anderson *et al.*, 2005). Data for 123 species of mammals (including *Mandrillus sphinx* and *M. leucophaeus*) are shown in Figure 10.9. Given that the sperm midpiece houses the mitochondria that provide a major source of energy for sperm, larger midpieces may be adaptive in terms of increased mitochondrial loading,

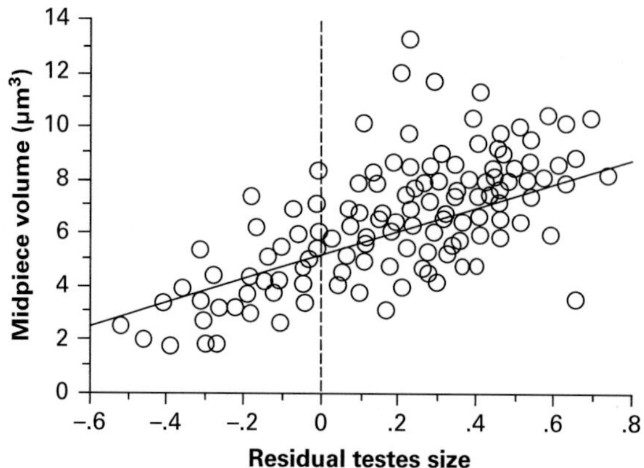

Figure 10.9. Correlation between sperm midpiece volume and relative testes size in 123 species of mammals. (From Dixson, 2009.)

and enhanced sperm motility (Anderson and Dixson, 2002; Anderson *et al.*, 2007). Primates that have single-male mating systems tend to have significantly smaller sperm midpieces. This is the case, for example, in gibbons and humans.

There are no published data on rates of spermatogenesis in *Mandrillus*, but one may speculate that rates are likely to be faster than those of many polygynous or monogamous primates, given that sperm production is known to be significantly more rapid in those mammals that engage in sperm competition (Ramm and Stockley, 2010).

Turning to a discussion of the accessory reproductive glands, it was noted in Chapter 3 that both the mandrill and drill have very large, lobulated seminal vesicles, as well as a large prostate gland. Recent work on a variety of mammalian species has shown that the weights of the prostate gland and seminal vesicles are positively correlated with relative testes sizes (Anderson and Dixson, 2009; Ramm *et al.*, 2005). Large seminal vesicles are also characteristic of many primates in which females mate with multiple partners, and in which sperm competition pressures are greatest. Examples include those species that have multimale–multifemale mating systems, such as mangabeys, baboons, macaques and chimpanzees. Mandrills and drills also belong to this group (Figure 10.10). Primates that have dispersed mating systems, including many of the non-gregarious nocturnal prosimians (e.g. galagos and mouse lemurs), also tend to have large seminal vesicles as they have large testes, and engage in multiple partner matings. Conversely, the seminal vesicles are much smaller in those taxa where polygynous, one-male units occur (as in the gorilla), or which form monogamous family groups (e.g. gibbons and owl monkeys: Figure 10.10).

In the majority of primates, seminal fluid undergoes varying degrees of rapid coagulation once ejaculation has occurred. Proteins secreted by the seminal vesicles (semenogelin 1 and semenogelin 2) are involved in this coagulation process, which is catalysed by an enzyme (vesiculase) produced by the cranial lobe of the prostate gland. The coagulation of semen is pronounced in both species of *Mandrillus*, so that a firm white coagulum forms rapidly once mating has occurred. Coagulum formation is

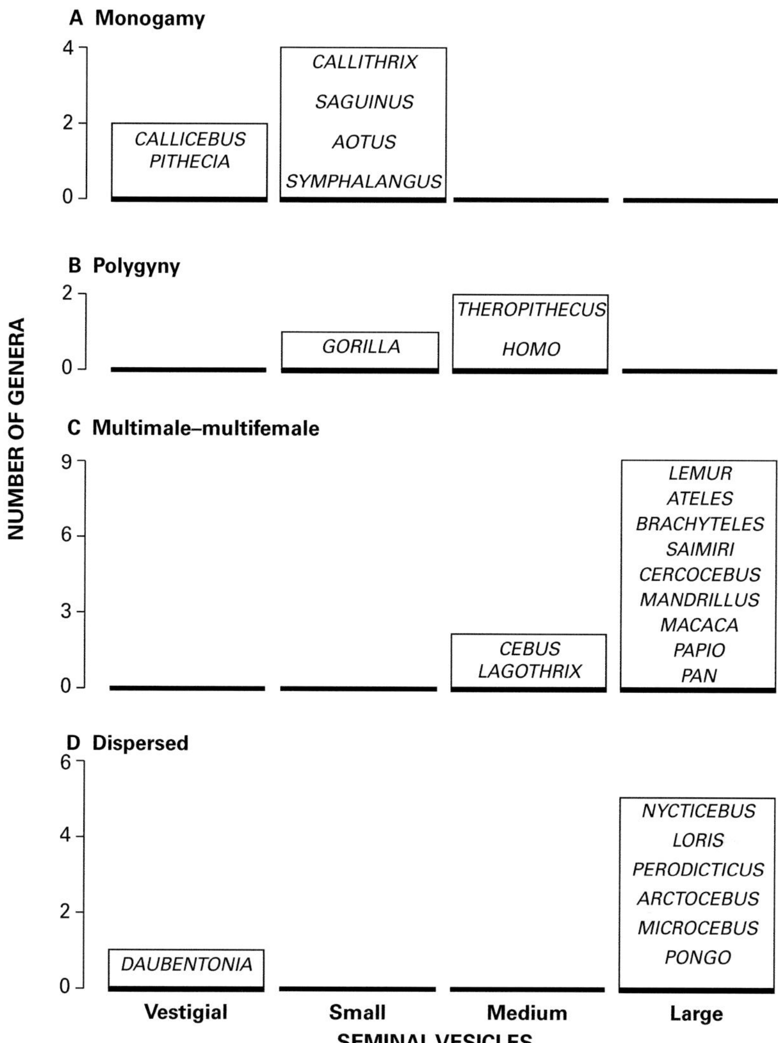

Figure 10.10. Seminal vesicle sizes and mating systems in 27 primate genera, including *Mandrillus*. Seminal vesicle sizes are rated on a four-point scale as being: 1. Vestigial. 2. Small. 3. Medium. 4. Large. Mating systems are classified as: A: Monogamous. B: Polygynous. C: Multimale–multifemale. D: Dispersed. Numbers of genera in each category are shown as open bars, with the genus names included within each bar. Seminal vesicles are significantly larger in multimale–multifemale and dispersed mating systems, than in monogamous or polygynous systems ($P < 0.001$). (From Dixson, 1998b.)

also characteristic of those primate species that engage in sperm competition (Dixson and Anderson, 2002). Comparative data on semen coagulation ratings for 28 primate genera (including *Mandrillus*) are shown in Figure 10.11. Mandrills and drills are rated as three on a four-point scale for degree of semen coagulation. A score of four refers to species in which a compact, rubbery copulatory plug is formed after ejaculation (as, for example, in ring-tailed lemurs, muriquis and chimpanzees: Dixson, 2012).

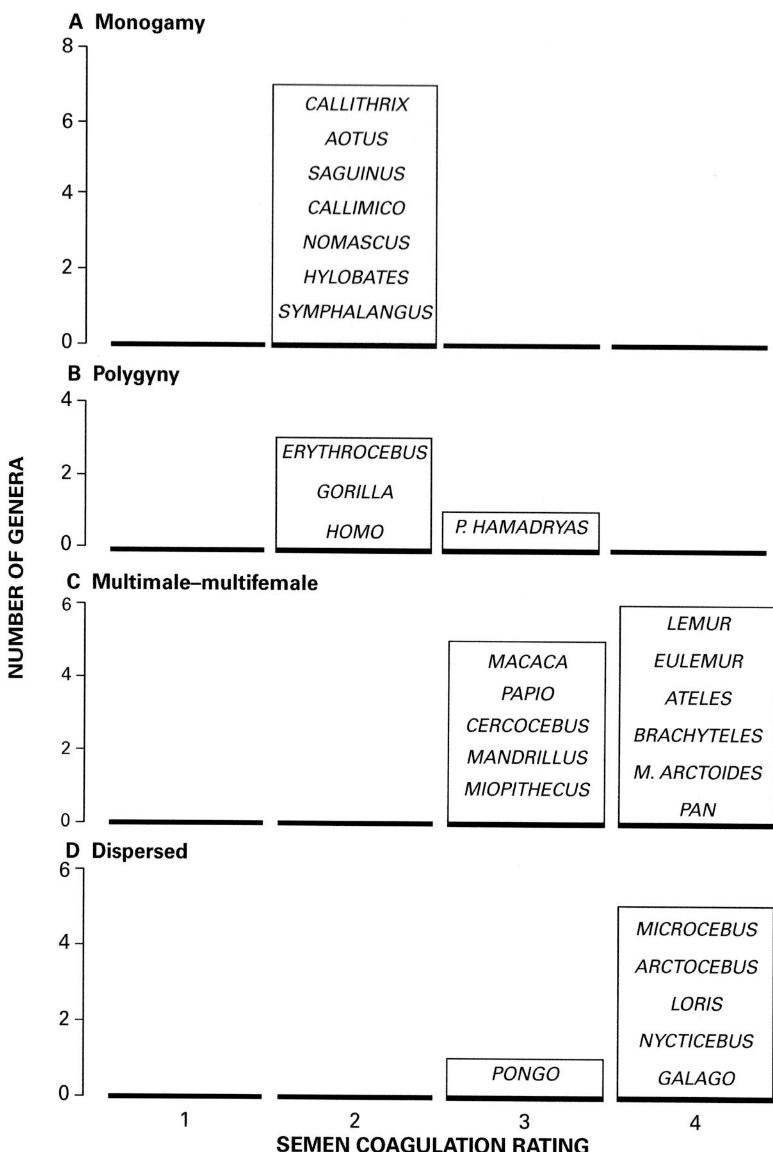

Figure 10.11. Semen coagulation ratings for 27 primate genera, including *Mandrillus*. Coagulation of semen occurs rapidly after ejaculation in most primates, but is more pronounced (coagulation ratings 3 and 4) in genera that have multimale–multifemale or dispersed mating systems, as compared to those having monogamous or polygynous mating systems ($P < 0.001$). (From Dixson and Anderson, 2002.)

Sperm competition may have selected for the evolution of larger seminal vesicles and marked seminal coagulation in mandrills, as in macaques and other multimale–multifemale species, because the coagulum assists sperm survival, and reduces loss of semen from the female's vagina after mating has occurred (Dixson, 2012). In the black-handed spider monkey, for example, Hernandez-Lopez *et al.* (2008) reported that the male's

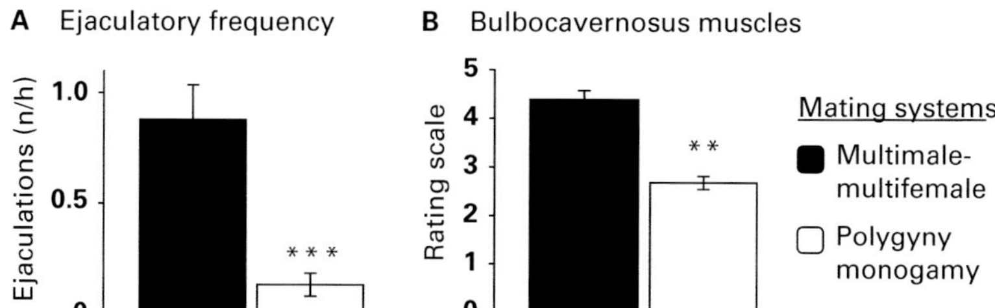

Figure 10.12. **A:** Ejaculatory frequencies are significantly higher in primate genera (including *Mandrillus*) that have multimale–multifemale mating systems than in genera that are monogamous or polygynous. *** = P < 0.001. **B:** The size of the penile bulbocavernosus muscles varies according to the mating systems of anthropoids. These muscles are larger in genera (including *Mandrillus*) that have multimale–multifemale mating systems than in genera that are monogamous or polygynous. ** = P < 0.01. (**A**. From Dixson, 1995; **B**. From Dixson, 2012).

coagulum buffers vaginal pH and raises the temperature of the vagina, as well as promoting the passage of highly motile, linearly moving sperm through the cervix. The male mandrill's coagulum may also serve to increase sperm survival and facilitate sperm transport through the lower portion of the female reproductive tract.

Moving now to consider some possible effects of sexual selection upon behaviour, sperm competition has been linked to the evolution of higher copulatory frequencies in mammals and birds (Birkhead *et al*., 1987; Dixson, 1995; Møller and Birkhead, 1989). Information on ejaculatory frequencies in non-human primates is limited, particularly for animals living in natural conditions. However, an analysis of data on 20 primate species (including the mandrill) indicates that copulations are more frequent in multimale–multifemale mating systems than in pair-living or polygynous systems. Hourly frequencies of ejaculation in those genera where multimale–multifemale mating systems occur average 0.879, as compared to 0.131/h in polygynous and monogamous forms. This difference is statistically highly significant (Figure 10.12A).

We have seen that mating frequencies are sometimes very high in mandrills, especially in alpha males, which sometimes copulate two to three times per hour during the annual mating season. Field studies have, likewise, revealed that copulatory frequencies are high in species such as the southern muriqui (*Brachyteles arachnoides*: Milton, 1985), yellow baboon (*Papio cynocephalus*: Bercovitch, 1989) and chimpanzee (*Pan troglodytes*: Tutin and McGinnis, 1981), all of which have multimale–multifemale mating systems. Despite their large testes sizes, sperm reserves might become depleted owing to repeated copulations by males of such species; as for example in the mandrill, given that its mating season spans five months of the year. Few data exist on the effects of repeated ejaculations upon sperm counts in primates. However, in the chimpanzee it has been shown that repeated ejaculations (six times in a five-hour period) cause total sperm numbers in the ejaculate to decrease from an average of 1278 million to 587 million (Marson *et al*., 1989). Males within this multimale–multifemale mating system are capable of maintaining much higher sperm counts than a monogamous/polygynous primate such as *Homo sapiens*. Thus, men who agreed to donate semen samples (by masturbation) two to three times daily over a ten-day period, exhibited a profound decline in their sperm counts, which averaged just 16.1% of their pre-depletion levels (Freund, 1963).

A number of proximate mechanisms might have provided the basis for the evolution of high frequencies of copulation by males in multimale–multifemale mating systems. Firstly, attractive signals from females, such as sexual skin swellings, are arousing to males and may promote higher copulatory rates. We have seen that the number of adult and sub-mature male mandrills entering supergroups during the mating season correlates with the number of females having swellings (see Figure 8.27, page 126). The sight of other males mating has been shown experimentally to stimulate copulation and to shorten the refractory period that follows ejaculation in male monkeys (Busse and Estep, 1984). Male mandrills, for example, sometimes remount a female within a few minutes of a previous ejaculation, in response to sexual interest shown by a second male. The same is true of Sulawesi black macaques (*Macaca nigra*), talapoins and patas monkeys (Dixson, 2012).

Is it possible that a male monkey might vary the sperm content of the ejaculate to maximise his potential for success in sperm competition?

Such 'prudent allocation' of sperm is known to occur in domestic fowl; cockerels ejaculate greater numbers of sperm if paired with a new hen (the Coolidge effect) or one with a larger comb on the head (females with larger combs have greater reproductive success: Pizzari *et al.*, 2003). The equivalent of the Coolidge effect doubtless occurs in mandrill groups, as dominant male mandrills will switch from guarding a female, and copulate with a new partner if she comes into maximum swelling. We do not know, however, whether male mandrills, or males of any other primate species, might engage in prudent allocation of sperm so as to maximize sperm numbers in the ejaculate during the female's fertile period. It is possible that increased sexual arousal in males, owing to enlargement of the female's swelling, might lead to greater sperm numbers in the ejaculate. No experiments have been performed yet on any primate species to determine whether males' sperm counts alter in response to changes in female attractiveness.

One further anatomical specialization that occurs in association with high copulatory frequencies by males in multimale–multifemale mating systems concerns the muscles at the base of the penis. The ischiocavernosus and bulbocavernosus muscles of the mandrill and drill were described briefly in Chapter 3. There it was noted that the bulbocavernosus muscles control the expulsion of semen during ejaculation, and that they are unusually large in both species of *Mandrillus*. Indeed, the bulbocavernosus muscles are significantly larger in those anthropoid genera that have multimale–multifemale mating systems, as compared to those that are polygynous or monogamous (Figure 10.12B). This may be linked to effects of sexual selection, via sperm competition, upon ejaculatory frequencies during the evolution of multimale–multifemale forms, including the mandrill and drill.

Because most of the reproductive traits discussed earlier occur in both male mandrills and male drills, one can infer with some confidence that they would have been present in their common ancestor, and that it would also have had a multimale–multifemale mating system. Barton (2000) has pointed out that multimale–multifemale mating systems are especially well represented among the

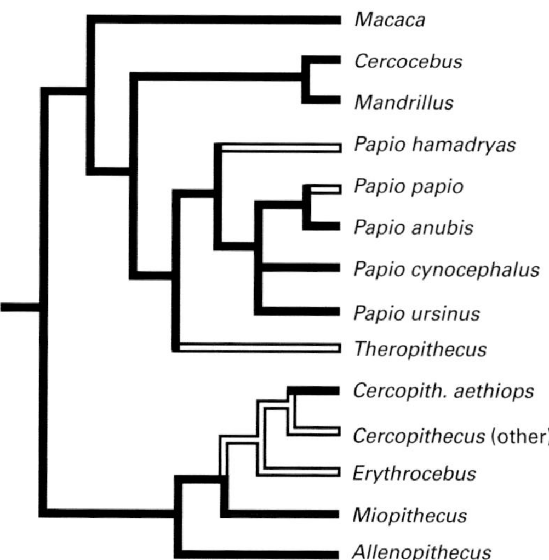

Figure 10.13. The occurrence of multimale–multifemale mating systems is likely to represent the ancestral condition for cercopithecine monkeys, as indicated by the closed bars of the phylogeny. Secondarily, division of larger groups into spatially separate one-male units has occurred in the ancestors of *Cercopithecus* and *Erythrocebus*, while groups consisting of nested one-male units are typical of *Papio hamadryas, P. Papio* and *Theropithecus gelada* (as indicated by the open bars of the phylogeny). *Mandrillus* is shown here as having retained the ancestral condition, as it has a multimale–multifemale mating system, albeit with some additional specializations. (From Dixson, 2012, redrawn and modified from Barton, 2000).

cercopithecines, including the mangabeys, baboons, macaques and related forms. This type of mating system is thus likely to represent the ancestral condition for the family Cercopithecinae as a whole (Figure 10.13). The tendency for polygynous one-male units to arise within what were originally multimale–multifemale groups has occurred several times during evolution of the Cercopithecinae (e.g. in *Papio hamadryas* and *Theropithecus gelada*). The emergence of multilevel societies and polygynous mating systems in these cases has been associated with reductions in relative testes size, seminal vesicles size, and other reproductive traits indicative of low sperm competition pressures. We have seen that these arguments do not apply to the mandrill, because it retains the reproductive traits typical of a multimale–multifemale mating system and engages in frequent multipartner matings, indicative of sexual selection by sperm competition. Thus, in adopting Barton's scheme for the evolution of mating systems in cercopithecine monkeys (Figure 10.13), I have made only one alteration; the genus *Mandrillus* is included among the multimale–multifemale forms. Without question, the mandrill's mating system is most unusual, as it involves very high levels of inter-male competition, interspersed with long periods of solitary foraging by adult males. Yet the available evidence indicates that the mandrill's mating system is derived from the ancestral multimale–multifemale model, which it shares with the mangabeys and with most other members of the Papionini.

Environmental endocrinology and reproductive success

> The physiological mechanisms that control energy balance are reciprocally linked to those that control reproduction, and, together these mechanisms optimize reproductive success under fluctuating metabolic conditions. Thus, it is difficult to understand the physiology of energy balance without understanding its link to reproductive success.
>
> Schneider, 2004

Strong linkages between intra-sexual competition for social rank, the neuroendocrine control of reproduction, and the physiology of nutrition make it essential to consider how these processes may have acted synergistically during the evolution of the mandrill. We have seen that mandrill groups range over large areas to acquire sufficient food, that they are seasonal breeders, and that maturing and adult males emigrate from supergroups for extended periods in order to forage alone. The large body size of adult males and the huge collective biomass of females and their offspring in supergroups may be presumed to engender a high level of feeding competition in this species. It is regrettable that we still know so little about behavioural interactions between the members of wild mandrill supergroups. Moreover, very little is known about the behaviour of solitary males living under completely natural conditions.

Studies of the semi-free ranging mandrill groups at the CIRMF in Gabon have provided some valuable insights concerning these gaps in knowledge, although there is still much to be learned. 'Clumped resources', such as the fruits and vegetables that are used to provision these groups, result in greater proximity between group members, with resultant increases in aggression. Higher-ranking females and members of their matrilines have priority of access to food. Female rank and reproductive success are also positively correlated; the infants of high-ranking females grow faster during the pre-weaning and post-weaning stages, and dominant females have greater lifetime reproductive success. Thus, it is likely that high rank confers metabolic and other advantages, which increase female fertility and help to sustain lactation once an infant is born. It will be worthwhile to discuss how dietary and social factors might interact at the neuroendocrine level to bring about such effects. In the absence of the necessary physiological studies of mandrills, however, I must beg the reader's indulgence at this point, as it is necessary to digress briefly, in order to consider some results of relevant work on other mammals.

It is now well established that the hypothalamic gonadotrophin releasing hormone (GnRH) 'pulse generator' is the primary regulator of fertility and, as was discussed in Chapter 7, GnRH pulsatile activity is modulated by peptidergic inputs (via kisspeptin and gonadotrophin inhibitory hormone neurons). These inputs convey information derived from an animals's external and internal environments, including its metabolic status as well as the seasonal and social stimuli that affect reproduction (see pages 84–86, and Figure 7.6). It has been demonstrated experimentally that luteinizing hormone (LH) pulse frequencies can be modulated by changes in nutrition (e.g. in male rhesus monkeys: Cameron and Nosbisch, 1991; and in rams: Blache *et al.*, 2003). In Merino rams, for example, the frequency of LH pulses increases two- to three-fold in just six to eight hours, if the animals are fed a highly nutritious supplement (lupin grain) rather than a basic maintenance diet. Acute effects on LH pulse frequency of withholding food, or re-feeding animals, are likely to be mediated via a common stimulus, the availability of oxidizable metabolic fuels (including free fatty acids and glucose: Schneider, 2004). However, the existence of adequate fat reserves likely provides a buffer against such short-term challenges to reproductive function. Fat tissue secretes leptin, a peptide hormone that acts on the brain to regulate both food intake and energy expenditure. It is argued that leptin provides 'a barometer of body condition' (Clarke and Henry, 1999) rather than exerting acute effects on reproduction in adulthood, or by triggering the onset of puberty (Clarke and Henry, 1999; Plant and Witchel, 2006).

Emotional ('stressful') factors and 'temperament' can also modulate the effects of diet upon the GnRH pulse generator, and thus alter gonadotrophin secretion and affect fertility. These points may be illustrated with reference to some interesting experiments on sheep. Blache and Bickell (2011) have shown that Merino ewes differ in their responses to social isolation and in their reactivity to humans; moreover, these differences in temperament are heritable. It proved possible to breed two lines of sheep, 'nervous' ewes that showed the greatest negative responses during isolation and arena tests, and 'calm' ewes that consistently scored lower on these measures. Diet is known to affect the number of ovulations in ewes (Scaramuzzi *et al.*, 2006), and it transpired that calm ewes achieved higher rates of multiple pregnancies than nervous ewes when all the animals were maintained on the same diet (Hart *et al.*, 2008).

Returning now to the subject of social rank, nutrition and reproduction in female mandrills, let us revisit some of the events that occur during the annual mating season, as discussed in Chapters 7 and 8. Most of the female mandrills in the CIRMF rainforested enclosures gave birth between the months of November and March, and were still lactating when the annual mating season began in June. Lactation is a metabolically costly process, which is known to inhibit the GnRH pulse generator and to cause lactational amenorrhoea in Old World monkeys, apes and humans (McNeilly, 2006). We may hypothesize that high-ranking female mandrills may be better able to cope with such metabolic challenges, given that rank confers priority of access to food and also, perhaps, because dominant females have built up greater fat reserves in the months prior to giving birth. Unfortunately, nothing is known about the physiology of fat metabolism in mandrills; for example, fat reserves and leptin levels have not been measured, and this should be a priority for future research. It is known, however, that high-ranking females are less likely to require multiple cycles in order to conceive, and that they have shorter interbirth intervals than lower-ranking females. Might it be the case that low rank compromises the ability of some females to maintain adequate energy balance, and thus to ovulate during the earlier part of the mating season? For those females that cycle towards the end of the mating season, it is likely that fertility is further compromised by changing environmental conditions, so that reproduction may have to be deferred until the following year.

It is interesting that all females in the CIRMF colony, irrespective of their ranks, tended to have long follicular phases (maximum duration recorded = 57 days). This might be partly a consequence of metabolic constraints upon gonadotrophin secretion and follicular development. Females in which metabolic reserves have been depleted by long periods of pregnancy, lactation and infant care may be challenged to ovulate. Oestrogen presumably remains at lower levels during long follicular phase cycles, as indicated by protracted periods of partial sexual skin tumescence. Measurements of circulating gonadal hormones and serial blood sampling to measure LH and follicle stimulating hormone (FSH) pulses are required in order to explore these questions. What is clear, is that the sexual skin swelling of the female mandrill is much more likely to have evolved as a graded signal of female reproductive condition (Nunn, 1999; Setchell and Wickings, 2004b) rather than representing a reliable indicator of individual long-term reproductive success (Domb and Pagel, 2001).

Studies to date have failed to reveal any statistical correlation between female rank and the duration of the follicular phase of the menstrual cycle in mandrills (see Figure 8.20, page 116). This question deserves more detailed study, however. In baboons, for example, females show significantly longer follicular phases in response to stressors, such as receipt of aggression or removal from their social group (Rowell, 1970). It is interesting that lower-ranking female mandrills sometimes exhibit interrupted follicular phases, involving transitory detumescence of their swellings, followed by resumption of the cycle (see Figure 8.9, page 103). These events may reflect transient reductions in GnRH because of social or metabolic stress. Whether changes in adrenal function and elevations in cortisol might play some role in this context has not been studied in the mandrill.

Any future work in this field might benefit from a broader approach to assessing social relationships among mandrills, rather than relying solely upon measurements of agonistic behaviour and dominance rank. Given that sheep differ in temperament (as was discussed earlier), and that such differences can affect reproductive success in ewes (Blache and Bickell, 2011), it is certainly valid to enquire whether similar effects might occur in monkeys. At an anecdotal level, it is apparent that some low-ranking female mandrills are extremely 'nervous' in social situations, as evidenced by higher levels of avoidance, fleeing, fear grimacing and crouch presentations during socio-sexual encounters. Yet, other low-ranking individuals react more 'calmly' and appear to cope better with the proximity of conspecifics during feeding and sexual interactions. A good example of this calmer behaviour was seen in female no. 12 in the CIRMF mandrill colony. This female was of medium rank, but she was highly successful socially and reproductively. For example, she managed to avoid aggressive interactions, and groomed infants and other relatives much more frequently than any other female. How far such individual differences in temperament might result in different reproductive outcomes for the females concerned is not known.

Do social factors coordinate the menstrual cycles of mandrills, so that synchrony, or asynchrony, of cycles might occur among a group's females during the annual mating season? Menstrual cycle synchrony, or reduction of irregular cycle lengths, reportedly occurs in a number of Old World anthropoids, including long-tailed macaques (Wallis *et al.*, 1986), hamadryas baboons (Zinner *et al.*, 1994) and captive chimpanzees (Wallis, 1985). However, asynchronous cycles were reported to occur in free-ranging chimpanzees by Matsumoto-Oda *et al.* (2007). Tight coordination in the timing of conceptions has been inferred for certain species that have very short birth seasons (e.g. vervet monkeys: Schapiro, 1985; squirrel monkeys: Milton and Johnston, 1984). In Milton and

Johnston's study of squirrel monkeys (*Saimiri sciureus*), 80% of births occurred over a ten-day period during the wet season. In the mandrill, although births also occur during the wettest months, we have seen that they are much less tightly grouped than in the squirrel monkey example (83% of female mandrills give birth between November and March: see Figure 7.1, page 78).

Despite the lack of distinct patterns of mating seasonality in the apes and in humans, there have been reports that menstrual synchrony occurs in women. McClintock, (1971) was the first to describe such synchrony, among women who shared a college dormitory. A number of later studies claimed to confirm this, although the effect was reported to be greatest among friends and room-mates (Weller and Weller, 1993; Weller *et al.*, 1995). Synchronization of cycles in response to olfactory cues alone has also been reported (by using axillary secretions collected from donor females at various stages of the menstrual cycle: Preti *et al.*, 1986; Stern and McClintock, 1998).

Serious doubts have emerged, however, concerning the existence of menstrual cycle synchrony among the Old World monkeys, apes and humans (Schank, 2000, 2006; Strassmann, 1997, 1999). It transpires that the 'synchronies' reported in many earlier studies were probably artefacts, resulting from statistical problems. For example, the use of more rigorous statistical tests has failed to confirm that menstrual cycle synchrony, or asynchrony, occurs in captive hamadryas baboons (Tobler, 2008; Tobler *et al.*, 2010). In my own work on mandrills, I initially thought that synchrony of cycles might occur during the annual mating season, and especially among closely related females (Dixson, 1998a). However, more extensive studies of the CIRMF mandrills have shown that synchrony of menstrual cycles does not occur, even among members of the same matriline (Setchell *et al.*, 2011a). Setchell *et al.* caution that the reason that synchronies have been reported in some primate studies might be, in part, a result of 'an evolved human tendency to detect patterns in meaningless noise'. This may very well be true! It seems highly unlikely, for example, that axillary secretions could exert pheromonal effects upon the timing of the menstrual cycle in women (Preti *et al.*, 1986; Stern and McClintock, 1998). Humans, like the apes and Old World monkeys, lack a functional vomeronasal organ and accessory olfactory system (Evans, 2003; Maier, 2000). This route for chemical communication, which plays such a vital role in mediating pheromonal effects upon reproduction in many vertebrates, has been lost during the evolution of the Old World anthropoids. By contrast, the retention of a functional vomeronasal system in many New World primates and prosimians renders it far more likely that pheromonal communication might continue to play a part in the neuroendocrine regulation of ovarian cycles in species such as ring-tailed lemurs or marmosets (Dixson, 2012).

Coordination of reproductive functions between the sexes is essential if fertile matings are to occur; this is especially so in species like the mandrill, where mating is constrained by seasonal factors. Social stimuli, as well as changes in climate and diet, are thus likely to play a significant role in this context. Social disruption can certainly alter the timing of the mandrill's mating season. Thus, when a second mandrill group was formed at the CIRMF, by transferring some animals to an adjacent enclosure, the timing of onset of the females' menstrual cycles was disrupted, so that longer mating seasons occurred during subsequent years (see Figure 7.4, page 83).

Under natural conditions, as increasing numbers of female mandrills develop their sexual skin swellings, larger numbers of males join supergroups, in order to mate (Abernethy *et al.*, 2002). The possible physiological effects of the onset of females' cycles upon males have been studied in semi-free ranging groups of rhesus monkeys (Vandenbergh, 1969; Vandenbergh and Drickamer, 1974). Treatment of ovariectomized female rhesus monkeys with oestradiol during the non-mating season brought about marked activation of sexual behaviour among the adult males, as well as androgen-dependent reddening of the males' sexual skin. Further work confirmed the expected rises in plasma testosterone in male rhesus monkeys, when exposed to sexually attractive females in a social group (Rose *et al.*, 1972). Mandrills are not such strictly seasonal breeders as the rhesus monkey, and their mating season is unlikely to be affected by photoperiodic cues, for the reasons discussed in Chapter 7. However, in addition to being visually attractive to male mandrills, females' swellings may have some effects upon neuroendocrine responsiveness. Thus, males may show some increases in reddening of the facial sexual skin and higher faecal androgen concentrations in the presence of females that have sexual swellings. How much of this increase in androgen might have been testicular in origin and how much was contributed by the adrenal glands is uncertain (Setchell *et al.*, 2008). I make this point because it is known that faecal glucocorticoid levels in male mandrills also increase during the annual mating season, indicating that the adrenal cortex is more active under these stressful conditions, and especially so in dominant males (Setchell *et al.*, 2010b).

Frequent confrontations must occur between male mandrills in the wild, as they begin to enter supergroups during the mating season; this represents a huge departure from their relatively non-gregarious existence during the rest of the year. Large changes in circulating testosterone may occur under such unstable conditions, as exemplified by the changes in hormone levels that follow rank disputes between males in semi-free ranging groups (see Figure 8.26, page 124). These fluctuations in testosterone are the result of rank changes and associated stressful interactions, not the cause of them. Comparative studies have shown that, where rank orders are well established, basal levels of testosterone do not correlate with dominance in captive Japanese macaques (Eaton and Resko, 1974b), rhesus macaques (Gordon *et al.*, 1976), baboons (Sapolsky, 1982) or vervets (Raleigh and McGuire, 1990). If plasma testosterone is measured before group formation, then hormone levels are not predictive of the ranks attained by males once they are introduced into the group environment (rhesus monkeys: Rose *et al.*, 1975; squirrel monkeys: Mendoza *et al.*, 1978; talapoins: Eberhart and Keverne, 1979). It may be inferred that the same is true for male mandrills; the huge size and impressive secondary sexual adornments of dominant males secure advantages for them once they enter supergroups, but individual differences in aggressiveness are unlikely to be predictable simply by measuring testosterone levels prior to the start of the mating season. This should not surprise us, because there is no simple cause and effect relationship between testosterone and aggression in male primates; effects exist, but they are much more subtle than are the effects of testosterone upon sexual behaviour (Dixson, 1980).

The situation is somewhat different, however, when one considers unstable hierarchies and the interplay between aggression and circulating testosterone in male primates.The volatile nature of inter-male relationships and associated changes in testosterone in male mandrills at the commencement of the mating season may usefully be discussed in relation to Wingfield's challenge hypothesis. As a result of his studies on territorial aggression in birds, John Wingfield proposed that associations between elevated testosterone levels and aggressive behaviour occur primarily when males challenge one another, as when male song sparrows are setting up territories or guarding the first set of offspring (Wingfield 2005; Wingfield *et al.*, 1990, 2001). Results of experimental studies using male California mice are consistent with the challenge hypothesis. Thus, testosterone injections that produce transient elevations in hormone levels result in increased aggression between males when they are given encounter tests the following day (Trainor *et al.*, 2004). In several species of monkeys, it has been shown that under unstable social conditions, dominant, aggressive males tend to exhibit elevated testosterone levels, whereas hormone levels decrease in subordinates (in rhesus macaques: Bernstein *et al.*, 1974; Rose *et al.*, 1972, 1975; baboons: Sapolsky, 1993; mandrills: Setchell and Dixson, 2001b; and possibly in geladas: Bergman *et al.*, 2009; Pappano and Beehner, 2014). Perhaps in some primates, as well as in birds or mice, the experience of 'winning' an aggressive encounter is somehow reinforced by transient elevations in testosterone, increasing the likelihood that a male will be confident and dominant during subsequent encounters. These ideas have yet to be fully tested experimentally in primates.

The intense nature of inter-male competition in the mandrill has led to selection for alternative reproductive strategies and associated physiological specializations in some males. The occurrence of a continuum of morphological variants, ranging from brightly adorned fatted males at one extreme to leaner individuals with muted sexual skin colouration at the other, was discussed in Chapter 8. Dominant, fatted males exhibit higher concentrations of circulating testosterone than do non-fatted males, in which there is a leaner body composition. Fatted and non-fatted male mandrills do not differ significantly in body weight or canine size, however, although crown-rump lengths are greatest among the non-fatted males. Perhaps pubertal development proceeds more slowly in males that are destined to become non-fatted, so that their skeletal growth continues for longer. More detailed studies of hormonal and skeletal development during puberty will be needed to address this question. We do know that a young male mandrill's high rank is positively correlated with earlier pubertal onset and emigration from his natal group (Setchell and Dixson, 2002). However, it is not clear whether such males, that emigrate early, are more likely to become fatted as adults. To date, nothing is known concerning development of adolescent males in the wild, once they emigrate from their groups. It might be the case, for example, that some males remain in a 'sub-mature' state for extended periods, only gradually transitioning to full physical development and entering supergroups for progressively longer periods during annual mating seasons. We do know that some sub-mature males, defined by Abernethy *et al.*, (2002) as being aged between six and nine years, re-enter groups during the mating season, but we have no idea how long these 'visits' might last.

Fatted male mandrills may be better able to cope with the metabolic challenges that occur throughout the long mating season, during which time they follow and mate-guard females for extended periods. As their fat reserves decrease, so these males revert to

Table 10.2. Some further examples of suppression/delay of physical development in male primates: possible effects of inter-male competition

Species	Traits affected	Primary mating system
Prosimians		
Galagoides demidoff	Body weight of subordinates is 75% of that of dominant males	Dispersed
Galago moholi	'B' males weigh less and have smaller sternal glands than 'A' males	Dispersed
New World monkeys		
Cacajao calvus	Muscular temporal 'bulges' on the head develop more slowly. Testicular growth is retarded	Multimale–multifemale
Alouatta caraya	Brown (juvenile) pelage colour is retained	Multimale–multifemale
Old World monkeys		
Nasalis larvatus	Delayed growth of the adult male's large nose	Polygyny
Cercopithecus neglectus	Retention of juvenile (russet-grey) pelage	Polygyny
Macaca fascicularis	Attainment of adult body weight delayed until males emigrate and forage as solitaries	Multimale–multifemale
Great Apes		
Pongo spp.	Delayed development of cheek flanges, vocal sac, hair length and adult body mass	See text

From Dixson (2012), and based upon sources cited therein.

foraging alone, and many of them eventually emigrate from supergroups at the end of the mating season. It may be much more difficult for non-fatted males to remain attached to supergroups throughout the mating season, however. They may be more likely to suffer from metabolic imbalances that cause suppression of gonadotrophin secretion, as was discussed previously in relation to the possible combined effects of social rank and poor nutrition on reproduction in female mandrills.

Suppression (or partial suppression) of the development of secondary sexual adornments and other physical traits, owing to intra-sexual competition between males, occurs in a number of primate species; some examples are provided in Table 10.2. Indeed, I think it is likely that such reproductive strategies may be more common among primates than is currently realised or documented (see Dixson, 2012, for a review). As regards the behavioural and physiological mechanisms that underpin such effects, some interesting comparisons may be made between the fatted and non-fatted reproductive strategies of male mandrills and the occurrence of 'flanged' versus 'non-flanged' strategies in male orang-utans. Dominant adult male orang-utans have fully developed cheek-flanges, a throat sac, and other secondary sexual traits, which are subject to developmental suppression in lower-ranking males (Figure 10.14). Dominant males occupy individual home ranges that overlap those of a number of females. Subordinate non-flanged males

Figure 10.14. Adult male Bornean orang-utans (*Pongo pygmaeus*). **A:** A dominant 'flanged' male exhibiting full development of the cheek flanges and laryngeal sac. **B:** A smaller 'non-flanged' male, in which secondary sexual traits are only partially developed. Photographs taken in the Gunung Palung National Park. Copyright Tim Laman. (From Dixson, 2012, after Knott and Kahlenberg, 2007.)

live a more peripatetic existence, avoiding flanged males and mating opportunistically (and sometimes coercively) with females (Knott and Kahlenberg, 2007). In captivity, male orang-utans reach puberty and begin to produce sperm at between six and seven years of age. However, their cheek flanges and other secondary sexual adornments are not fully expressed until they are nine or ten years old (Dixson *et al.*, 1982). The presence of a dominant (i.e. fully developed) adult male in the same captive environment can suppress the growth of secondary sexual characteristics in younger males for up to seven years (Kingsley, 1982, 1988). The same is true in the wild, where some adult males remain in this suppressed condition for many years, and where both flanged and non-flanged males are known to sire offspring (Utami Atmoko and van Hooff, 2004).

Endocrine studies of captive orang-utans have shown that non-flanged males have lower levels of luteinizing hormone, testosterone, dihydrotestosterone and growth hormone than do fully developed males (Maggioncalda *et al.*, 1999, 2000). The growth hormone data are intriguing because non-flanged males are much smaller than dominant orang-utans. This is not the case in non-fatted male mandrills, however, as their body weights are comparable to those of fatted males. No measurements of growth hormone are available for mandrills, but it seems unlikely that fatted and non-fatted males will be found to differ in this respect. Non-fatted males are longer and leaner in the body and limbs, whereas fatted males are stocky, and adipose tissue is concentrated in the rump and flanks. These major physiological differences between mandrills and orang-utans imply that the underlying neuroendocrine mechanisms mediating reproductive suppression are likely to differ in significant ways between the two species. The division between the two orang-utan morphs is more extreme, for example, and suppression probably lasts for longer under natural conditions in male orang-utans. What is common to both species is that developmental arrest is reversible, so that testosterone levels increase and secondary sexual traits develop in subordinates if dominant males are removed (orang-utan: Kingsley, 1988; mandrill: Setchell and Dixson, 2001b).

The precise nature of the cues that mediate inter-male suppression is not known. Given that male orang-utans lead a largely solitary existence, their 'long-calls' may play some role in this respect, as well as their striking appearance and display tactics during direct encounters. Male mandrills also spend long periods foraging alone, but they mark trees with their sternal glands, and fatted males retain their striking colouration throughout the year. Continued communication between males is thus possible in theory, so that 'solitary' males should not be thought of as being totally isolated from one another or from supergroups.

Clearly, much remains to be learned about proximate mechanisms that cause, and maintain, arrested physical development in male mandrills, orang-utans and males of other primate species listed in Table 10.2. Because 'suppressed' males may be sexually active and fertile, it is possible that specific androgen-sensitive peripheral tissues, such as the mandrill's red sexual skin or the orang-utan's cheek flanges, might be selectively desensitized to effects of androgens. Hormone receptor concentrations might be reduced, or perhaps the enzymatic conversion of testosterone to oestrogen (by aromatization) or DHT (by 5-alpha-reductase) might be compromised. Some results of experiments on male rhesus monkeys by Rhodes *et al.* (1997) are relevant to this discussion. These workers showed that red sexual skin colouration decreased after treatment of male rhesus monkeys with fadrazole (an aromatase inhibitor), but was maintained by either testosterone or oestradiol. DHT failed to maintain the red colour, however. As testosterone (but not DHT) may be aromatized to oestrogen, these authors concluded that testosterone acts as a prohormone in the control of sexual skin colouration. If regional differences in aromatase activity were to occur in sexual skin, this might explain, for example, why muting of red colouration on the nose and genitalia of subordinate male mandrills is often greater than that involving the perianal field.

Sexual selection and behaviour: a critical discussion

> The operational distinction between male competition and female choice is not always obvious, since they are often complexly interrelated. For example, a female might behave in a manner that incites competition between males with the result that the female acquires a competitively superior mate.
>
> Because the pairing of the sexes may result from a number of causes, such as male competition or random encounters, it is necessary to make detailed observations of the behavior of both sexes prior to mating.
>
> Ryan, 1985

This quote is taken from Michael Ryan's superb book *The Tungara Frog: A Study in Sexual Selection and Communication*. His choice of this unassuming little animal as a study species (these frogs are only about 3 cms long) was made after careful deliberation, and it was a wise decision. Male Tungara frogs on Barro Colorado Island, which is situated in Gatun Lake in the Panama Canal, attract females by calling; the advertisement call consists of a 'whine', which is sometimes followed by a short series of 'chuck' sounds. Female frogs are attracted by calls given by large aggregations of males. Each female chooses her mate, however, as she approaches and contacts a specific caller, at which point the male initiates amplexus (the clasping response that precedes the shedding of ova and sperm). Via a series of ingenious experiments, Ryan demonstrated that female Tungara frogs are more likely to choose males that include chuck sounds in their calls, and they show a preference for lower-frequency chucks, which are often given by larger males. These preferred, complex advertisement calls incur a serious 'cost' for the males concerned, however, as their major predator, the bat *Trachops cirrhosus*, is most likely to attack male frogs that are giving complex calls.

I have begun this discussion with the example of male advertisement, predation risk, and female mate choice in the Tungara frog, because the results of Ryan's painstaking work are so much more complete and compelling than anything that has been achieved by studying primates. In general, the role played by sexual selection in female mate choice has been much more extensively researched in birds, reptiles, amphibians and fish than in primates and other mammals, as reflected by the content of authoritative textbooks that deal with animal behaviour (e.g. Alcock, 2013; Davies *et al.*, 2012). By necessity, therefore, the following discussion includes results of work carried out on birds and other vertebrates, as well as on mandrills and other primates.

In his *Descent of Man*, Darwin (1871) noted that: 'With mammals the male appears to win the female more through the law of battle than through the display of his charms'. Where the mandrill is concerned, this appears to be the case, with the caveat that not all battles necessarily involve physical conflict between the participants. In the male mandrill, intra-sexual selection has resulted in the evolution of a diverse battery of traits, some of which enhance fighting ability, while others function indirectly as conspicuous badges of status. Thus, we have seen that males are three times larger than adult females, and they have huge canine teeth. They engage in potentially damaging fights only rarely, however; instead, their bright colouration, two-phase grunting vocalizations, and scent-marks serve as distance signals of social status that also help them to avoid unnecessary proximity with each other. This latter point is important, as visual communication is limited among the dense vegetation, and in the deep shade of the rainforest floor, where the mandrill and its forebears have evolved over the millennia.

It might be argued that the striking colouration of the male mandrill should make it more conspicuous to leopards and other predators, as in the case of the male Tungara frog's complex advertisement calls and predation by bats. As was discussed earlier, it is possible that natural selection as well as sexual selection might have contributed to the evolution of extreme body weight and canine size dimorphism in male mandrills. Long periods of solitary foraging by adult males place them at greater risk of attacks by leopards, but any leopard that risked attacking an adult male mandrill might be severely injured as a result.

I consider it unlikely that the male mandrill's red sexual skin and other extraordinary adornments could have evolved primarily as a result of female mate choice. Indeed, as was stressed in Chapter 8, direct observations of the pre-copulatory behaviour of mandrills provide only weak evidence for effects of female mate choice. Males initiate mate-guarding, and the vast majority of copulations are likewise male-initiated. Females, moreover, will mate with multiple partners, to the extent that dominant male mandrills exhibit various genital and behavioural adaptations indicative of the effects of sexual selection via sperm competition.

As has been emphasized in earlier chapters, I am not implying that female mandrills mate indiscriminately; males are sometimes invited, avoided or refused. Yet, the complex social and sexual lives of mandrills, and of most anthropoid primates, are very different from those of frogs, birds or fish, in which masculine secondary sexual adornments and courtship displays play such well-defined roles in determining which animals will mate with each other. However, as pointed out by Ryan (1985), if a female behaves in ways that increase inter-male competition, then the result may be 'that the female acquires a competitively superior mate.' This is likely to be the case where the mandrill is concerned. When females develop their highly attractive sexual skin swellings during the annual mating season, males enter supergroups in increasing numbers to compete for copulations. Female attractiveness stimulates intra-sexual competition between males.

Swellings are widespread among the Old World anthropoids, including at least some members of all genera that belong to the Papionini (Figure 10.15). It is likely that sexual skin has arisen at least five times during the evolution of the Old World monkeys and apes (Dixson, 2012). At the morphological level, the origins of the phenomenon probably reside in the slight oedema and pinkness of the vulva caused by oestrogen around the time of ovulation, as occurs in various prosimian primates. Among the diurnal anthropoids, the existence of excellent colour vision, and the propensity of males to find such female cues visually attractive, has given rise to runaway sexual selection for increasingly elaborate pink or reddish swellings (Dixson, 1983a), which function as graded signals of female fertility (Nunn, 1999). This applies to a number of species that have multimale–multifemale mating systems, including the mandrill, talapoin and chimpanzee. In such cases, swellings may enhance the ability of females to attract multiple potential mating partners (Clutton-Brock and Harvey, 1976). Indeed swellings might encourage sperm competition, if a female mates with a series of partners during her fertile period (Harvey and May, 1989).

Thus, it is possible in the mandrill that heightened female sexual attractiveness, and graded visual signals of impending ovulation, have been the main drivers of male–male competition. Intra-sexual competition between male mandrills at both the pre-copulatory level (for access to females), and post-copulatory level (via sperm competition) is at its most intense when sexual skin turgescence is greatest, and females are most likely to conceive. Although female mandrills may not actively choose males on the basis of their

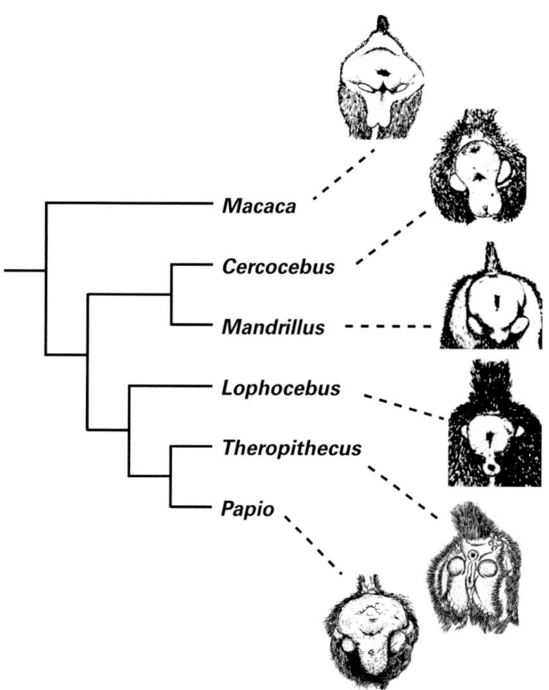

Figure 10.15. Phylogenetic variations in the morphology of sexual skin among the various genera of the Papionini (excluding *Rungwecebus*). (Author's drawings.)

colourful adornments, they are more likely to mate with males of high rank, and such males tend to be the most colourfully adorned.

Females may also compete with one another in ways that affect their access to mates and, ultimately, their reproductive success. Female mandrills establish rank orders that are measurable in terms of agonistic interactions, approach–avoidance behaviour, displacement of conspecifics from feeding areas and socio-sexual relationships. High-ranking females have preferential access to food, and nutritional status may have important effects upon fertility. Competition for access to mating partners among females is thus achieved less by direct aggression than by the competitive ability to display sexually attractive adornments. It should also be kept in mind that the ratio of adult females to adult males (the socionomic sex ratio) is very high in mandrills. The most attractive females are likely to have greater success in arousing the interest of males and eliciting competition between them.

Modern studies of sexual selection and mate choice often focus upon the hypothetical 'benefits' (whether material resources or genetic benefits) that possessors of attractive traits might confer upon their partners or offspring. Because females usually invest so much more than males in the physiological processes that govern reproduction, mate choice is usually exerted by the female sex. Males, by contrast, are more likely to attempt to 'charm females' or to follow 'the law of battle' by competing among themselves for access to mating partners (Darwin, 1871). As regards mate choice, Darwin expressed the view that females might possess some aesthetic preference for striking masculine adornments, as for example in peafowl and birds of paradise. He commented, however, that 'I was well aware that I thus laid myself open to ridicule.' In modern terms, however, Darwin's idea is not ridiculous, if one substitutes the explanation that females might have some pre-disposition to prefer certain male attributes (visual, auditory or olfactory) owing to the way that their sense organs and nervous systems operate. Such sensory biases could make the favoured traits (and the female preferences for them) subject to runaway sexual selection (Fisher, 1930). In Fisher's model, both the genes affecting female mate choice and the genes for preferred male attributes are thus linked. Runaway selection favours the evolution of progressively more pronounced male adornments, as well as progressively greater female preferences.

Do the masculine traits that initiate such a runaway process necessarily have to be linked to beneficial qualities, such as the greater ability of males to provide resources to females, or 'better genes' such as those affecting resistance to parasites? Mathematical modelling of Fisherian runaway sexual selection (Kirkpatrick, 1982; Lande, 1981) indicates that this is not necessarily the case. In theory at least, selection might occur for a male attribute that has no benefit to those females that choose it. As Richard Prum (2010) has pointed out, the Lande–Kirkpatrick models thus provide researchers with 'a null model of evolution by inter-sexual selection'. This is useful and preferable to the usual approach, which assumes a priori that female choice for particular male traits must be adaptive in terms of good genes or resources associated with possession of those traits.

Some examples may be given of mate choice for traits possessed by the opposite sex simply because they are attractive, with no additional genetic or resource benefits. However, most of these examples relate to artificial situations. For example, in the zebra finch (*Taeniopygia guttata*), females prefer males fitted with red leg-bands, whereas males prefer black and pink-banded females (Burley *et al.* (1982). Female zebra finches also find males fitted with an artificial white crest on top of the head more attractive, but do not favour red or green crests (Burley and Symanski, 1998). Male chacma baboons may be sexually aroused by the sight of an ovariectomized female fitted with an abnormally coloured, (black) artificial sexual skin swelling (Bielert *et al.*, 1989). To cite another example derived from work on primates, certain novel artificial odours can activate mounting behaviour by male rhesus monkeys, when applied to the rumps of ovariectomized females (Goldfoot, 1981).

One example where Prum's (2010) idea can be applied to a natural situation concerns mate choice by female bowerbirds. Male bowerbirds attract females by building decorative arrangements of twigs, berries and other colourful objects. The best constructed bowers attract more females, and the males that build them secure significantly more matings and have greater reproductive success (spotted bowerbird: Madden, 2003; satin bowerbird: Reynolds *et al.*, 2007). Males play no part in providing resources for females or their offspring; the latter receive only the male's gametes, and hence his genes. There is some evidence that successful male satin bowerbirds might have genetic advantages as reflected by the quality of their plumage and greater resistance to ectoparasites (Doucet and Montgomerie, 2003), as well as by the attractiveness of their bowers.

Field experiments have also shown that successful males of this species may possess superior cognitive abilities (Keagy *et al.*, 2011). When we consider the remote evolutionary origins of bower construction, however, might the propensity of some males to collect randomly a few colourful objects have tapped into a sensory bias present in females, and thus secured a mating advantage? Might runaway sexual selection have sparked the evolution of complex bower construction fortuitously, rather than because of any initial link to male genetic advantages?

Turning now to a discussion of the costs imposed by the production and possession of secondary sexual adornments, an alternative hypothesis that attempts to account for the evolution of extravagant adornments, such as the peacock's tail, in relation to female mate choice is the handicap hypothesis (Zahavi, 1975; Zahavi and Zahavi, 1997). According to this hypothesis, males signal their underlying genetic quality by displaying traits that actually handicap them, for example by placing them at increased risk of predation. Theoretical modelling to test this proposition indicates either that it 'does not work' (Kirkpatrick, 1986), or that it might do so, but only under special conditions (Grafen, 1990a, 1990b).

The handicap hypothesis carries with it an element of plausible story telling, sometimes in the absence of objective data. As an example, in their book *The Handicap Principle: A Missing Piece of Darwin's Puzzle*, Zahavi and Zahavi (1997) seek to explain the evolution of the human beard as a handicap. They do so on the basis that 'it can make a man more vulnerable in a fight' because it might be grasped by an opponent! Yet, experimental studies of the beard's functions, indicate that it acts as a visual signal of masculine status and age, that probably evolved as such because it accentuates sex differences in the size of the jaw, chin and lower face (Dixson and Vasey, 2012; see also Barber, 1995). The beard itself is not perceived as being sexually attractive by women. However, given that somewhat older and socially more dominant males may be preferred as partners, the beard might have had some indirect effect upon mate choice, especially during earlier periods of human evolution.

If one were to adopt a handicap approach to interpreting the male mandrill's adornments, one might argue that, because they render a male highly conspicuous, they handicap him by attracting predators. The most brightly coloured males thus display their 'good genes' and superior survival abilities to potential mates. In reality, we have seen that, like the human beard, the male mandrill's adornments are badges of status, signalling high rank and competitive abilities. They are primarily the result of intrasexual selection. Females benefit by mating with the most dominant males, and their sexual skin swellings encourage pre-copulatory competition between males, as well as post-copulatory sperm competition when matings involve multiple partners. Nor is there any evidence that the sexual skin swellings of female mandrills act as handicaps. Unusual, or unsightly, as they may appear to the human eye, they are highly attractive to males of their own species, and sexual selection has favoured their evolution on that basis. Moreover, we have seen that morphologically very similar swellings occur in both the mandrill and drill, as well as in their closest phylogenetic relatives among the mangabeys (*Cercocebus* spp.). This means that similar swelling morphologies have probably been in existence in this branch of the Papionini for more than four million years. If they had handicapped females, and compromised survival, they would have undergone negative selection and ceased to exist long ago.

There is a physiological iteration of the handicap hypothesis, which is equally problematic in my view. This posits that testosterone has negative effects on the male immune system, and that males with high levels of testosterone (and well-developed secondary sexual traits) signal their ability to withstand testosterone's immunosuppressive effects. This immunocompetence-handicap hypothesis was originally proposed more than two decades ago in a frequently cited paper by Folstad and Karter (1992). The hypothesis has been evaluated mostly by studying birds, and only rarely by experiments on mammals. To the best of my knowledge, it has yet to be adequately tested in any primate species. A meta-analysis of published work, conducted by Roberts *et al.* (2004), produced only weak support for the proposition that testosterone reduces immunocompetence, or that it effects parasite loads. The results of some studies on birds have failed to support the immunocompetence-handicap hypothesis (e.g. red-winged blackbirds: Westneat *et al.*, 2003; house sparrows: Greenman *et al.*, 2005). Where the mandrill is concerned, the male's testosterone-dependent red nasal colouration is unrelated to parasite load, at least in semi-free ranging animals (Setchell *et al.*, 2009). Yet, even if it were possible to measure parasite loads, sexual skin colouration and testosterone levels accurately in completely free-ranging male mandrills, I suggest that any correlations would not necessarily provide convincing evidence in support of the immunocompetence-handicap hypothesis.

To place the immunocompetence-handicap hypothesis in broad evolutionary perspective, it is useful to consider just how ancient, and how successful are the 'sex hormones', with regard to their reproductive and associated functions. The following quote is taken from Bentley's (1976) textbook *Comparative Vertebrate Endocrinology*:

> The sex hormones show a remarkable uniformity; testosterone, progesterone and estradiol-17β are common throughout the vertebrates.
>
> This possibly reflects the "conservative" nature of the sexual process and the early evolution of a mechanism of such efficiency that little subsequent endocrine modification of the hormonal excitants could be advantageous.

Given that testosterone was already present at the dawn of vertebrate evolution, it would seem probable that the immune systems of extant vertebrate species should be well adapted to this hormone. Therefore, the immunocompetence-handicap hypothesis is unlikely to help us understand the evolution of colourful testosterone-dependent adornments.

Setting aside the possible trade-offs that might exist between fitness and hormonal regulation, there is now increasing interest in examining the role that other physiological factors, such as reactive oxygen species, might play in affecting life-history trade-offs (for a comprehensive review of this field, see Dowling and Simmons, 2009). Reactive oxygen species (ROS) are produced as by-products during a wide range of essential oxidative enzymatic reactions in animal and plant cells. When ROS are produced at chronically high levels, however, they become costly in a physiological sense, as the resulting oxidative stress can damage cellular functions, including gene expression. It is of considerable interest, therefore, that trade-offs may exist between production of ROS and the fitness costs incurred during reproductive competition. For example, in mandrills, it has been shown that high-ranking males exhibit significant increases in markers of oxidative stress, owing to ROS, during the annual mating season (Beaulieu *et al.*, 2014). This, of course, is the period when males enter supergroups and engage in intense intra-sexual competition in order to guard and mate with females. Beaulieu *et al.* also reported that female mandrills, irrespective of social rank, tended to show increased levels of oxidative stress during the mating season. This is interesting, in view of the findings discussed earlier (in Chapter 8) concerning the high levels of metabolic stress that are likely to occur at this time, and the very long follicular phases that are commonly seen in females of all social ranks.

Darwin proposed that sexual selection concerns those traits that have been favoured by evolution, not because they necessarily enhance an individual's chances of survival, but because they improve reproductive success. However, the physiology of reproduction, including its hormonal control, were subjects that Darwin could not consider in relation to his theories about sexual selection; the required knowledge of these subjects simply did not exist during the mid-nineteenth century. Had Darwin been aware of these matters, he would undoubtedly have discussed them at some length. Hormones such as testosterone and oestradiol play crucial roles in reproduction, and they also affect the expression of secondary sexual traits in animals and in humans. First and foremost, therefore, we should consider whether the major benefit conferred by a secondary sexual adornment, such as hormone-dependent colourful sexual skin in mandrills, is that it signals possession of a healthy, functional reproductive system. The health of an individual's reproductive system may, in turn, vary depending upon many factors, including its genetic constitution, the functioning of its immune system (including genes of the major histocompatibility complex: Setchell *et al.* 2010a), its ability to resist parasites and diseases (Hamilton and Zuk, 1982), its resistance to oxidative stress (Dowling and Simmons, 2009) and so forth. However, such factors are likely to represent trade-offs and secondary regulators of sexual selection for many adornments that are hormone dependent. The primary drivers are the linkages between adornments, the neuroendocrine system and reproductive fitness.

Small initial degrees of sexual dimorphism in those traits that depend upon sex hormones for their expression may have undergone sexual selection precisely because they provide clues about an individual's reproductive potential or its social rank (or both). A male mandrill's brilliant androgen-dependent red sexual skin, on the face, rump and genitalia, signals his physical condition and his likely rank, as well as his reproductive potential. Other males may avoid conflicts with such brightly adorned individuals precisely for these reasons. Adequate basal levels of testosterone are required for the functioning of the male's reproductive organs; more brightly coloured male mandrills have higher levels of testosterone, but they also have larger testes, and are thus very likely to have higher sperm counts as a result. If, indeed, sexual selection by female choice has had any influence upon the evolution of the male mandrill's red colouration, it may be because the most brightly adorned males are potentially more fertile, as well as being more dominant. Where the female mandrill's oestrogen-dependent sexual skin swelling is concerned, it is clearly sexually attractive, and it provides even more information about fertility, including the approximate timing of the fertile period (but not the exact timing of ovulation, as we have seen). Note, however, that female swellings do not signal long-term qualities of reproductive success; instead, they provide graded information about current fertility.

There is a crucial difference between those Old World primate species, in which females exhibit marked changes in sexual skin morphology and sexual attractiveness during the menstrual cycle, and humans, as women lack such traits. Does this mean that oestrogen has had no effect upon the evolution of sexual attractiveness in women? That such is not the case is shown by results of modern research on female facial attractiveness and reproductive status. Thus, men judge the faces of women who have the highest oestrogen levels during their menstrual cycles as being most attractive (Law-Smith *et al.*, 2006). Women with narrow waists and large breasts also have higher follicular phase levels of oestrogen and greater reproductive potential (Jasieńska *et al.*, 2004).The distribution of fat in the female body, from puberty onwards, is affected by oestrogen. Young women in their prime reproductive years, who have shapely 'hourglass' figures with a low waist-to-hip ratio, have a healthier body fat distribution and are more likely to be fertile. Men, belonging to diverse cultures around the world, judge images of such shapely women to be most sexually attractive and marriageable (Dixson *et al.*, 2010a, 2010b; Singh, 2006; Singh *et al.*, 2010). During the course of human evolution, these feminine traits may have been advantageous in acquiring and retaining long-term partners, and may thus have been moulded by sexual selection as well as by natural selection. Note that male preferences for these attractive female traits are based upon clear linkages between reproductive physiology (hormone levels), sexually attractive morphological cues (facial and bodily traits) and greater fertility and reproductive potential.

I suggest that the quest to understand mate choice in primates, for traits that might signal 'better resources' or 'better genes' is likely to be most successful when focusing on species where a single long-term partner is chosen for reproductive purposes. The choice of one mate, rather than copulating with multiple partners, places a greater premium on selecting the best possible option. Monogamous mate choices are thus likely to be highly discriminating, not withstanding the possibility that extra-pair copulations might occur, as is now known to be the case in some monogamous primates as well as in many bird species (Birkhead and Møller, 1992).

To cite an ornithological example, king penguin pairs are monogamous during an individual breeding cycle. Both partners cooperate closely for 14 months to incubate the single egg and rear their chick to the stage where it becomes independent. Both sexes thus invest heavily in rearing their single offspring. These birds are monomorphic, but both sexes exhibit three adornments that are thought to play important roles in mate choice. These are an auricular patch of dull yellow or orange feathers, yellow or rusty-brown breast feathers, and a beak spot on the lower mandible (Pincemy *et al.*, 2009). The beak spot is unusual as it reflects ultraviolet (UV) light; varnishing over this spot, so that UV reflectance is diminished, significantly increases the time taken by male king penguins to pair with females (Jouventin *et al.*, 2009). In their studies, Pincemy *et al.*

(2009) showed that reducing the size of the auricular patch, or changing its colour and the colour of the breast patch from yellow to white, also significantly lengthened pairing times for males (but not for females). Females therefore exerted the strongest mate choices, at least in this particular penguin colony where the sex ratio was strongly male biased. The quality of the adornments that affect mate choice in king penguins may also correlate with an individual's immunocompetence (male breast patch colour: Nolan *et al.*, 2006), body condition (male beak spot UV reflectance: Dobson *et al.*, 2008), and strength of territorial defence (auricular patch size in both sexes: Viera *et al.*, 2008).

It has long been known that some primates form pairs and live in small family groups. These relationships endure for much longer than those of king penguins, however, as the latter usually choose a new partner during each breeding cycle. Given the long-term nature of pairings that occur in primates such as owl monkeys, titi monkeys and gibbons, selection may have favoured the evolution of strong patterns of mate choice by both sexes. Males, as well as females, invest a huge amount in these monogamous relationships, and paternal care of the resulting offspring is much more prevalent (as, for example, in owl monkeys, marmosets and tamarins: Dixson and Fleming, 1981; Dixson and George, 1982) than in those primates that have multimale–multifemale or polygynous mating systems. Extra-pair copulations certainly occur in a number of pair-living primates, as in the siamang (Palombit, 1994), white-handed gibbon (Reichard, 1995) and the indris (Bonadonna *et al.*, 2014). Yet, it is probable in such cases that the great majority of each female's offspring are sired by a single, long-term partner. This is the case, for example, in the white-handed gibbon (*Hylobates lar*), in which genetic studies have shown that 38 out of 41 (i.e. 92.7%) offspring in free-ranging groups had been sired by females' long-term partners (Barelli *et al.*, 2013).

Monogamous primates are usually sexually monomorphic, and in some species, both males and females possess specialized cutaneous adornments. This is the case in the emperor tamarin (*Saguinus imperator*), in which both sexes have a magnificent 'moustache', and in the aptly named cottontop tamarin (*S. oedipus*), in which males and females have a prominent white crest on the head. In some species, however, the adornments differ between the sexes, as for example in the white-cheeked gibbon (*Hylobates leucogenys*). In this species, pelage colour and facial markings are sexually dimorphic. In owl monkeys (genus *Aotus*), which are nocturnal, the two sexes are virtually identical in appearance, but their subcaudal scent-marking glands are histologically sexually dimorphic, and olfactory signals play important roles in social communication and sexual recognition (Dixson, 1983b; Hunter and Dixson, 1983). Long-term fieldwork in Argentina on family groups of *Aotus azarai* has shown that stable monogamous pairs achieve significantly greater reproductive success than pairings in which partners change, sometimes because one member is ousted by a solitary 'floater' in the surrounding population (Fernandez Duque and Huck, 2013). DNA analysis has confirmed that resident males in family groups of these monkeys sire 100% of the offspring (Huck *et al.*, 2014). The extent to which olfactory communication between the sexes might influence long-term mate choice in owl monkeys remains an unexplored question. Likewise, the role that visual adornments of certain monogamous primate species might play in mate choice has not been studied experimentally. Nor is it known whether better development of such traits might signal possession of 'better genes' or greater 'resource potential' in a prospective partner. Clearly, much remains to be learned about inter-sexual selection in monogamous primate species. Indeed, they may provide better models for research on female mate choice than primate species such as the mandrill that have multimale–multifemale mating systems.

Birth sex ratios and male infanticides

In the concluding section of this chapter, I shall deal briefly with two possible avenues of selection that have been much debated by primatologists: selection for skewing of the birth sex ratio, and sexual selection and infanticidal behaviour by adult males.

Sex ratio biases in a number of primate species have been reported. Attempts have been made both to correlate such biases with female dominance rank and to explain why mothers should have either more sons or more daughters than the 50:50 birth sex ratio expected by chance. Three hypotheses to account for biases in the birth sex ratio have been advanced:

1. The Trivers-Willard hypothesis.
2. The advantaged daughter hypothesis.
3. The local resource competition hypothesis.

1. Trivers and Willard (1973) postulated that females in good reproductive condition might benefit from having more sons than daughters, as successful sons are potentially able to sire very large numbers of offspring, and to produce more grandchildren than daughters can. Support for this hypothesis derives from studies of red deer (Clutton–Brock *et al.*, 1988) and feral horses (Cameron and Linklater, 2007), but the evidence pertaining to primates is weak, at best (e.g. in free-ranging spider monkeys: McFarland Symington, 1987).
2. Some studies of primates have produced the reverse result, that high-ranking females give birth to significantly more daughters than sons. This was the case, for example, in the captive groups of rhesus monkeys that were maintained for many years at the University of Cambridge, Sub-Department of Animal Behaviour (Nevison *et al.*, 1996; Simpson and Simpson, 1982). These findings are consistent with the advantaged daughter hypothesis, which posits that high-ranking females benefit from producing more daughters because this strengthens their matrilines. Unlike male rhesus monkeys, females are philopatric, and tend to inherit their mothers' ranks in the group. Further work on rhesus monkeys and some other species has failed to support the results of the Cambridge studies.
3. The third hypothesis (the local resource competition hypothesis) seeks to explain why lower-ranking females might bias the birth sex ratio towards production of sons, rather than daughters (Clark, 1978; Silk, 1983). Clark originally pointed out that female greater galagos (*Otolemur crassicaudatus*) tend to give birth to a higher than expected proportion of male offspring. This might be because female offspring tend to remain in the same areas as their mothers, whereas males emigrate as they attain

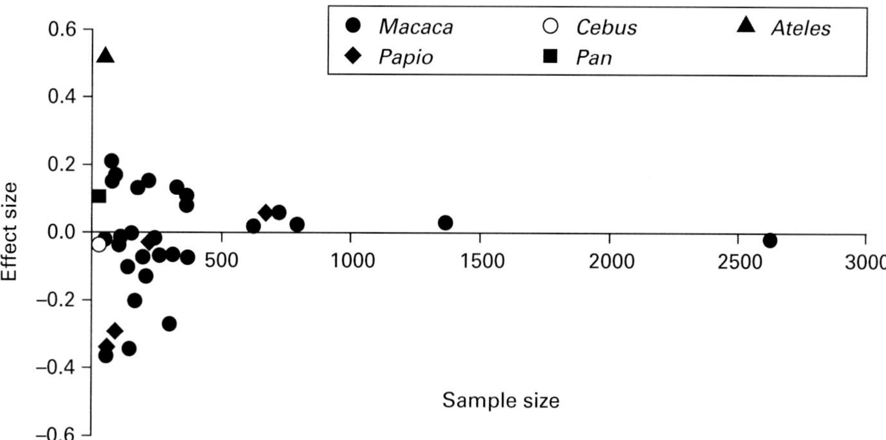

Figure 10.16. Differences in birth sex ratios among progeny of high-ranking and low-ranking females in five anthropoid genera, and effects of sample size. The effect size is equal to the proportion of male offspring produced by high-ranking females minus the proportion of male offspring produced by low-ranking females.Thus, when the effect size is zero, high- and low-ranking females produce equal proportions of male infants. Sample size is the total number of offspring produced. As sample size increases, effect sizes decline towards zero. (From Silk and Brown, 2004.)

sexual maturity. By giving birth to more sons, lower-ranking females might reduce local competition for resources between themselves and other females. Subsequent work on a number of species of monkeys has garnered some support to these ideas (Johnson, 1988; Silk, 1983, 1988).

A major problem with most tests of the hypotheses for skewing of primate birth sex ratios is that primates reproduce slowly. Data sets on the numbers and sexes of offspring are therefore often quite small. A crucial insight into the nature of this problem is due to Gillian Brown and Joan Silk (2002), who conducted a meta-analysis to examine relationships between skewing of birth sex ratios and sample sizes in primates. Their results are shown in Figure 10.16. What clearly emerges from this analysis is that skewing of birth sex ratios typically involves studies with small sample sizes, and that more males may be born to either high-ranking or low-ranking females. However, as sample sizes increase these effects diminish, and tend towards zero (Figure 10.16). In a subsequent review of this field, Silk and Brown (2004) emphasized that there are no significant effects of female rank upon proportions of sons, at least for the 35 data sets they were able to examine.

Brown and Silk's work did not include any information on the mandrill. It remains the case that data on female rank and birth sex ratios in mandrills are insufficient to conduct accurate tests of the three hypotheses discussed here. I venture to suggest that it is unlikely that future studies will reveal any meaningful connection between skewing of birth sex ratios and social rank in female mandrills.

Turning to the question of male infanticides, these have only been recorded by direct observations in approximately 6% of all extant primate species, and mostly in a handful of cases (Dixson, 2012). A few species are exceptional, however, as substantial numbers of male infanticides have now been observed directly in free-ranging groups of Hanuman langurs (*Semnopithecus entellus*), red howler monkeys (*Alouatta seniculus*), as well as several other species (see reviews by Palombit, 2012 and van Schaik, 2000a). It was fieldwork on Hanuman langurs that initially gave rise to the sexual selection hypothesis of male infanticide, which posits that male langurs kill (non-related) infants, especially after group takeovers, in order that their mothers will begin to cycle again and engage in conceptive matings with them. Such behaviour might thus improve the reproductive success of the males concerned. Subsequent fieldwork on Hanuman langurs has confirmed that infanticide is an important reproductive strategy for males of this species (e.g. Borries and Koenig, 2000; Borries *et al.*, 1999; Sommer 1993). Considerable evidence has also been obtained concerning sexual selection in relation to infanticidal behaviour in male red howler monkeys (Crockett 2003; Crockett and Janson 2000).

Problems with the sexual selection hypothesis of male infanticide soon emerged when some primatologists began to encourage its application to the Order Primates as a whole. This was, in my view, a scientifically hazardous exercise, given that male infanticides are rare events that have never been observed in the vast majority of primate species. Yet, a good many authors now treat as established fact the proposition that sexual selection by male infanticide has played a pervasive role in the evolution of all manner of traits. Consider the following examples: the evolution of bright natal coat colours in infants (Treves, 2000); the evolution of multiple partner matings and mating during pregnancy as paternity confusion tactics (van Noordwijk and van Schaik, 2000); the evolution of sexual skin swellings and copulatory vocalizations (Pradhan *et al.*, 2006), and the evolution of monogamy (Opie *et al.*, 2013).

I do not think that the available field evidence would justify reaching any of these conclusions. In the absence of direct observations of males killing young infants, some authors have made use of data on inferred infanticides, so that the death or disappearance of an infant in suspicious circumstances is taken as evidence that it was a result of infanticide. This assumption is highly problematic, as male infanticide is acknowledged as being a rare event. Predation, disease, and accidents are much more common causes of infant mortality. Thus, using Occam's razor, surely the presumption in cases where infants disappear from groups should be that infanticide was most unlikely to be the cause of death. I remember many years ago, during a census of mountain gorillas, coming across a dead infant; it had been left by its mother in her night nest. The infant had not sustained any external injuries, and appeared not to have died owing to violent causes. Mountain gorilla groups not infrequently interact with each other, or with lone silverback males. Had I been observing such behaviour involving this group of mountain gorillas, but unaware of this particular nest site, should I have inferred that the infant had disappeared because of male infanticide?

Inclusion of data on inferred infanticides in some studies is one thing, but indirect measures of infanticide risk are even more problematic, because these do not require evidence of infant mortality. For example, Carel Van Schaik (2000b) has posited that the ratio of gestation length/lactation duration in a primate species should provide a measure of potential risk of infanticide by males of that species. Females that exhibit longer periods of lactational amenorrhoea are thus presumed to be at greater risk of having their infants killed, as this would result in an earlier resumption of their ovarian cycles, and enhance mating opportunities for infanticidal males. This rationale was used by Opie *et al.* (2013), as part of their comparative analysis of infanticide in relation to the evolution of monogamy in primates. They concluded that avoidance of male infanticide has played a pivotal role in the origins of primate monogamy, more important even than paternal care or female ranging patterns.

There are good reasons to question such conclusions, however (Dixson, 2013; Lucas and Clutton-Brock, 2013; Lucas and Huchard, 2014), one of them being the unwarranted assumption that longer lactation periods really do equate to higher incidences of male infanticides for the species concerned. The mandrill and drill are both listed as being at ‘high risk’ of male infanticides in Opie *et al.*’s data set, despite the fact that males have not been observed to kill infants in either species. In a peculiar reversal of the normal practice, the mandrill and drill have thus been presumed guilty until proven innocent. I should like to offer some words in their defence. To my knowledge, the only published account claiming ‘evidence for sexually selected infanticide’ in the drill (Böer and Sommer, 1992) concerns an incident at Hanover Zoo where a male drill threatened an orphaned infant in an adjacent cage; the Zoo authorities were attempting to introduce this infant to the established group. Where the mandrill is concerned, Setchell *et al.*, (2006c) state that ‘strong evidence for infanticide exists in the CIRMF mandrill colony’, but later in the same paper they note that ‘infanticide was not observed during the study’. Three suspected infanticides had apparently occurred at other times, but here again, there was only circumstantial evidence.

The complex sexual and social lives of mandrills in the wild, where males enter and exit groups, and where sperm competition plays an important role in reproductive success, make it unlikely that males could have ‘certainty of paternity’. A sexually active male might easily risk killing his own offspring, therefore, if he engaged in infanticide. Indeed, in a comparative analysis of the evolution of male infanticides across mammalian societies, Lucas and Huchard (2014) found that, in lineages where large relative testes sizes and sperm competition occur, infanticides tend to be absent. We have seen that the mandrill is a member of such a lineage in the Papionini.

Palombit (2012) has listed nine other adaptive hypotheses, besides the sexual selection hypothesis, to account for occurrences of male infanticides in primates. If future fieldwork should bring to light incidences of male infanticide in wild mandrills or drills, it would be prudent to explore all the possible causes (adaptive and non-adaptive) for such rare events, rather than assuming that they must be because of sexual selection.

11 Epilogue: conservation status of the genus *Mandrillus*

Forests precede civilizations and deserts follow them.

Chateaubriand

All things pass.

Arab proverb

As human populations continue to increase, and the planet becomes ever more crowded, many wonderful species are rapidly dwindling in numbers and becoming extinct. As the largest of all the monkeys that depend for their survival upon extensive tracts of undisturbed rainforest, both the mandrill and drill are already in difficulties, and their long-term future is precarious at best. We have seen that many questions concerning the biology of wild mandrills and drills remain to be answered. Yet the opportunity to conduct the necessary field studies diminishes with each passing year, as wild populations of these magnificent animals are forced ever closer to extinction. Thus, although this book is concerned primarily with behavioural biology, reproduction and evolution, it would surely be remiss of me to conclude matters without devoting some attention to the vital subject of conservation.

Extinction is an insidious process. It often involves the progressive loss of local habitats and populations, the emergence of a patchwork of increasingly isolated subpopulations, and finally the total disappearance of a species. This process is already well advanced where the drill is concerned. For we now know that it is extinct, or virtually so, throughout many areas of its former range in Cameroon, Nigeria and on the Island of Bioko. Thus, it may be useful to consider firstly what has befallen drill populations, and then proceed to a discussion of the conservation status of the mandrill.

Drills on Bioko Island (Equatorial Guinea)

The volcanic Island of Bioko (2020 km^2 in area) forms part of Equatorial Guinea (formerly Rio Muni), which was a Spanish colony until 1969. Bioko lies 32 km off the coast of West Africa, from which it became separated by rising sea levels at the end of the last ice age. Seven endangered monkey taxa are found in its rainforests, including a subspecies of the drill (*Mandrillus leucophaeus poensis*) that occurs nowhere else. More than 20 years ago, Butynski and Koster (1994) conducted a survey of the distribution and conservation status of Bioko's primate populations. They reported

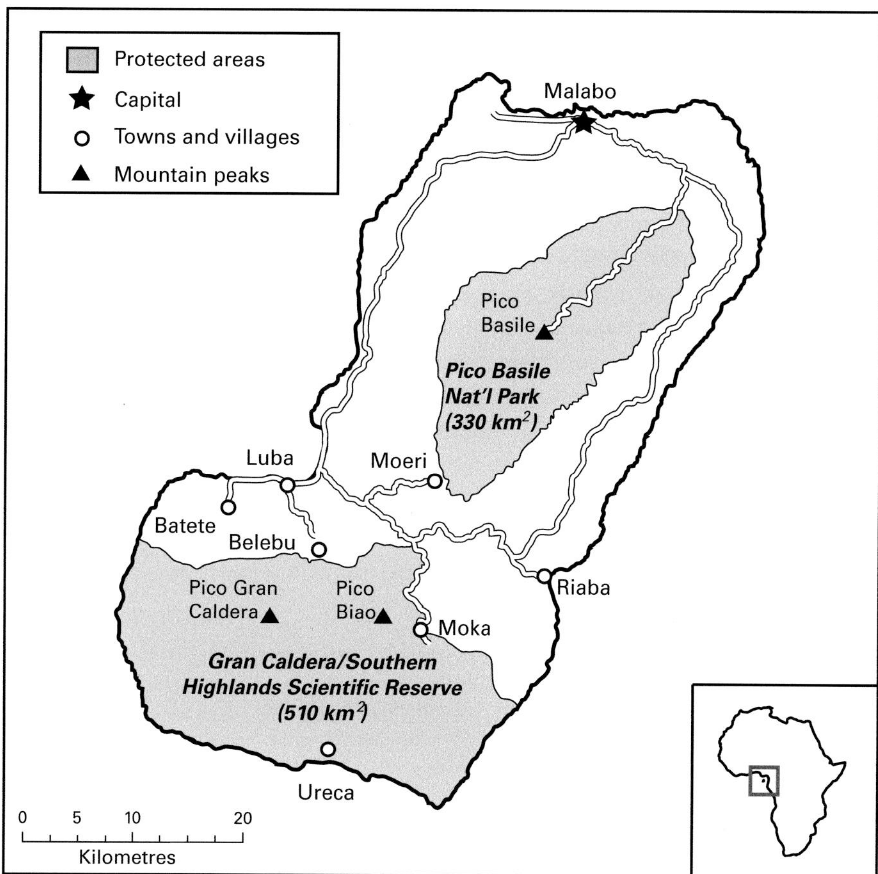

Figure 11.1. A map of Bioko Island (Equatorial Guinea), including its designated protected areas.

that drills were more plentiful in the southern quarter of the island (in what is now the Gran Caldera Southern Highlands Scientific Reserve) and to the north (in what is currently the Pico Basilé National Park) (Figure 11.1). Butynski and Koster noted, however, that drills were much sought after by hunters, and 'probably much more so than any other animal on the island'.

Unfortunately, this situation has became progressively worse over the years, as has been documented by various authors who have monitored bushmeat sales in Malabo, the capital of Equatorial Guinea (Butynski *et al.*, 2009; Fa *et al.*, 2000; Hearn *et al.*, 2006). The Bioko Biodiversity Protection Progam, founded by Gail Hearn, has conducted long-term bushmeat surveys in Malabo. As a result, we know that between the years 1997 and 2010, more than 197,000 animals were traded via this bushmeat market, including at least 35,000 monkeys.

Bushmeat provides only a minor source of protein for the population of Malabo, indeed it is too expensive for most people, and is mainly purchased as a delicacy by wealthier members of the community. Demand has fueled a steady increase in prices, and in commercial hunting, especially using shotguns as well as snares. Commercial hunters are depleting even remote forests of primates, duikers, giant pouched rats and other prized species.

Of course, further research by conservation biologists and primatologists is always useful; for example, much more information is needed about the current status of wild populations of drills and other primates on Bioko. However, research alone cannot save endangered species such as the drill. Ultimately, the power to enforce effective conservation policies rests with the government and its agencies, and in the hearts and minds of the people of Equatorial Guinea. Yet, theirs is a poor nation, heavily dependent upon aid despite its natural resources. Until quite recently, life expectancy was just 49 years (up from 37 years in 1960), and there were approximately 25 doctors for every 100,000 people. Historically, public education has been woefully neglected, corruption is endemic and uncontrolled destruction of the rainforest has been extensive.

Experienced conservation biologists are, understandably, pragmatic in the face of such problems, as the following quotation shows. It is taken from a paper dealing with a field survey of primate populations in a remote area of southeastern Bioko.

> Our results illustrate an increasingly desperate situation. Conservation strategies, beginning with legitimate enforcement of existing legislation, could greatly improve the current status of Bioko's primates. An island – wide reduction of hunting is necessary, with a focus on government – supported conservation efforts (e.g. training of national staff, sensitization of government, police, and military personnel, and conservation education programs). (Cronin *et al.*, 2013.)

Drill populations in Nigeria

Drills are found in Nigeria only in certain forested areas of the Cross River State, which is situated in the southeastern part of the country, adjacent to its border with Cameroon. More than 90% of Nigeria's original forests have been destroyed, so that Cross River State is one of very few areas where substantial blocks of rainforest still remain.

The most important conservation area is the Cross River National Park, approximately 4000 km^2 in area, which consists of a northern section (Okwango) and a southern section (Oban Hills). As can be seen in Figure 11.2, the Oban Hills section of the Park shares its eastern border with Korup National Park in neighbouring Cameroon. Drills occur in both these Parks. The Okwango Division also borders Cameroon, where it is contiguous with the Takamanda National Park. As well as drills, these Parks contain Cross River gorillas (*Gorilla g. diehli*), perhaps 200–250 of which may still survive (Robbins, 2007).

Unfortunately, neither division of the Cross River National Park is secure. Its primate populations are under constant pressure, owing to illegal logging and poaching, as well as encroachment by farms, and by coffee and rubber plantations around the Park's borders. Nigeria is a densely populated country, beset by numerous social and political problems including widespread corruption. Although drills are protected by law in Cross

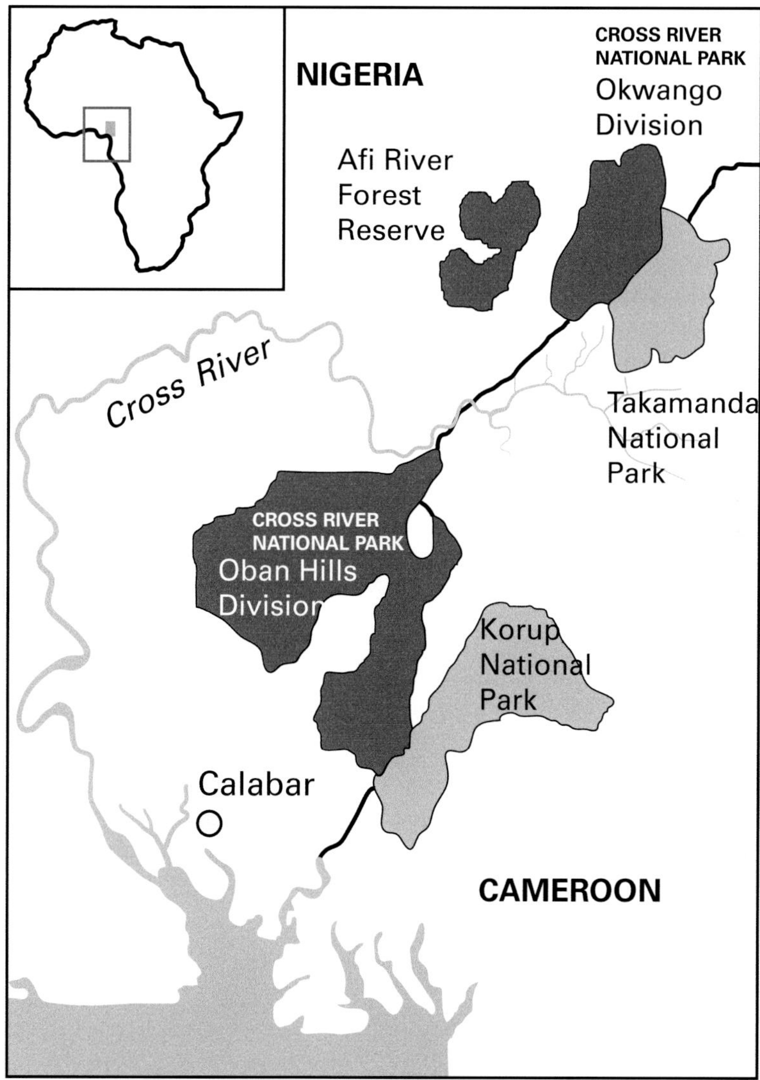

Figure 11.2. A map of eastern Nigeria (Cross River State) and western Cameroon, to show the positions of the various National Parks and protected areas discussed in the text.

River State, they are often killed by hunters, and orphaned infants may be kept as pets. Pandrillus, which is a non-governmental organization, acts as a rescue centre, and since 1991 it has developed a large breeding population of drills, housed in semi-free ranging groups. Pandrillus is situated on Afi Mountain, 100 km^2 of which was officially declared a Wildlife Sanctuary in 2000 (Figure 11.2).

How many drills might remain in the wild in Nigeria is not known. Their numbers are unlikely to be high, however, and their continued survival will depend upon effective conservation and management of protected areas, in just the same way as stated earlier with regard to the drill populations of Bioko Island.

Conservation prospects for the drill in Cameroon

If the drill is to have any long-term prospect of survival, much will depend upon its conservation in Cameroon, as this encompasses 80% of its historic distribution range. The most up to date assessment of the current distribution and conservation outlook for drills in Cameroon (Morgan *et al.*, 2013) reported the results of field surveys and associated research carried out between 2002 and 2009. I shall give a brief account of this project here; interested readers may wish to refer to the original publication for more detailed information.

For the purposes of this exercise, we divided the historic distribution range of the drill (approximately 46,000 km^2 in total) into 52 survey units. These units included a number of important protected areas, such as the Korup and Takamanda National Parks; these have been mentioned as being contiguous with the Cross River National Park in neighbouring Nigeria. Figure 11.3 shows the extent of survey area, and Figure 11.4 shows the positions and names of all 52 survey units.

The current status of drills in each survey unit was evaluated using four criteria: direct evidence of drill presence; hunters' perceptions of drill status; primate groups encountered/day, and occurrences of forest elephant/leopards and chimpanzees in the same area. A five-point scale was used for each criterion, and the composite scores (maximum drill status score = 20) were used to map drill status. A simplified version of this map is shown in Figure 11.5. Drills were confirmed as occupying 16 of the survey units, based

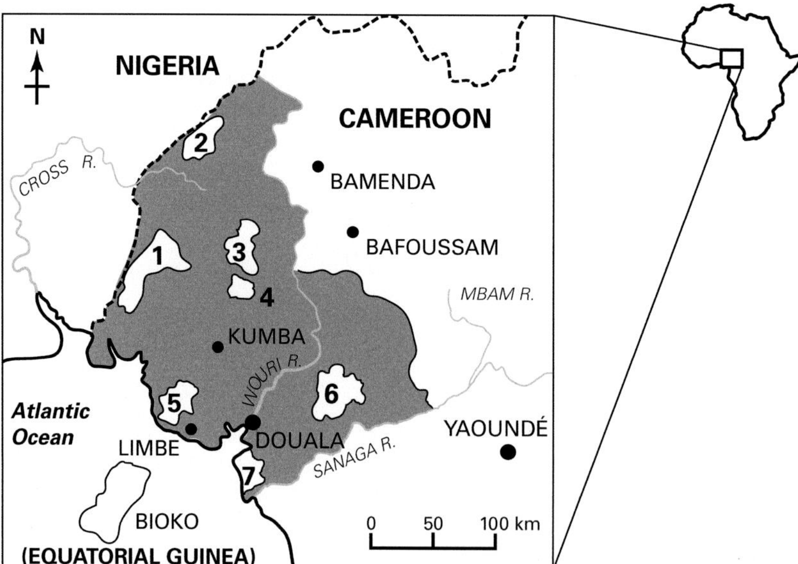

Figure 11.3. Drill survey area (shaded area on map) in Cameroon, and selected protected areas: **1.** Korup National Park (NP); **2.** Takamanda NP; **3.** Banyang Mbo Wildlife Sanctuary; **4.** Bakossi NP; **5.** Mt. Cameroon NP; **6.** Proposed Ebo NP; **7.** Douala-Edea Forest Reserve. (Redrawn and Modified from Morgan *et al.*, 2013.)

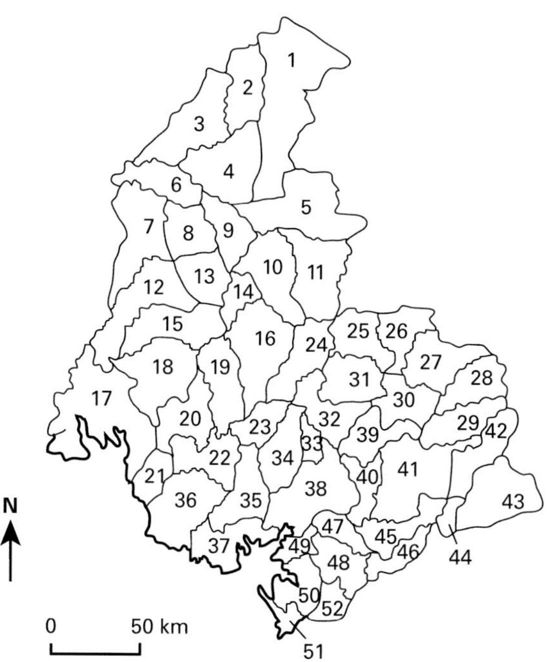

Figure 11.4. Names and positions of the 52 drill survey units in Cameroon. **1.** Benakuma; **2.** Akwaya; **3.** Takamanda; **4.** Mone; **5.** North Fontem; **6.** Cross River; **7.** Ejagham; **8.** Kembong; **9.** Nta Ali; **10.** Banyang Mbo; **11.** South Fontem; **12.** Korup; **13.** Nkwende Hills; **14.** Mungo Ndor; **15.** Libangenie; **16.** Bakossi Mountains; **17.** Bakassi; **18.** Rumpi Hills; **19.** Butu; **20.** Ekondo-Titi; **21.** Mokoko; **22.** Bakundu; **23.** Mungo River Forest Reserve; **24.** Kupe-Manenguba; **25.** Nkonjock West; **26.** Nkonjock East; **27.** Tongo; **28.** Ndokbou North; **29.** Ndokbou South; **30.** Makombe; **31.** Nlonako; **32.** Bakaka; **33.** Dibombe Mabombe; **34.** West Sawa; **35.** Mungo River Plain; **36.** Mount Cameroon; **37.** Tiko; **38.** East Sawa; **39.** Yingui; **40.** Dibamba; **41.** Ebo; **42.** Ngambe; **43.** Ndom; **44.** Massock; **45.** Loungame; **46.** Sanaga; **47.** Massoumbou; **48.** Dizangue; **49.** Douala; **50.** Douala coastal forest; **51.** Douala-Edea; **52.** Lake Mboli. (Redrawn from Morgan *et al.*, 2013.)

upon direct evidence (sightings of animals or hearing their vocalizations) and indirect evidence of their presence (foraging signs, faeces). However, the species was found to be extinct, or virtually so, in at least 24 of the 52 areas surveyed. Figure 11.5 shows that there are two distinct clusters of survey units (darkly shaded) where drill status is higher than elsewhere in Cameroon. The first of these is in the southeast, and consists of the Ndokbou South, Yingui and Ebo Forests; Ebo being the site of a proposed National Park. The second, and larger cluster, lies to the west and includes the Korup National Park and Ejagham forests, adjacent to the border with Nigeria, as well as six other units with slightly lower ratings (Mungo, Kembong, Nta Ali, Nkwende Hills, Libangenie and Rumpi Hills).

The future conservation outlook for drill populations in Cameroon was assessed by rating each of the 52 survey units on nine additional criteria. These criteria dealt with degree of forest cover, logging concessions, impacts of hunting, law enforcement, population densities and other matters that crucially affect whether long-term

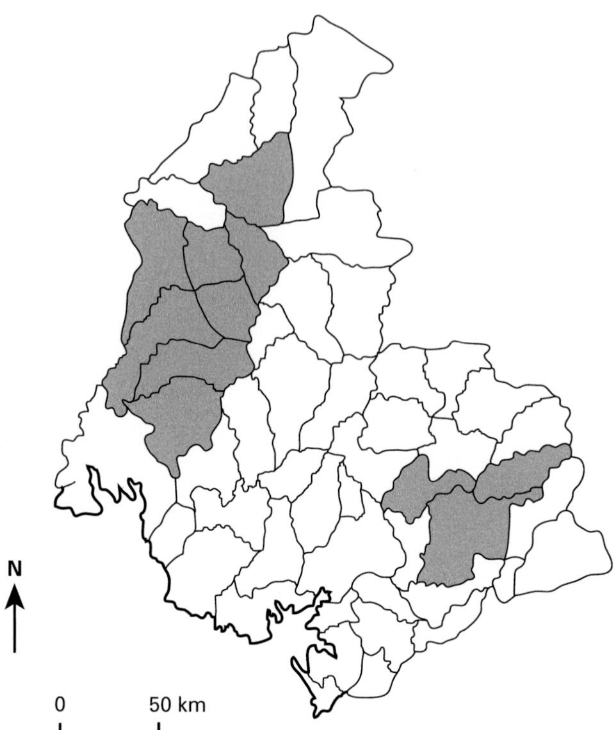

Figure 11.5. Drill status in the survey units shown in Figure 11.4. Dark-shaded areas are the units with higher scores (range, 9–20). Other (unshaded) areas have lower scores (range, 0–8), and the drill is effectively extinct in some of these (see text for further details). (Redrawn and modified from Morgan *et al.*, 2013.)

conservation is likely to succeed at a given site. On this basis, five survey units achieved very high scores (Ebo, Korup, Takamanda, Akwaya and Banyang Mbo), while eight others had moderately high scores. It was clear that all survey units have been affected, to a greater or lesser extent, by logging. There has been a worrying trend in Cameroon to reclassify what were 'forest reserves' as logging concessions. Extensions, or planned extensions of large-scale commercial plantations threaten some areas where forests and viable drill populations still exist (e.g. to the west of the Ebo Forest where a 210,000-hectare oil palm plantation has been proposed). Drills are widely hunted, including commercial hunting for the bushmeat trade (MacDonald *et al.*, 2011). Hunting with dogs is a major problem, as drill groups take to the trees when pursued by dogs, fitted with 'drill bells' (Figure 11.6), and hunters may then shoot the monkeys in large numbers. The drill is protected by law (it is ranked as a 'Class A' species, and is totally protected in the same way as the great apes and forest elephants). In practice, the law is rarely enforced.

Hope, it is said, 'springs eternal in the human breast'. Yet, as stated earlier in this chapter, conservation biologists are obliged to be pragmatic. So let me conclude by

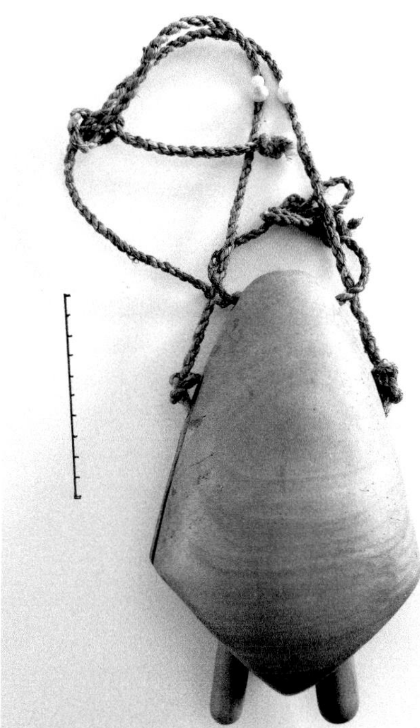

Figure 11.6. A traditional wooden 'drill bell' worn by hunting dogs. This one was made in Bakossiland, Cameroon. The scale is 10 cms long. (Author's photograph.)

quoting Bethan Morgan, who is an expert in this field, regarding the future of conservation efforts, and the possible survival of the drill in Cameroon:

> There are some promising signs in drill conservation, such as the persistence of the species in Korup National Park, the creation of the Takamanda and Bakossi National Parks (2008), the recent Gazettement of the Mount Cameroon National Park (2010), and the planned creation of Ebo National Park. However, if Cameroon is to retain its dwindling and increasingly isolated drills, the newly created and planned protected areas need to be supported by a renewed commitment to applied conservation, providing the management of these parks with both the financial resources and – crucially – the appropriately trained (and motivated) personnel. (Morgan *et al.*, 2013)

With these thoughts in mind, let us now consider what is known about the current status of the mandrill.

The mandrill in Gabon

I shall concentrate here upon Gabon because the majority of mandrills live in its vast rainforests. Mandrills are also found in neighbouring Equatorial Guinea, but destruction of the rainforest there has been so extensive that conservation prospects are very poor. The same is likely to be the case in most parts of southern Cameroon where, historically, mandrills occurred in forests to the south of the Sanaga River. The status of mandrills in Congo Brazzaville, where they are restricted to forests to the north of the Congo River, is uncertain.

The prospects for long-term conservation of mandrills in Gabon are more promising, although significant problems exist owing to increasing exploitation of its forests and mineral resources. Gabon is an unusual country, in the sense that its human population density is very low. Approximately 1.5 million people inhabit an area of 270,000 km^2. This is quite similar to the land area of the United Kingdom, but the UK has a population 42 times greater than that of Gabon. Moreover, one third of all the Gabonese live in Libreville, the capital city on the west coast. Gabon's rainforests contain an extraordinarily diverse fauna and flora, which, until very recently, has remained largely intact, especially within the remote interior of the country.

From 1989 to 1992, I lived with my family in Gabon, and worked in Franceville, in the Southeastern Province of Haute Ogooué. The International Medical Research Centre (CIRMF), which maintains the semi-free ranging mandrill groups discussed in this book, is situated close to Franceville. In addition to its captive primate colonies, the CIRMF established a Field Station in the Lopé Reserve (now the Lopé National Park), focusing on the ecology and conservation of gorillas and chimpanzees. This programme was initiated after a country-wide census of gorilla and chimpanzee populations had been completed in the early 1980s. This census had produced a most encouraging result, for at that time there were estimated to be approximately 35,000 gorillas and 64,000 chimpanzees in Gabon (Tutin and Fernandez, 1984). Caroline Tutin and Michel Fernandez noted that 'Gabon's large areas of undisturbed primary forest offer exceptional potential for conservation, not only of gorillas and chimpanzees, but also of the intact tropical rainforest ecosystems which they inhabit.'

The Lopé Reserve was chosen as the site to establish a Field Station because it contained substantial populations of gorillas and chimpanzees, as well as forest elephants, buffalo, duiker, red river hogs, mandrills and a host of other species. The Field Station gradually became a centre for research on a wide range of projects, including the first detailed studies of the mandrill's extraordinary supergroups (Abernethy *et al.*, 2002; Rogers *et al.*, 1996; White *et al.*, 2010).

It is fruitless to discuss the conservation of mandrills, and the rainforests upon which they depend, without some brief consideration of the economic, social and political challenges that occur in Gabon. Development of Gabon's natural resources has been a mixed blessing for its people, and has had a negative impact on its wildlife. Traditionally, the economy of Gabon relied heavily upon exploitation of its forests for timber, but since the 1970s, oil revenues have been of paramount importance. The Gross National Income (GNI) per head of the population is currently the third highest in sub-Saharan Africa. Yet 80% of Gabon's people receive little or nothing of this largesse; most of the population lives in poverty. Corruption, greed, and exploitation by international companies are issues in Gabon, as elsewhere.

Travel throughout Gabon has always been difficult, and until quite recently, timber extraction and mining operations were limited by the remoteness of the forests and the difficult nature of the terrain. Many roads are almost impassable during the wet season

Figure 11.7. In Gabon, dirt roads in the interior of the country often become impassable during wet seasons. (Author's photograph.)

(Figure 11.7). Traditionally, logs were transported down to the coast by river. This method sufficed for certain species (e.g. Okoumé) that produce lighter timber, but not for many valuable hardwoods, including ebony and mahogany. Matters changed dramatically, however, with the construction of the Trans-Gabon Railway, which was completed in 1987; 649 km of track now connected Owendo (a coastal port near Libreville) in the west to Franceville in the southeast of the country.

The creation of the railroad doubtless gladdened the hearts of those international companies that acquired further timber or mining concessions as a result. However, the The Financial Times World Desk Reference expressed a less sanguine point of view when it noted that 'the Trans-Gabon Railroad sliced through one of the world's finest virgin rainforests and opened the interior to indiscriminate exploitation of rare woods'. I travelled on the railroad a number of times in the early years after its completion. It was dispiriting to see people lining the station platforms, deep in the forest, carrying the carcasses of monkeys, duiker, dwarf crocodiles, and sacks of bushmeat of all kinds. As the logging companies and their dirt roads advanced into the wilderness, so too did the hunters (Figure 11.8). As well as supplying logging camps with meat, commercial hunting provisioned markets in towns along the railway's route. Lest we judge such hunting operations too harshly, it is as well to keep in mind the low income and lack of alternative sources of protein that characterizes many rural communities. We need also to consider the failure of logging operations to provide adequate supervision or provisions for their own workers.

Figure 11.8. Bushmeat. Mandrills are often hunted in Gabon, either for local consumption, or for sale in the markets. (Photograph by Dr Pierre Moisson.)

Many years have passed since I lived in Gabon and travelled on its railroad. Now there is no shortage of detailed scientific publications that quantify the progressive expansion of hunting and the consumption of wildlife in Gabon. Such practices remain relatively uncontrolled and, in the long run, completely unsustainable (Coad *et al.*, 2010; Fa *et al.*, 2003; Foerster *et al.*, 2012; Laurance *et al.*, 2006; Thibault and Blaney, 2003).

More than 50% of Gabon's rainforests have been leased to timber companies. Fortunately, however, 10% of the country has been incorporated into a National Parks system. Thirteen National Parks were set up in 2002 owing to a collaboration between the Gabonese government and the Wildlife Conservation Society (WCS: formerly the Bronx Zoo) in the USA. Figure 11.9 shows a map of Gabon and the positions of these Parks. Crucial to the long-term success of any National Park is genuine political commitment to applied conservation programmes, financial support, and provision of trained and motivated staff. These are the same factors that were referred to earlier, in relation to the support required to ensure effective management of protected areas, and conservation of the drill in Cameroon and elsewhere.

An acid test of whether a country's National Parks are operating effectively may be made by examining how its authorities deal with conflicts of interest or external threats. For example, what happens if it is discovered that a National Park, or its surrounding 'buffer zone', contains important mineral deposits, or other valuable resources? A chilling illustration of the worst that may happen is provided by events in the Minkébé National Park in northeast Gabon (Figure 11.9; Park No. 7). The Minkébé Park encompasses a huge forest, some 7570 km^2 in area. Until 2004, it also contained an estimated

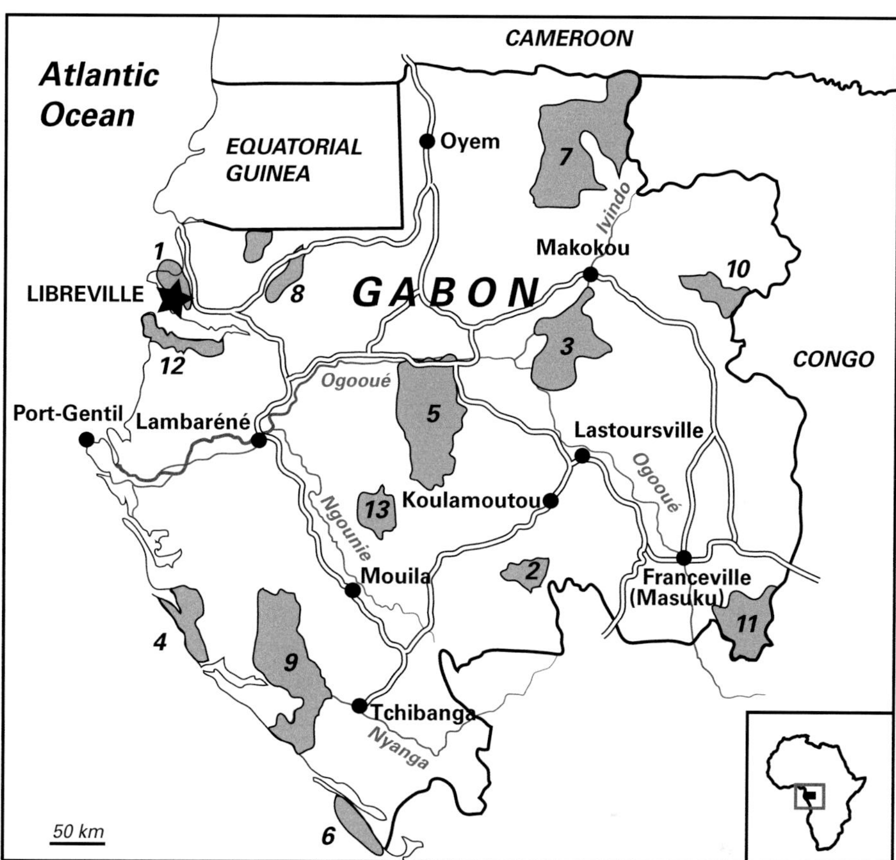

Figure 11.9. Map of Gabon to show the positions of its National Parks: **1.** Akanda; **2.** Birougou; **3.** Ivindo; **4.** Loango; **5.** Lopé; **6.** Mayumba; **7.** Minkébé; **8.** Monts de Cristal; **9.** Moukalaba–Doudou; **10.** Mwagné; **11.** Plateau Batéké; **12.** Pongara; **13.** Waka.

21,000 elephants; this was considered to be the largest intact population of forest elephants remaining in Africa (MIKE, 2005). Historically, the remoteness of Minkébé had largely shielded its forests and animals from human interference. However, the discovery of gold, and subsequent illegal mining operations, produced an influx of thousands of people into the area (many of whom had crossed the border from Cameroon). The growth of the mines and burgeoning human population between 2007 and 2011 was associated with widespread hunting inside the National Park, both for food and to poach ivory. In 2011, the Gabonese military intervened, and 6000 miners were expelled from the Park. It was too late, however, as more than half the elephants had been killed. According to Parks Gabon, in 2013 approximately 6900 elephants were thought to remain in the southern section of Minkébé National Park. There are logging concessions situated adjacent to the boundaries of Minkébé, as is the case for most of the National Parks that were created in 2002 (e.g. Ivindo, Lopé and Waka: Rayden and Essame Essono, unpublished report). The long-term viability and effectiveness of the National Parks system clearly should not be taken for granted.

Throughout this chapter I have resisted the temptation to estimate how many mandrills, or drills, might remain in any particular area of forest. The fact is that nobody knows for certain, as it is technically very difficult to conduct censuses of these monkeys. The drill is surely an endangered species, as its numbers are clearly declining throughout much of its historic distribution range in Cameroon, as elsewhere (Morgan *et al.*, 2013). There has been no comparable survey of the mandrill's current distribution and conservation status. Although the mandrill is not yet classed as being in danger of extinction (e.g by the International Union for the Conservation of Nature: IUCN), I believe that readers will understand that its situation is far from secure. Sadly, if matters do not improve, it is likely that first the drill and then the mandrill will become extinct.

Yet, with proper management and support of the National Parks, this dire prediction need not become a reality. For example, the Lopé National Park (4970 km^2) offers excellent prospects for long-term conservation of the mandrill, as well as forest elephants, gorillas, chimpanzees and many other wonderful species. In Cameroon, the Korup National Park and the proposed Ebo National Park have the potential to shelter huge numbers of rainforest species, of which the drill is just one example. Indeed, if properly managed, the National Parks of Gabon and Cameroon could ensure the survival of much of these countries' exceptional biodiversity in perpetuity. Future generations would doubtless be grateful, and Gabon and Cameroon might then benefit from increased tourist revenues. As the majority of the rainforests are logged, and wildlife disappears over the next 50–100 years, the National Parks are likely to become the last havens for many animals, including the mandrill and drill.

During their long evolutionary history, mandrills and drills have adapted and evolved throughout multiple cycles of rainforest contraction and expansion, resulting from climate change. Naturally occurring forest refuges have made it possible for them to survive in the past. Now, as human populations relentlessly expand, and we continue to cut down the rainforests, will we ensure that some secure refuges remain, to shelter the remarkable animals that depend upon them? The omens are not promising in this respect. In global terms, vertebrate diversity continues to decline inexorably, despite every effort made by conservation organizations and by dedicated individuals (Hoffmann *et al.*, 2010). Will our descendants have cause to curse us for allowing the last wild places to be destroyed? If so, then the Latin phrase *litera scripta manet* (the written word remains) will seem apposite. For, although zoological treatises such as this one may remain in serried ranks upon dusty library shelves, yet the animals themselves will no longer exist.

Appendix Behavioural scoring system

The communicatory behaviour of the mandrill is described in detail in Chapter 5. During observations of the semi-free ranging mandrill groups at the CIRMF, the various behavioural patterns were scored using a shorthand system composed of 50 codes. The mandrills were fitted with ear tags bearing their ID numbers. Thus, it was possible to recognize them individually and to record sequences of behavioural interactions.

As an example: the code for facial threat = TH; and for fleeing = FL.

So: 2 TH 12, 12 FL 2 = female no. 2 threatens female no. 12, and female no. 12 flees from no. 2. Adding 'Y' to a code means that the animal concerned receives the behaviour; e.g. 12 THY 2 = female no. 12 is threatened by female no. 2.

Codes for social behaviour

Allogrooming (GR); self-grooming (SG); grinning (Z); head shake (SH); piloerection of the scalp crest (CR); lip smacking (SM); play face (PF); chasing play (CHPL); rough-and-tumble play (RT); mouth to mouth: in greeting and/or as an olfactory inspection (MM); two-phase grunting (2PG).

Codes for sexual behaviour

Male

Follow female (FO); look at sex skin (XS); sniff genital area (SN); mount attempt (ATT); half mount (HM); hip touch (HT); full mount (FM); penile erection (ER); pelvic thrusts, pre-intromission (counted) (T); pelvic thrusts, during intromission (counted) (IT); ejaculation (EJAC); mate-guarding (MG); masturbation (MAST).

Female

Sexual presentation (XPR); eye contact and sexual presentation (EX); approach and withdraw ('parading') (AW); running from the male after copulation has ended (RM).

Derived measures

Male initiated mount (MIM); female initiated mount (FIM); female refusal of male's mount attempt (RF); female terminates mount prior to ejaculation (FT).

Codes for socio-sexual behaviour

Presentation (male–male or female–female) (PR); mounts (male–male or female–female) (MT).

Codes for agonistic behaviour (including avoidance and displacement)

Non-contact aggression

Stare (ST); head bob (HB); ground slap (GS); lunge (LU); chase (CH).

Contact aggression

Hit or grab (RA); bite (BI).

Submissive behaviour

Flee (FL); squeal (SQ); grimace (GM); duck face (DK).

Avoidance and displacement

Avoid (AV); displace (DI).

Other coded patterns

Branch shaking display (BR); 'Tension' yawn (Y); chest-rub (scent-marking with the sternal gland) (CU).

References

Abernethy, K. and White, L.J.T. (2013). *Mandrillus sphinx*: the mandrill. In *Mammals of Africa, Vol.11, Primates*, eds. T.M Butynski, J. Kingdon and J. Kalani. London: Bloomsbury Publishing, pp. 192–6.

Abernethy, K.A., White, L.J.T. and Wickings, E.J. (2002). Hordes of mandrills (*Mandrillus sphinx*): extreme group size and seasonal male presence. *Journal of Zoology, London*, 258, 131–7.

Alcock, J. (2013). *Animal Behavior*, Tenth edn. Sunderland, MA: Sinauer Associates Inc.

Alexander, R.D., Hoogland, J.L., Howard, R.D., Noonan, K.M. and Sherman, P.W. (1979). Sexual dimorphisms and breeding systems in pinnipeds, ungulates, primates and humans. In *Evolutionary Biology and Human Social Behavior*, eds. N.A. Chagnon and W. Irons. North Scituate, MA: Duxbury Press, pp. 402–35.

Ally-Mwamende, K. (2009). *Social Organization, Ecology and Reproduction in the Sanje Mangabey (Cercocebus sanjei) in the Udzungwa Mountains National Park, Tanzania*. MSc Thesis. Wellington, NZ: Victoria University of Wellington.

Anderson, J.R. (2010). Non-human primates: a comparative developmental perspective on yawning. In *Mystery of Yawning in Physiology and Disease*, ed. O. Walusinski. Basel: Karger, pp. 63–76.

Anderson, M.J. and Dixson, A.F. (2002). Sperm competition: motility and the midpiece. *Nature, London*, 416, 496.

Anderson, M.J. and Dixson, A.F. (2009). Sexual selection affects the sizes of the mammalian prostate gland and seminal vesicles. *Current Zoology*, 55, 1–8.

Anderson, M.J., Chapman, S.J., Videan, E.N., *et al.* (2007). Functional evidence for differences in sperm competition in humans and chimpanzees. *American Journal of Physical Anthropology*, 36(3), 369–75.

Anderson, M.J., Dixson, A.S. and Dixson, A.F. (2006). Mammalian sperm and oviducts are sexually selected: evidence for coevolution. *Journal of Zoology, London*, 270, 682–6.

Anderson, M.J., Hessel, J.K. and Dixson, A.F. (2004). Primate mating systems and the evolution of immune response. *Journal of Reproductive Immunology*, 61, 31–8.

Anderson, M.J., Nyholt, J. and Dixson, A.F. (2005). Sperm competition and the evolution of sperm midpiece volume in mammals. *Journal of Zoology, London*, 267, 135–42.

Andrew, R.J. (1963). The origin and evolution of the calls and facial expressions of the primates. *Behaviour*, 20, 1–109.

Astaras, C. (2009). *Ecology and Status of the Drill* (Mandrillus leucophaeus) *in Korup National Park, Southwest Cameroon: Implications for Conservation.* PhD Thesis. Göttingen: Faculty of Mathematics and Natural Sciences, Georg-August-University of Göttingen.

Bancroft, J. (1989). *Human Sexuality and its Problems*. Second edn. Edinburgh: Churchill Livingstone.

Bancroft, J. (2009). *Human Sexuality and its Problems*. Third edn. Edinburgh: Churchill Livingstone.

Barber, N. (1995). The evolutionary psychology of physical attractiveness: sexual selection and human morphology. *Ethology and Sociobiology*, 16, 395–424.

Barelli, C., Matsudaira, K., Wolf, T., *et al.* (2013). Extra-pair paternity confirmed in white-handed gibbons. *American Journal of Primatology*, 75, 1185–95.

Bartholinus, T. (1671–1672). Anatome cercopitheci mamonet dicti. In *T. Bartholini Acta Medica et Philosophica Hafniensia*.

Barton, R. (1985). Grooming site preferences in primates and their functional implications. *International Journal of Primatology*, 6(5), 519–32.

Barton, R.A. (2000). Socioecology of baboons: the interaction of male and female strategies. In *Primate Males: Causes and Consequences of Variation in Group Composition*, ed. P.M. Kappeler. Cambridge: Cambridge University Press, pp. 97–107.

Baum, M.J., Keverne, E.B., Everitt, B.J. and Herbert, J. (1976). Reduction in sexual interaction in rhesus monkeys by a vaginal action of progesterone. *Nature, London*, 263, 606–8.

Beach, F.A. (1976). Sexual attractivity, proceptivity and receptivity in female mammals. *Hormones and Behavior*, 7, 105–38.

Beasley, D.E.A., Bonisoli-Aliquati, A. and Mousseaux, T.A. (2013). The use of fluctuating asymmetry as a measure of environmentally induced developmental instability: a meta-analysis. *Ecological Indicators*, 30, 218–26.

Beaulieu, M., Mboumba, S., Williams, E., Kappeler, P.M. and Charpentier, M.J.E. (2014). The oxidative cost of unstable social dominance. *Journal of Experimental Biology*, 217, 2629–32.

Benneton, C. and Noë, R. (2004). Reproductive tactics of adult male sooty mangabeys in TaÏ National Park, Ivory Coast. *Folia Primatologica*, 75, 169.

Bentley, P.J. (1976). *Comparative Vertebrate Endocrinology*. Cambridge: Cambridge University Press.

Bercovitch, F.B. (1988). Coalitions, cooperation and reproductive tactics among adult male baboons. *Animal Behaviour*, 36, 1198–209.

Bercovitch, F.B. (1989). Body size, sperm competition, and determinants of reproductive success in savanna baboons. *Evolution*, 43, 1507–21.

Bercovitch, F.B. (1992). Estradiol concentrations, fat deposits, and reproductive strategies in male rhesus macaques. *Hormones and Behavior*, 26(2), 272–82.

Bercovitch, F.B. (1996). Testicular function and scrotal coloration in patas monkeys. *Journal of Zoology, London*, 239, 93–100.

Bercovitch, F.B. (2000). Behavioral ecology and socioecology of reproductive maturation in cercopithecine monkeys. In *Old World Monkeys*, eds. P.F. Whitehead and C.J. Jolly. Cambridge: Cambridge University Press, pp. 298–320.

Bergman, T.J., Ho, L. and Beehner, J.C. (2009). Chest color and social status in male geladas (*Theropithecus gelada*). *International Journal of Primatology*, 30, 791–806.

Bernstein, I.S. (1970). Primate status hierarchies. In *Primate Behavior, Vol.1*, ed. L.A. Rosenblum. New York, NY: Academic Press, pp. 71–109.

Bernstein, I.S., Rose, R.M. and Gordon, T.P. (1974). Behavioral and environmental events influencing primate testosterone levels. *Journal of Human Evolution*, 3, 517–25.

Bielert, C. (1986). Sexual interactions between captive male and female chacma baboons (*Papio ursinus*) as related to the female's menstrual cycle. *Journal of Zoology, London*, 209, 521–36.

Bielert, C. and Busse, C. (1983). Influences of ovarian hormones on food intake and feeding of captive and wild chacma baboons (*Papio ursinus*). *Folia Primatologica*, 30, 103–11.

Bielert, C. and Van Der Walt, L.A. (1982). Male chacma baboon (*Papio ursinus*) sexual arousal: mediation by visual cues from female conspecifics. *Psychoneuroendocrinology*, 7, 31–48.

Bielert, C., Girolami, L. and Jowell, S. (1989). An experimental examination of the colour component in visually mediated sexual arousal of the male chacma baboon (*Papio ursinus*). *Journal of Zoology, London*, 219, 569–79.

Birkhead, T.R. and Møller, A.P. (1992). *Sperm Competition in Birds*. London: Academic Press.

Birkhead, T.R. and Møller, A.P. (1998). Eds. *Sperm Competition and Sexual Selection*. San Diego, CA: Academic Press.

Birkhead, T.R., Atkin, L. and Möller, A.P. (1987). Copulation behavior in birds. *Behaviour*, 101, 101–38.

Blache, D. and Bickell, S. (2011). External and internal modulators of sheep reproduction. *Reproductive Biology*, 11(3), 61–77.

Blache, D., Zhang, S. and Martin, G.B. (2003). Fertility in male sheep: modulators of the acute effects of nutrition on the reproductive axis of male sheep. *Reproduction*, 61, 387–402.

Böer, M. and Sommer, V. (1992). Evidence for sexually selected infanticide in captive *Cercopithecus mitis, Cercocebus torquatus*, and *Mandrillus leucophaeus*. *Primates*, 33(4), 557–63.

Boinski, S. (1987). Mating patterns in squirrel monkeys (*Saimiri oerstedii*): implications for seasonal sexual dimorphism. *Behavioral Ecology and Sociobiology*, 21, 13–21.

Boinski, S. (1998). Monkeys with inflated sex appeal. In *The Primate Anthology: Essays on Primate Behavior, Ecology, and Conservation from Natural History*, eds. R.L. Ciochon and R. A. Nisbett. Englewood Cliffs, NJ: Prentice Hall, pp. 174–9.

Bonadonna, G., Torti, V., Randrianarison, R.M., *et al.* (2014). Behavioral correlates of extra-pair copulation in *Indri indri*. *Primates*, 55(1), 119–23.

Borries, C. and Koenig, K. (2000). Infanticide in hanuman langurs: social organization, male migration and weaning age. In *Infanticide by Males and its Implications*, eds. C.P. Van Schaik and C.H. Janson. Cambridge: Cambridge University Press, pp. 99–122.

Borries, C., Launhardt, K., Epplen, C., Epplen, J.T. and Winkler P. (1999). DNA analyses support the hypothesis that infanticide is adaptive in langur monkeys. *Proceedings of the Royal Society of London, Series B*, 266, 901–4.

Bossi, T. (1991). *Sozial – und Sexualverhalten von Männlichen Mandrills (*Mandrillus sphinx*)*. Zurich: Diplomarbeit, Anthropologisches Institut der Universität Zurich, Schweiz.

Bout, N. and Thierry, B. (2005). Peaceful meaning for the silent bared-teeth displays of mandrills. *International Journal of Primatology*, 26(6), 1215–28.

Bret, C., Sueur, C., Ngoubangoye, B., *et al.* (2013). Social structure in a semi-free ranging group of mandrills (*Mandrillus sphinx*): a social network analysis. *PLoS ONE*, 8(12), e83015. DOI: 10.1371/journal.pone.0083015.

Bronson, F.H. (1989). *Mammalian Reproductive Biology*. Chicago, IL: University of Chicago Press.

Brown, G.R. and Silk, J.B. (2002). Reconsidering the null hypothesis: is maternal rank associated with birth sex ratios in primate groups? *Proceedings of the National Academy of Sciences, USA*, 99, 11252–5.

Buffon, G.L. (1766). *Histoire Naturelle Générale et Particulière, 14. Orangs, Catarrhini.* Paris: L'Imprimerie du Roi.

Burley, N.T. and Symanski, R. (1998). A taste for the beautiful: latent aesthetic mate preferences for white crests in two species of Australian grassfinches. *American Naturalist*, 152(6), 792–802.

Burley, N., Krantzberg, G. and Radman, P. (1982). Influence of colour-banding on the conspecific preferences of zebra finches. *Animal Behaviour*, 30(2), 44–55.

Busse, C.D. and Estep, D.Q. (1984). Sexual arousal in male pigtailed monkeys (*Macaca nemestrina*): effects of serial matings by two males. *Journal of Comparative Psychology*, 98, 227–31.

Butynski, T.M. (1988). Guenon birth seasons and correlates with rainfall and food. In *A Primate Radiation: Evolutionary Biology of the African Guenons*, eds. A. Gautier-Hion, F. Boulière, J.P. Gautier and J. Kingdon. Cambridge: Cambridge University Press, pp. 284–322.

Butynski, T.M. and Koster, S.H. (1994). Distribution and conservation status of primates in Bioko Island, Equatorial Guinea. *Biodiversity and Conservation*, 3, 893–909.

Butynski, T.M., de Jong, Y. and Hearn, G.W. (2009). Body measurements of monkeys on Bioko Island, Equatorial Guinea. *Primate Conservation*, 24, 99–105.

Cameron, E.Z. and Linklater, W.L. (2007). Extreme sex ratio variation in relation to change in condition around conception. *Biology Letters*, 3, 395–7.

Cameron, J.L. and Nosbisch, C. (1991). Suppression of pulsatile luteinizing hormone and testosterone secretion during short term food restriction in the adult male rhesus monkey (*Macaca mulatta*). *Endocrinology*, 128, 1532–40.

Chambers, K.C. and Phoenix, C.H. (1987). Differences among ovariectomized female rhesus monkeys in the display of sexual behavior with and without estradiol treatment. *Behavioral Neuroscience*, 101, 303–8.

Charlton, B.D. (2013). Experimental tests of mate choice in nonhuman mammals: the need for an integrative approach. *Journal of Experimental Biology*, 216, 1127–30.

Charpentier, M.J.E., Mbouba, S., Ditsoga, C. and Drea, C.M. (2013). Nasopalatine ducts and flehmen behavior in the mandrill: reevaluating olfactory communication in Old World primates. *American Journal of Primatology*, 75(7), 703–14.

Charpentier, M., Peignot, P., Hossaert-McKey, M., *et al.* (2005). Constraints on control: factors influencing reproductive success in male mandrills (*Mandrillus sphinx*). *Behavioral Ecology*, DOI: 10/1093/beheco/ari034.

Chevalier-Skolnikoff, S. (1974). Male–female, female–female, and male–male sexual behavior in the stumptail monkey, with special attention to the female orgasm. *Archives of Sexual Behavior*, 3, 95–116.

Clark, A.B. (1978). Sex ratio and local resource competition in a nocturnal primate. *Science*, 210, 163–5.

Clarke, I.J. (2011). Control of GnRH secretion: one step back. *Frontiers in Neuroendocrinology*, 32(3), 367–75.

Clarke, I. and Henry, B.A. (1999). Leptin and reproduction. *Reviews of Reproduction*, 4, 48–55.

Clutton-Brock, T.H. (1991). The evolution of sex differences and consequences of polygyny in mammals. In *Development and Integration of Behaviour: Essays in Honour of Robert Hinde*, ed. P. Bateson. Cambridge: Cambridge University Press, pp. 229–53.

Clutton-Brock, T.H. and Harvey, P.H. (1976). Evolutionary rules and primate societies. In *Growing Points in Ethology*, eds. P. Bateson and R.A. Hinde. Cambridge: Cambridge University Press, pp. 195–237.

Clutton-Brock, T.H. and Harvey, P.H. (1977). Primate ecology and social organization. *Journal of Zoology, London*, 183(1), 1–39.

Clutton-Brock, T.H., Albon, S.D. and Guiness, F.E. (1988). Reproductive success in red deer. In *Reproductive Success*, ed. T.H. Clutton-Brock. Chicago, IL: University of Chicago Press, pp. 325–43.

Clutton-Brock, T.H., Harvey, P.H. and Rudder, B. (1977). Sexual dimorphism, socionomic sex ratio and body weight in primates. *Nature, London*, 269, 797–800.

Coad, L., Abernethy, K., Balmford, A., Manica, A., Airey, L. *et al.* (2010). Distribution and use of income from bushmeat in a rural village, central Gabon. *Conservation Biology*, 24(6), 1510–18.

Colmenares, F. (1990). Greeting behaviour in male baboons, 1: communication, reciprocity and symmetry. *Behaviour*, 113, 81–116.

Colmenares, F. (1991). Greeting behaviour between male baboons: oestrous females, rivalry and negotiation. *Animal Behaviour*, 41, 49–60.

Crockett, C.M. (2003). Re-evaluating the sexual selection hypothesis for infanticide by *Alouatta* males. In *Sexual Selection and Reproductive Competition in Primates: New Perspectives and Directions*, ed. C.B. Jones. Norman, OK: American Society of Primatologists, pp. 327–65.

Crockett, C.M. and Janson, C.H. (2000). Infanticide in red howlers: female group size, male membership, and a possible link to folivory. In *Infanticide by Males and its Implications*, eds. C. P. Van Schaik and C.H. Janson. Cambridge: Cambridge University Press, pp. 75–98.

Cronin, D.T., Riaco, C. and Hearn, G.W. (2013). Survey of threatened monkeys in the Iladyi River valley region, southwestern Bioko Island, Equatorial Guinea. *African Primates*, 8, 1–8.

Crook, J.H. (1966). Gelada baboon herd structure and movement: a comparative report. *Symposia of the Zoological Society of London*, 18, 237–58.

Curtis, R.F., Ballantine, J.A., Keverne, E.B., Bonsall, R.W. and Michael, R.P. (1971). Identification of primate sexual pheromones and properties of synthetic attractants. *Nature, London*, 232, 396–8.

Cuvier, F. (1807). *Annals du Musée d'Histoire Naturelle, Paris*, 9, 477–82.

Cuvier, F. (1833). In *Histoire Naturelle des Mammifères, Vol. 1*, eds. Geoffroy, E. and Cuvier, F. Paris: A. Belin.

Darwin, C. (1871). *The Descent of Man and Selection in Relation to Sex*. London: John Murray.

Darwin, C. (1872). *The Expression of the Emotions in Man and Animals*. London: John Murray.

Darwin, C. (1876). Sexual selection in relation to monkeys. *Nature, London*, Nov. 2, 18.

Davenport, T.R.B., Stanley, W.T., Sargis, E.J., *et al.* (2006). A new genus of African monkey, *Rungwecebus*: morphology, ecology and molecular phylogenetics. *Science*, 312(5778), 1378–81.

Davies, N.B., Krebs, J.R. and West, S.A. (2012). *An Introduction to Behavioural Ecology*. Fourth edn. Chichester: Wiley–Blackwell.

Davis, J.A. (1976). The ascent of mandrill. *Animal Kingdom*, 76(6), 4–11.

Deaner, R.O., Khera, A.V. and Platt, M.L. (2005). Monkeys pay per view: adaptive valuation of social images by rhesus macaques. *Current Biology*, 15, 543–8.

Denton, C.H. (1999). Cenozoic climate change. In *African Biogeography, Climate Change and Human Evolution*, eds. T.G. Bromage and F. Schrenk. New York, NY: Oxford University Press.

Deschner, T., Heistermann, M., Hodges, K. and Boesch, C. (2004). Female sexual swelling size, timing of ovulation, and male behavior in wild West African chimpanzees. *Hormones and Behavior*, 46, 204–15.

DeVore, I. (1964). Ed. *Primate Behavior: Field Studies of Monkeys and Apes*. New York, NY: Holt, Rinehart and Winston.

Devreese, L. and Gilbert, C.C. (2013). Phylogenetic relationships within the *Cercocebus* clade as indicated by craniodental morphology. *Folia Primatologica*, 84(3–5), 268.

De Waal, F. (1982). *Chimpanzee Politics*. London: Jonathan Cape.

De Waal, F.B.M. (1989). *Peacemaking Among Primates*. Cambridge, MA: Harvard University Press.

Disotell, T.R. (2000). The molecular systematics of the *Cercopithecidae*. In *Old World Monkeys*, eds. P.F. Whitehead and C.J. Jolly. Cambridge: Cambridge University Press, pp. 29–56.

Disotell, T.R., Honeycutt, R.L. and Ruvolo, M. (1992). Mitochondrial phylogeny of the Old World monkey tribe Papionini. *Molecular Biology and Evolution*, 9, 1–13.

Dixson, A.F. (1977). Observations on the displays, menstrual cycles and behaviour of the "black ape" of Celebes. *Journal of Zoology, London*, 182, 63–84.

Dixson, A.F. (1980). Androgens and aggressive behavior in primates: a review. *Aggressive Behavior*, 6, 37–67.

Dixson, A.F. (1983a). Observations on the evolution and behavioral significance of "sexual skin" in female primates. *Advances in the Study of Behavior*, 13, 63–106.

Dixson, A.F. (1983b). The owl monkey. In *Reproduction in New World Primates*, ed. J.P. Hearn. Lancaster: MTP Press, pp. 69–113.

Dixson, A.F. (1987). Observations on the evolution of the genitalia and copulatory behaviour in male primates. *Journal of Zoology, London*, 213, 423–43.

Dixson, A.F. (1991). Sexual selection, natural selection and copulatory patterns in male primates. *Folia Primatologica*, 57, 96–101.

Dixson, A.F. (1995). Sexual selection and ejaculatory frequencies in primates. *Folia Primatologica*, 64, 146–52.

Dixson, A.F. (1998a). *Primate Sexuality: Comparative Studies of the Prosimians, Monkeys, Apes, and Human beings*. First edn. Oxford: Oxford University Press.

Dixson, A.F. (1998b). Sexual selection and the evolution of the seminal vesicles in primates. *Folia Primatologica*, 69, 300–6.

Dixson, A.F. (2009). *Sexual Selection and the Origins of Human Mating Systems*. Oxford: Oxford University Press.

Dixson, A.F. (2010). Homosexual behaviour in primates. In *Animal Homosexuality: A Biosocial Perspective*, by A. Poiani. Cambridge: Cambridge University Press, pp. 381–400.

Dixson, A.F. (2012). *Primate Sexuality: Comparative Studies of the Prosimians, Monkeys, Apes, and Humans*. Second edn. Oxford: Oxford University Press.

Dixson, A.F. (2013). Male infanticide and primate monogamy. *Proceedings of the National Academy of Sciences, USA*, 110(51), E4937.

Dixson, A.F. and Anderson, M.J. (2002). Sexual selection, seminal coagulation and copulatory plug formation in primates. *Folia Primatologica*, 73, 63–9.

Dixson, A.F. and Fleming, D. (1981). Parental behaviour and infant development in owl monkeys (*Aotus trivirgatus griseimemba*). *Journal of Zoology, London*, 194, 25–39.

Dixson, A.F. and George, L. (1982). Prolactin and parental behaviour in a male New World primate. *Nature, London*, 299, 551–3.

Dixson, A.F. and Herbert, J. (1974). The effects of testosterone on the sexual skin and genitalia of the male talapoin monkey. *Journal of Reproduction and Fertility*, 38, 217–19.

Dixson, A.F. and Mundy, N.I. (1994). Sexual behavior, sexual swelling and penile evolution in chimpanzees (*Pan troglodytes*). *Archives of Sexual Behavior*, 23, 267–80.

Dixson, A.F., Bossi, T. and Wickings, E.J. (1993). Male dominance and genetically determined reproductive success in the mandrill (*Mandrillus sphinx*). *Primates*, 34, 525–32.

Dixson, A.F., Dixson, B.J. and Anderson, M. (2005). Sexual selection and the evolution of visually conspicuous sexually dimorphic traits in male monkeys, apes, and human beings. *Annual Reviews of Sex Research*, 15, 1–19.

Dixson, A.F., Harvey, N., Patton, M.L. and Setchell, J.M. (2003). Behaviour and Reproduction. In *Reproduction and Integrated Conservation Science*, eds. W.V. Holt, A.R. Pickard, J. Rodger and D.E. Wildt. Cambridge: Cambridge University Press, pp. 24–41.

Dixson, A.F., Knight, J., Moore, H.D.M. and Carman, M. (1982). Observations on sexual development in male orang-utans (*Pongo pygmaeus*). *International Zoo Yearbook*, 22, 222–7.

Dixson, A.F., Scruton, D.M. and Herbert, J. (1975). Behaviour of the talapoin monkey (*Miopithecus talapoin*) studied in groups, in the laboratory. *Journal of Zoology, London*, 176, 177–210.

Dixson, B.J. and Vasey, P.L. (2012). Beards augment perceptions of men's age, social status, and aggressiveness, but not attractiveness. *Behavioral Ecology*, 23(3), 481–90.

Dixson, B.J., Dixson, A.F., Bishop, P.J. and Parish, A. (2010a). Human physique and sexual attractiveness in men and women: a New Zealand–U.S. comparative study. *Archives of Sexual Behavior*, 39, 798–806.

Dixson, B.J., Li, B. and Dixson, A.F. (2010b). Female waist-to-hip ratio, body mass index and sexual attractiveness in China. *Current Zoology*, 56, 175–81.

Dobroruka, L.J. (1966). Kleine notizen über baumpaviane, *Papio leucophaeus* (F. Cuvier 1807) und *Papio sphinx* (Linnaeus 1758). *Revue de Zoologie et de Botanique Africaines*, 73, 155–8.

Dobson, F.S., Nolan, P.M., Nicolaus, M., *et al.* (2008). Comparison of color and body condition between early and late breeding king penguins. *Ethology*, 114, 925–33.

Domb, L. and Pagel, M. (2001). Sexual swellings advertise female quality in wild baboons. *Nature, London*, 410, 204–6.

Doucet, S.M. and Montgomerie, R. (2003). Multiple sexual ornaments in satin bowerbirds: ultraviolet plumage and bowers signal different aspects of male quality. *Behavioral Ecology*, 14, 503–9.

Dowling, D.K. and Simmons, L.W. (2009). Reactive oxygen species as universal constraints in life-history evolution. *Proceedings of the Royal Society of London, Series B*, 276, 1737–45.

Dubuc, C., Allen, W.I., Maestripieri, D. and Higham, J.P. (2014a). Is male rhesus macaque red color ornamentation attractive to females? *Behavioral Ecology and Sociobiology*, DOI: 10.1007/s00265-014–1732-9.

Dubuc, C., Winters, S., Allen, W.L., *et al.* (2014b). Sexually selected skin colour is heritable and related to fecundity in a non-human primate. *Proceedings of the Royal Society of London, Series B*, 281, 20141602.

Dunbar, R.I.M. (1984). *Reproductive Decisions: an Economic Analysis of Gelada Baboon Social Strategies*. Princeton, NJ: Princeton University Press.

Dunbar, R.I.M. and Dunbar, E.P. (1975). *Social Dynamics of Gelada Baboons*. Basel: Karger.

Dunkel, L., Van Noordwijk, M., Cadilek, M. and Mardianah, N. (2013). Geographic variation in orangutan males' monopolization potential is driven by female choice and, ultimately, ecology. *Folia Primatologica*, 84, 271.

Eaton, G.G. and Resko, J.A. (1974a). Ovarian hormones and behavior of *Macaca nemestrina*. *Journal of Comparative and Physiological Psychology*, 86, 919–25.

Eaton, G.G. and Resko, J.A. (1974b). Plasma testosterone and male dominance in a male Japanese macaque (*Macaca fuscata*) troop compared to repeated measures of testosterone in laboratory males. *Hormones and Behavior*, 5, 251–9.

Eberhard, W.G. (1985). *Sexual Selection and Animal Genitalia*. Cambridge, MA: Harvard University Press.

Eberhard, W.G. (1996). *Female Control: Sexual Selection by Cryptic Female Choice*. Princeton, NJ: Princeton University Press.

Eberhard, W.G. (2009). Postcopulatory sexual selection: Darwin's omission and its consequences. *Proceedings of the National Academy of Sciences, USA*, 106(1), 10025–32.

Eberhart, J.A. and Keverne, E.B. (1979). Influences of the dominance hierarchy on LH, testosterone and prolactin in male talapoin monkeys. *Journal of Endocrinology*, 83, 42–3.

Emery, M.A. and Whitten, P.L. (2003). Size of sexual swellings reflects ovarian function in chimpanzees (*Pan troglodytes*). *Behavioral Ecology and Sociobiology*, 54, 340–51.

Enstam, K.L. and Isbell, L.A. (2007). The guenons (genus *Cercopithecus*) and their allies. In *Primates in Perspective*, eds. C.J. Campbell, A. Fuentes, K.C. MacKinnon, M. Panger and S.K. Bearder. Oxford: Oxford University Press, pp. 252–74.

Epple, G., Belcher, A.M., Kuderling, I., *et al.* (1993). Making sense out of scents: species differences in scent glands, scent-marking behaviour, and scent-mark composition in the *Callitrichidae*. In *Marmosets and Tamarins: Systematics, Behaviour and Ecology*, ed. A.B. Rylands. Oxford: Oxford University Press, pp. 123–51.

Escalante, A.A., Cornejo, O.E., Freeland, D.E., *et al.* (2005). A monkey's tale: the origin of *Plasmodium vivax* as a human malaria parasite. *Proceedings of the National Academy of Sciences, USA*, 102, 1980–5.

Escalante, A.A., Freeland, D.E., Collins, W.E. and Lal, A.A. (1998). The evolution of primate malaria parasites based on the gene encoding cytochrome b from the linear mitochondrial genome. *Proceedings of the National Academy of Sciences, USA*, 95, 8124–9.

Estaquier, J., Peeters, M., Bedjabaga, L., *et al.* (1991). Prevalence and transmission of simian immunodeficiency virus and simian T-cell leukemia virus in a semi-free range breeding colony of mandrills in Gabon. *AIDS*, 5(11), 1385–6.

Estes, R.D. (1972). The role of the vomeronasal organ in mammalian reproduction. *Mammalia*, 36, 315–41.

Evans, C. (2003). *Vomeronasal Chemoreception in Vertebrates: A Study of the Second Nose*. London: Imperial College Press.

Fa, J.E., Currie, D. and Meeuwig, J. (2003). Bushmeat and food security in the Congo Basin: linkages between wildlife and peoples' future. *Environmental Conservation*, 1, 71–8.

Fa, J.E., Yuste, J.E.G. and Castello, R. (2000). Bushmeat markets on Bioko Island as a measure of hunting pressure. *Conservation Biology*, 14(6), 1602–13.

Feistner, A.T.C. (1989). *The Behaviour of a Social Group of Mandrills (*Mandrillus sphinx*)*. PhD Thesis. Stirling: University of Stirling.

Feistner, A.T.C. (1991). Scent marking in mandrills, *Mandrillus sphinx*. *Folia Primatologica*, 57, 42–7.

Fernandez-Duque, E. and Huck, M. (2013). Till death (or an intruder) do us part: intrasexual competition in a monogamous primate. *PLoS One*, 8(1), e53724, DOI: 10.1371/journal.pone.0053724.

Fisher, R.A. (1930). *The Genetical Theory of Natural Selection*. Oxford: Clarendon Press.

Fleagle, J.G. and McGraw, W.C. (1999). Skeletal and dental morphology supports diphyletic origin of baboons and mandrills. *Proceedings of the National Academy of Sciences, USA*, 96, 1157–61.

Fleagle, J.G. and McGraw, W.C. (2002). Skeletal and dental morphology of African papionins: unmasking a cryptic clade. *Journal of Human Evolution*, 42, 267–92.

Foerster, S., Wilkie, D.S., Morelli, G.A., *et al.* (2012). Correlates of bushmeat hunting among remote rural households in Gabon, Central Africa. *Conservation Biology*, 26(2), 335–44.

Folstad, I. and Karter, A.J. (1992). Parasites, bright males, and the immunocompetence handicap. *American Naturalist*, 139, 603–22.

Frei, E. (1991). *Analyse von Präsentationen und Besteigungen bei Mandrills (*Mandrillus sphinx*)*. Zurich: Diplomarbeit, Zoologisches Institut der Universität Zurich, Schweiz.

Freund, M. (1963). The effect of frequency of emission on semen output and an estimate of daily sperm production in man. *Journal of Reproduction and Fertility*, 6, 269–86.

Furuichi, T. (1987). Sexual swelling, receptivity and grouping of wild pigmy chimpanzee females at Wamba, Zaire. *Primates*, 28, 309–18.

Galbany, J., Romero, A., Mayo-Alesón, M., *et al.* (2014). Age-related tooth wear differs between forest and savanna primates. *PLoS One*, DOI: 10.1371/journal.pone. 0094938.

Galdikas, B.F.M. (1983). The orang-utan long call and snag crashing at Tanjung Puting Reserve. *Primates*, 34, 371–84.

Galdikas, B.F.M. (1995). Social and reproductive behavior of wild adolescent female orangutans. In *The Neglected Ape*, eds. R.D. Nadler, B.F.M. Galdikas, L.K. Sheeran and N. Rosen. New York, NY: Plenum Press, pp. 163–82.

Gartlan, J.S. (1970). Preliminary notes on the ecology and behavior of the drill *Mandrillus leucophaeus, Ritgen, 1824*. In *Old World Monkeys: Evolution, Systematics and Behavior*, eds. J. Napier and P. Napier. New York, NY: Academic Press, pp. 445–75.

Gerald, M.S. (2001). Primate colour reveals social status and predicts aggressive outcome. *Animal Behaviour*, 61, 559–66.

Gerald, M.S. (2003). How color may guide the primate world: possible relationships between sexual selection and sexual dichromatism. In *Sexual Selection and Reproductive Competition in Primates: New Perspectives and Directions*, ed. C. Jones. Norman, OK: American Society of Primatologists, pp. 141–71.

Gesner, C. (1551–1558). *Medici Tigurini, Historiae Animalium*, Libr. 1–1V. Tiguri: Folio.

Gesner, C. (1606). *Das Tierbuch*. Zurich: N. Ochsembach.

Girolami, L. and Bielert, C. (1987). Female perineal swelling and its effects on male sexual arousal: an apparent sexual releaser in the chacma baboon (*Papio ursinus*). *International Journal of Primatology*, 8, 651–61.

Goldfoot, D.A. (1981). Olfaction, sexual behavior, and the pheromone hypothesis in rhesus monkeys: a critique. *American Zoologist*, 21, 153–64.

Goldfoot, D.A., Essock-Vitale, S.M., Asa, C.S., Thornton, J.E. and Leshner, A.I. (1978). Anosmia in male rhesus monkeys does not alter copulatory activity with cycling females. *Science*, 199, 1095–6.

Goldfoot, D.A., Kravetz, M.A., Goy, R.W. and Freeman, S.K. (1976). Lack of effect of vaginal lavages and aliphatic acids on ejaculatory responses in rhesus monkeys: behavioral and chemical responses. *Hormones and Behavior*, 7, 1–27.

Goodall, J. (1986). *The Chimpanzees of Gombe: Patterns of Behavior*. Cambridge, MA: Belknap Press of Harvard University Press.

Goodman, M., Porter, C.A., Czelusniak, J., *et al.* (1998). Toward a phylogenetic classification of primates based on DNA evidence complemented by fossil evidence. *Molecular Phylogenetics and Evolution*, 9, 585–98.

Gordon, T.P., Rose, R. and Bernstein, I.S. (1976). Seasonal rhythm in plasma testosterone levels in the rhesus monkey (*Macaca mulatta*): a three-year study. *Hormones and Behavior*, 7, 229–43.

Grafen, A. (1990a). Biological signals as handicaps. *Journal of Theoretical Biology*, 144, 517–46.

Grafen, A. (1990b). Sexual selection unhandicapped by the Fisher process. *Journal of Theoretical Biology*, 144, 473–516.

Greenman, C.G., Martin, L.B. and Hau, M. (2005). Reproductive state, but not testosterone, reduces immune function in male house sparrows (*Passer domesticus*). *Physiological and Biochemical Zoology*, 78(1), 60–8.

Groves, C.P. (2000). The phylogeny of the Cercopithecoidea. In *Old World Monkeys*, eds. P.F. Whitehead and C.J. Jolly. Cambridge: Cambridge University Press, pp. 77–98.

Grubb, P. (1973). Distribution, divergence and speciation of the drill and mandrill. *Folia Primatologica*, 20, 161–77.

Grubb, P. (1982). Refuges and dispersal in the speciation of African Forest mammals. In *Biological Diversification in the Tropics*, ed. G.T. Prance. New York, NY: Columbia University Press, pp. 537–53.

Grubb, P. (1990). Primate geography in the Afro-tropical forest biome. In *Vertebrates in the Tropics*, eds. C. Peters and R. Hutterer. Bonn: Museum Alexander Koenig, pp. 187–214.

Guevara, E.E. and Steiper, M.E. (2014). Molecular phylogenetic analysis of the Papionina using concatenation and species tree methods. *Journal of Human Evolution*, 66, 18–28.

Hamilton, A.C. (1988). Guenon evolution and forest history. In *A Primate Radiation: Evolutionary Biology of the African Guenons*, eds. A. Gautier-Hion, F. Boulière, J.P. Gautier and J. Kingdon. Cambridge: Cambridge University Press, pp. 13–34.

Hamilton, A.C. and Taylor, D. (1992). History of climate and forests in tropical Africa during the last 8 million years. In *Tropical Forests and Climate*, ed. N. Myers. Dordrecht, Netherlands: Kluwar Academic Publishers.

Hamilton, W.D. and Zuk, M. (1982). Heritable true fitness and bright birds: a role for parasites? *Science*, 218, 384–7.

Harcourt, A.H. and Wood, M.A. (2012). Rivers as barriers to primate distributions. *International Journal of Primatology*, 33, 168–83.

Harcourt, A.H., Harvey, P.H., Larson, S.G. and Short, R.V. (1981). Testis weight, body weight and breeding system in primates. *Nature, London*, 293, 55–7.

Harcourt, A.H., Purvis, A. and Liles, L. (1995). Sperm competition: mating system, not breeding season, affects testes size of primates. *Functional Ecology*, 9, 468–76.

Harper, M.J.K. (1994). Gamete and zygote transport. In *The Physiology of Reproduction, Vol. 1*, Second edn., eds. E. Knobil and J.D. Neill. New York, NY: Raven Press, pp. 123–87.

Harrison, M.J.S. (1988). The mandrill in Gabon's rain forest: ecology, distribution and status. *Oryx*, 22(4), 218–28.

Hart, K., Van Lier, E., Viñoles, C., Paganoni, B. and Blache, D. (2008). Calm merino ewes have more multiple pregnancies than nervous merino ewes due to higher ovulation rate. *Reproduction in Domestic Animals*, 43(3), 88.

Harvey, P.H. and May, R.M. (1989). Out for the sperm count. *Nature, London*, 337, 508–9.

Hausfater, G. and Takacs, G. 1987). Structure and function of hindquarter presentations in yellow baboons (*Papio cynocephalus*). *Ethology*, 74, 297–319.

Hearn, G.W., Morra, W.A. and Butynski, T.M. (2006). Monkeys in trouble: the rapidly deteriorating conservation status of the monkeys on Bioko Island, Equatorial Guinea. *Report of the Bioko Biodiversity Protection Program, Acadia University, Glenside, PA*, http://www.bioko.org/conservation/2006MonkeysInTrouble8.pdf.

Henschel, P., Abernethy, K.A. and White, L.J.T. (2005). Leopard food habits in the Lopé National Park, Gabon, Central Africa. *African Journal of Ecology*, 43(1), 21–8.

Herbert, J. (1966). The effect of oestrogen applied directly to the genitalia upon the sexual attractiveness of the female rhesus monkey. *International Congress Series, Excerpta Medica*, 111, 212.

Herbert, J. (1970). Hormones and reproductive behaviour in rhesus and talapoin monkeys. *Journal of Reproduction and Fertility*, 11, 119–40.

Hernández-López, L., Cerda-Molina, A.L., Páez-Ponce, D.L. and Mondragón-Ceballos, R. (2008). The seminal coagulum favours passage of fast-moving sperm into the uterus in the black-handed spider monkey. *Reproduction*, 136, 411–21.

Higham, J.P., MacLarnon, A.M., Ross, C., Heistermann, M. and Semple, S. (2008). Baboon sexual swellings: information content of size and color. *Hormones and Behavior*, 53, 452–62.

Hill, W.C.O. (1954). Sternal glands in the genus *Mandrillus*. *Journal of Anatomy, London*, 88, 582.

Hill, W.C.O. (1955). A note on integumental colours with special references to the genus *Mandrillus*. *Säugetierkd Mitt.*, 3, 145–51.

Hill, W.C.O. (1970). *Primates, Comparative Anatomy and Taxonomy, Vol.8, Cynopithecinae: Papio, Mandrillus, Theropithecus*. Edinburgh: Edinburgh University Press.

Hill, W.C.O. (1974). *Primates, Comparative Anatomy and Taxonomy, Vol.7, Cynopithecinae: Cercebus, Macaca, Cynopithecus*. Edinburgh: Edinburgh University Press.

Hoffmann, M., Hilton-Taylor, C., Angulo, A., *et al.* (2010). The impact of conservation on the status of the world's vertebrates. *Science*, 330, 1503–9.

Hongo, S. (2014). New evidence from observations of progressions of mandrills (*Mandrillus sphinx*): a multilevel or non-nested society? *Primates*, DOI: 10.1007/s10329-014–0438-y.

Horr, D.A. (1972). The Bornean Orang-utan: population structure and dynamics in relation to ecology and reproductive strategy. *In Primate Behavior, Vol. 4*, ed. L.A. Rosenblum. New York, NY: Academic Press, pp. 25–80.

Hoshino J. (1985). Feeding ecology of mandrills (*Mandrillus sphinx*) in Campo Animal Reserve, Cameroon. *Primates*, 26, 248–73.

Hoshino, J., Mori, A., Kudo, H. and Kawai, M. (1984). Preliminary report on the grouping of mandrills (*Mandrillus sphinx*) in Cameroon. *Primates*, 25, 295–307.

Hotchkiss, J. and Knobil, E. (1994). The menstrual cycle and its neuroendocrine control. *In The Physiology of Reproduction, Vol. 2*, Second edn., eds. E Knobil and J.D. Neill. New York, NY: Raven Press, pp. 711–49.

Huchard, E., Benavides, J.A., Setchell, J.M., *et al.* (2009). Studying shape in sexual signals: the case of primate sexual swellings. *Behavioral Ecology and Sociobiology*, 63, 1231–42.

Huck, M., Fernandez-Duque, E., Babb, P. and Schurr, T. (2014). Correlates of genetic monogamy in socially monogamous mammals: insights from Azara's owl monkeys. *Proceedings of the Royal Society of London, Series B*, 281, DOI: 10.1098/rspb.2014.0195.

Hunter, A.J. and Dixson, A.F. (1983). Anosmia and aggression in male owl monkeys (*Aotus trivirgatus*). *Physiology and Behavior*, 30, 875–9.

I'Anson, H., Foster, D.L., Foxcroft, G.R. and Booth, P.J. (1991). Nutrition and reproduction. *Oxford Reviews of Reproductive Biology*, 13, 239–311.

Islam, K. and Crawford, J.A. (1998). Comparative displays among four species of tragopans and their derivation and function. *Ethology, Ecology and Evolution*, 10(1), 17–32.

Jack, K.M. (2007). The cebines: towards an explanation of variable social structure. In *Primates in Perspective*, eds. C.J. Campbell, A. Fuentes, K.C. MacKinnon, M. Panger and S.K. Bearder. Oxford: Oxford University Press, pp. 107–23.

Jasieńska, G., Lipson, S.F., Ellison, P., Thune, I. and Ziomkiewicz, A. (2006). Symmetrical women have higher potential fertility. *Evolution and Human Behavior*, 27, 390–400.

Jasieńska, G., Ziomkiewicz, A., Ellison, P., Lipson, S.F., and Thune, I. (2004). Large breasts and narrow waists indicate high reproductive potential in women. *Proceedings of the Royal Society of London, Series B*, 271, 1213–17.

Jeannin, A. (1936). *Les Mammifères sauvages du Cameroun*. Paris: Lechevalier.

Johnson, C.N. (1988). Dispersal and sex ratio at birth in primates. *Nature, London*, 332, 726–8.

Jolly, C.J. (2007). Baboons, mandrills and mangabeys: Afro – papionin socioecology in a phylogenetic perspective. In *Primates in Perspective*, eds. C.J. Campbell, A. Fuentes, K.C. MacKinnon, M. Panger and S.K. Bearder. Oxford: Oxford University Press, pp. 240–51.

Jones, T.E., Ehardt, C.L., Butynski, T.M., *et al.* (2005). The highland mangabey *Lophocebus kipunji*: a new species of African monkey. *Science*, 308(5725), 1161–4.

Jooste, E., Pitman, R.T., Van Hoven, W. and Swanepoel, L.H. (2012). Unusual high predation on chacma baboons (*Papio ursinus*) by female leopards (*Panthera pardus*) in the Waterburg Mountains, South Africa. *Folia Primatologica*, 83, 353–60.

Jouventin, P. (1975a). Observations sur la socio-écologie du mandrill. *La Terre et la Vie*, 29, 493–532.

Jouventin, P. (1975b). Les rôles des colorations du mandrill (*Mandrillus sphinx*). *Zeitschrift für Tierpsychologie*, 39, 455–62.

Jouventin, P., Couchoux, C. and Dobson, F.S. (2009). UV signals in penguins. *Polar Biology*, 32, 513–14.

Keagy, J., Savard, J.F. and Borgia, G. (2011). Complex relationship between multiple measures of cognitive ability and male mating success in satin bowerbirds, *Ptilonorhynchus violaceus*. *Animal Behaviour*, 81, 1063–70.

Kingdon, J. (1997). *The Kingdon Field Guide to African Mammals*. London: Academic Press.

Kingdon, J.S. (1980). The role of visual signals and face patterns in African forest monkeys (guenons) of the genus *Cercopithecus*. *Transactions of the Zoological Society of London*, 35(4), 425–75.

Kingsley, S.K. (1982). Causes of non-breeding and the development of the secondary sexual characteristics in the male orang-utan: a hormonal study. In *The Orang-Utan: Its Biology and Conservation*, ed. L.E.M. de Boer. The Hague: Dr. W. Junk, pp. 215–29.

Kingsley, S.K. (1988). Physiological development of male orang-utans and gorillas. In *Orang-Utan Biology*, ed. J.H. Schwartz. New York, NY: Oxford University Press, pp. 123–31.

Kirkpatrick, M. (1982). Sexual selection and the evolution of female choice. *Evolution*, 36, 1–12.

Kirkpatrick, M. (1986). The handicap mechanism of sexual selection does not work. *American Naturalist*, 127, 222–40.

Klopp, E.B. (2012). Craniodental features in male *Mandrillus* may signal size and fitness. *American Journal of Physical Anthropology*, 147, 593–603.

Knobil, E. (1974). On the control of gonadotropin secretion in the rhesus monkey. *Recent Progress in Hormone Research*, 30, 1–46.

Knott, C.D. and Kahlenberg, S.M. (2007). Orangutans in perspective: forced copulations and female mating resistance. In *Primates in Perspective*, eds. C.J. Campbell, A. Fuentes, K.C. MacKinnon, M. Panger and S.K. Bearder. Oxford: Oxford University Press, pp. 290–305.

Krasnow, S.M. and Steiner, R.A. (2006). Physiological mechanisms integrating metabolism and reproduction. In *Knobil and Neill's Physiology of Reproduction, Vol 2*. Third edn., ed. J.D. Neill. St. Louis, MO: Elsevier, Academic Press, pp. 2553–625.

Kudo, H. (1987). The study of vocal communication of wild mandrills in Cameroon in relation to their social structure. *Primates*, 28, 289–308.

Kudo, H. and Mitani, M. (1985). New record of predatory behavior by the mandrill in Cameroon. *Primates*, 26, 161–7.

Kumar, A. and Kurup, G.U. (1985). Sexual behavior of the lion-tiled macaque (*Macaca silenus*). In *The Lion-tailed Macaque: Status and Conservation*, ed. P. Heltne. New York, NY: Alan R. Liss, pp. 109–30.

Kummer, H. (1968). *Social Organization of Hamadryas Baboons: A Field Study. Bibliotheca Primatologica*, 6, 1–189. Basel: Karger.

Kummer, H. (1971). *Primate Societies*. Chicago, IL: Aldine–Atherton.

Kummer, H. (1990). The social system of hamadryas baboons and its presumable evolution. In *Baboons: Behavior and Ecology, Use and Care*, eds. M.T. de Melo, A. Whitten and R.W. Byrne. Brasil: Brasilia, pp. 43–60.

Kummer, H. and Kurt, F. (1963). Social units of a free-living population of hamadryas baboons. *Folia Primatologica*, 1, 4–19.

Lahm, S.A. (1986). Diet and habitat preference of *Mandrillus sphinx* in Gabon: implications of foraging strategy. *American Journal of Primatology*, 11, 9–26.

Lande, R. (1981). Models of speciation by natural selection of polygenic traits. *Proceedings of the National Academy of Sciences, USA*, 78, 3721–5.

Laurance, W.F., Alonso, A., Lee, M. and Campbell, P. (2006). Challenges for forest conservation in Gabon, Central Africa. *Futures*, 36(4), 454–70.

Law-Smith, M.J.L., Perret, D.I., Jones, B.C., *et al.* (2006). Facial appearance is a cue to oestrogen levels in women. *Proceedings of the Royal Society of London, Series B*, 273, 135–40.

Little, A.C., Apicella, C.L. and Marlowe, F.W. (2007). Preferences for symmetry in human faces in two cultures: data from the UK and the Hadza, an isolated group of hunter gatherers. *Proceedings of the Royal Society of London, Series B*, 274, 3113–17.

Loireau, J.N. and Gautier-Hion, A. (1988). Olfactory marking behaviour in guenons and its implications. In *A Primate Radiation: Evolutionary Biology of the African Guenons*, eds. A. Gautier-Hion, F. Boulière, J.P. Gautier and J. Kingdon. Cambridge: Cambridge University Press, pp. 246–53.

Lucas, D. and Clutton-Brock, T.H. (2013). The evolution of social monogamy in mammals. *Science*, 341, 526–30.

Lucas, D. and Huchard, E. (2014). The evolution of infanticide by males in mammalian societies. *Science*, 346(6211), 841–4.

MacDonald, D.W., Johnson, P.J., Albrechtsen, L., *et al.* (2011). Association of body mass with price of bushmeat in Nigeria and Cameroon. *Conservation Biology*, 25(6), 1220–8.

Madden, J.R. (2003). Bower decorations are good predictors of mating success in the spotted bowerbird. *Behavioral Ecology and Sociobiology*, 53, 269–77.

Maestripieri, D., Leoni, M., Raza, S.S., Hirsch, E.J. and Whitham, J.C. (2005). Female copulation calls in Guinea baboons: evidence for postcopulatory female choice? *International Journal of Primatology*, 26, 737–58.

Maggioncalda, A.N., Czekala, N.M. and Sapolsky, R.M. (2000). Growth hormone and thyroid stimulating hormone concentrations in captive male orang-utans: implications for understanding developmental arrest. *American Journal of Primatology*, 50, 67–76.

Maggioncalda, A.N., Sapolsky, R.M. and Czekala, N.M. (1999). Reproductive hormone profiles in captive male orang-utans: implications for understanding developmental arrest. *American Journal of Physical Anthropology*, 109, 19–32.

Maier, W. (1997). The nasopalatine duct and the nasal floor cartilages in catarrhine primates. *Zeitschrift Fuer Morphologie und Anthropologie*, 81, 289–300.

Maier, W. (2000). Ontogeny of the nasal capsule in cercopithecoids: a contribution to the comparative and evolutionary morphology of the catarrhines. In *Old World Monkeys*, eds. P. F. Whitehead and C.J. Jolly. Cambridge: Cambridge University Press, pp. 99–132.

Main, M.B. and Du Toit, J.T. (2005). Sex differences in reproductive strategies affect habitat choice in ungulates. In *Sexual Selection in Vertebrates: Ecology of the Two Sexes*, eds. K.E. Ruckstuhl and P. Neuhaus. Cambridge: Cambridge University Press, pp. 148–61.

Malbrant, R. and Maclatchy, A. (1949). *Faune de L'Equateur Africain Français, Tome 111, Mammifères*. Paris: Paul Lechevalier.

Markiewicz, M., Asano, Y., Znoyko, S., *et al.* (2007). Distinct effects of gonadectomy in male and female mice on collagen fribrillogenesis in the skin. *Journal of Dermatological Science*, 47, 217–26.

Marson, J., Gervais, D., Cooper, R.W. and Jouannet, P. (1989). Influence of ejaculation frequency on semen characteristics in chimpanzees (*Pan troglodytes*). *Journal of Reproduction and Fertility*, 85, 43–50.

Martin, R.D., Willner, L.A. and Dettling, A. (1994). The evolution of sexual size dimorphism in primates. In *The Differences between the Sexes*, eds. R.V. Short and E. Balaban. Cambridge: Cambridge University Press, pp. 159–200.

Martin, W.C.L. (1841). *A General Introduction to the Natural History of Mammiferous Animals: Quadrumana*. London: Wright and Co.

Marty, J.S., Higham, J.P., Gadsby, E.L. and Ross, C. (2009). Dominance, coloration and social and sexual behavior in male drills, *Mandrillus leucophaeus*. *International Journal of Primatology*, 30, 807–23.

Matsumoto-Oda, A., Hamai, M., Hayaki, H., *et al.* (2007). Estrus cycle asynchrony in wild female chimpanzees, *Pan troglodytes schweinfurthii*. *Behavioral Ecology and Sociobiology*, 61, 661–8.

McClintock, M.K. (1971). Menstrual synchrony and suppression. *Nature, London*, 229, 244–5.

McFarland Symington, M. (1987). Sex ratio and maternal rank in wild spider monkeys: when daughters disperse. *Behavioral Ecology and Sociobiology*, 20, 421–5.

McNeilly, A.S. (2006). Suckling and the control of gonadotropin secretion. In *Knobil and Neill's Physiology of Reproduction, Vol. 2, Third edn.*, ed. J.D. Neill. St. Louis: Elsevier, Academic Press, pp. 2511–51.

Mellen, J.D., Littlewood, A.P., Barrow, B.C. and Stevens, V.J. (1981). Individual and social behavior in a captive group of mandrills (*Mandrillus sphinx*). *Primates*, 22, 206–20.

Mendoza, S.P., Coe, C.L., Lowe, E.L. and Levine, S. (1978). The physiological response to group formation in adult male squirrel monkeys. *Psychoneuroendocrinology*, 3, 221–9.

Michael, R.P. and Keverne, E.B. (1968). Pheromones and the communication of sexual status in primates. *Nature, London*, 218, 746–9.

Michael, R.P. and Keverne, E.B. (1970). Primate sex pheromones of vaginal origin. *Nature, London*, 225, 84–5.

Michael, R.P., Bonsall, R.W. and Kutner, M. (1975). Volatile fatty acids, "copulins," in human vaginal secretions. *Psychoneuroendocrinology*, 1, 153–63.

MIKE. (2005). Monitoring the illegal killing of elephants – Central African forests: final report on population surveys (2003–2004). Washington, DC: MIKE-CITES-WCS.

Milton, K.M. (1985). Mating patterns of woolly spider monkeys, *Brachyteles arachnoides*: implications for female choice. *Behavioral Ecology and Sociobiology*, 17, 53–9.

Milton, K. and Johnston, P. (1984). The relationship between diet and reproduction in New World Primates. *American Journal of Physical Anthropology*, 63(2), 175.

Møller, A.P. (1988). Ejaculate quality, testes size and sperm competition in primates. *Journal of Human Evolution*, 17, 479–88.

Møller, A.P. (1989). Ejaculate quality, testes size and sperm production in mammals. *Functional Ecology*, 3, 91–6.

Møller, A.P. (1991). Concordance of mammalian ejaculate features. *Proceedings of the Royal Society of London, Series B*, 246, 237–41.

Møller, A.P. (1992). Parasites differentially increase the degree of fluctuating asymmetry in secondary sexual characters. *Journal of Evolutionary Biology*, 5, 691–9.

Møller, A.P. (1993). Morphology and sexual selection in the barn swallow *Hirundo rustica* in Chernobyl, Ukraine. *Proceedings of the Royal Society of London, Series B*, 252, 51–7.

Møller, A.P. and Birkhead, T.R. (1989). Copulation behaviour of mammals: evidence that sperm competition is widespread. *Biological Journal of the Linnean Society*, 38, 119–31.

Morgan, B.J., Abwe, E.E., Dixson, A.F. and Astaras, C. (2013). The distribution, status, and conservation outlook of the drill (*Mandrillus leucophaeus*) in Cameroon. *International Journal of Primatology*, 34(2), 281–302.

Morley, R.J. and Kingdon, J. (2013). Africa's environmental and climatic past. In *Mammals of Africa, Vol. 1, Introductory Chapters and Afrotheria*, eds. J. Kingdon, D. Happold, M. Hoffmann, et al. London: Bloomsbury Publishing, pp. 43–56.

Napier, J.R. and Napier, P.H. (1967). *A Handbook of Living Primates: Morphology, Ecology and Behaviour of Nonhuman Primates*. London: Academic press.

Nerrienet, E., Amouretti, X., Müller-Trutwin, M.C., *et al.* (1998). Phylogenetic analysis of SIV and STLV Type 1 in mandrills (*Mandrillus sphinx*): indications that intracolony transmissions are predominantly the result of male-to-male aggressive contacts. *AIDS Research and Human Retroviruses*, 14(9), 785–96, DOI: 10.1089/aid.1998.14.785.

Nevison, C.M., Rayment, F.D.G. and Simpson, M.J.A. (1996). Birth sex ratios and maternal rank in a captive colony of rhesus monkeys (*Macaca mulatta*). *American Journal of Primatology*, 39, 123–38.

Nolan, P.M., Dobson, F.S., Dresp, B. and Jouventin, P. (2006). Immunocompetence is signalled by ornamental colour in king penguins, *Aptenodytes patagonicus*. *Evolutionary Ecology Research*, 8, 1325–32.

Nunn, C.L. (1999). The evolution of exaggerated sexual swellings in female primates and the graded-signal hypothesis. *Animal Behaviour*, 58, 229–46.

Nunn, C.L. (2012). Primate disease ecology in comparative and theoretical perspective. *American Journal of Primatology*, 74, 497–509.

Nunn, C.L. and Altizer, S. (2006). *Infectious Diseases in Primates*. Oxford: Oxford University Press.

Nunn, C.L., Gittleman, J.L. and Anthonovics, J. (2000). Promiscuity and the primate immune system. *Science*, 290, 1168–70.

Oh, J.-W., Chung, W.-J., Heo, K., *et al.* (2014). Biomimetic virus-based colourimetric sensors. *Nature Communications*, 5, DOI: 10.1038/ncomms.4043.

Opie, C., Atkinson, Q.D., Dunbar, R.I.M. and Shultz, S. (2013). Male infanticide leads to social monogamy in primates. *Proceedings of the National Academy of Sciences, USA*, 110(33), 13328–32.

Otovic, P., Partan, S.R., Bryant, J.B. and Hutchinson, E. (2014). Let's call a truce. . . .for now: the silent bared-teeth facial expression in mandrills(*Mandrillus sphinx*) during base-line and post-conflict conditions. *Ethology*, DOI: 10.1111/eth.12285.

Page, S.L., Chiu, C.H. and Goodman, M. (1999). Molecular phylogeny of Old World monkeys as inferred from Ɣ-globin DNA sequences. *Molecular Phylogenetics and Evolution*, 13, 348–59.

Palombit, R.A. (1994). Extra-pair copulations in a monogamous ape. *Animal Behaviour*, 47, 721–3.

Palombit, R.A. (2012). Infanticide: male strategies and female counter strategies. In *The Evolution of Primate Societies*, eds. J.C. Mitani, J. Call, P.M. Kappeler, R.A. Palombit and J. Silk. Chicago, IL: The University of Chicago Press, pp. 432–68.

Pappano, D.J. and Beehner, J.C. (2014). Harem-holding males do not rise to the challenge: androgens respond to social but not seasonal challenges in wild geladas. *Royal Society Open Science*, 1, 140081.

Parhar, I., Ogawa, S. and Kitahashi, T. (2012). RF-amide peptides as mediators in environmental control of GnRH neurons. *Progress in Neurobiology*, 98(2), 176–96.

Parker, G.A. (1970). Sperm competition and its evolutionary consequences in the insects. *Biological Reviews*, 45, 525–67.

Peignot, P., Charpentier, M.J.E., Bout, N., *et al.* (2008). Learning from the first release project of captive-bred mandrills (*Mandrillus sphinx*) in Gabon. *Oryx*, 42(1), 122–31.

Pennant, T. (1771). *Synopsis of Quadrupeds*. Chichester.

Pennant, T. (1781). *History of Quadrupeds*. London: B. White.

Pincemy, G., Dobson, S.F. and Jouventin, P. (2009). Experiments on colour ornaments and mate choice in king penguins. *Animal Behaviour*, 78(5), 1247–53.

Pinder, M. (1988). Loa loa – a neglected filaria. *Parasitology Today*, 4(10), 279–84.

Pizzari, T., Cornwallis, C.K., Løvlie, H., Jakobsson, S. and Birkhead, T.R. (2003). Sophisticated sperm allocation in a male fowl. *Nature, London*, 426, 70–4.

Plant, T.M. and Witchel, S.F. (2006). Puberty in non-human primates and humans. In *Knobil and Neill's Physiology of Reproduction, Vol.2, Third edn.*, ed. J.D. Neill. St. Louis, MO: Elsevier, Academic Press, pp. 2177-230.

Plavcan, J.M. and Van Schaik, C.P. (1992). Intrasexual competition and canine dimorphism in anthropoid primates. *American Journal of Physical Anthropology*, 87(4), 461–77.

Pradhan, G.R., Engelhardt, A. and Van Schaik, C.P. (2006). The evolution of female copulation calls in primates: a review and a new model. *Behavioral Ecology and Sociobiology*, 59, 333–43.

Preti, G., Cutler, W.B., Garcia, C.R., Huggins, G.R. and Lawley, H. (1986). Human axillary secretions influence women's menstrual cycles: the role of donor extract of females. *Hormones and Behavior*, 20, 474–82.

Price, J.S., Burton, J.L., Shuster, S. and Wolff, K. (1976). Control of scrotal colour in the vervet monkey. *Journal of Medical Primatology*, 5, 296–304.

Prum, R.O. (2010). The Lande-Kirkpatrick mechanism is the null model of evolution by inter-sexual selection: implications for meaning, honesty, and design in inter-sexual signals. *Evolution*, 64(11), 3085–100.

Prum, R.O. and Torres, R.H. (2004). Structural colouration of mammalian skin: convergent evolution of coherent scattering dermal collagen arrays. *Journal of Experimental Biology*, 207, 2157–72.

Qi, X.-G., Garber, P.A., Ji, W., *et al.* (2014). Satellite telemetry and social modeling offer new insights into the origin of primate multilevel societies. *Nature Communications*, 5:5296, DOI: 10.1038/ncomms6296.

Raleigh, M.J. and McGuire, M.T. (1990). Social influences on endocrine function in male vervet monkeys. In *Socioendocrinology of Primate Reproduction*, eds. T.E. Ziegler and F.B. Bercovitch. New York, NY: Wiley-Liss, pp. 95–111.

Ramaswamy, S., Guerriero, K.A., Gibbs, R.B. and Plant, T.M. (2008). Structural interactions between kisspeptin and GnRH neurons in the mediobasal hypothalamus of the male rhesus monkey (*Macaca mulatta*) as revealed by double immunofluorescence and confocal microscopy. *Endocrinology*, 149, 4387–95.

Ramm, S.A. and Stockley, P. (2010). Sperm competition and sperm length influence the rate of mammalian spermatogenesis. *Biology Letters*, 6, 219–21.

Ramm, S.A., Parker, G.A. and Stockley, P. (2005). Sperm competition and the evolution of male reproductive anatomy in rodents. *Proceedings of the Royal Society of London, Series B*, 272, 949–55.

Reichard, U. (1995). Extra-pair copulations in a monogamous gibbon (*Hylobates lar*). *Ethology*, 100, 99–112.

Renoult, J.P., Schaefer, H.M., Salle, B. and Charpentier, M.J.E. (2011). The evolution of the multicoloured face of mandrills: insights from the perceptual space of colour vision. *PLoS One* 6(12), e29117.

Reynolds, S.M., Dryer, K., Bollback, J., *et al.* (2007). Behavioral paternity predicts genetic paternity in satin bowerbirds (*Ptilonorhynchus violaceus*), a species with a non-resource-based mating system. *Auk*, 124, 857–67.

Rhodes, L., Argersinger, M.E., Gankert, L.T., *et al.* (1997). Effects of administration of testosterone, dihydrotestosterone, oestrogen and fadrazole, an aromatase inhibitor, on sex skin color in intact male rhesus macaques. *Journal of Reproduction and Fertility*, 111, 51–7.

Ritgen, F.F.A. (1824). *Natüliche Einteilung Säugetiere.*

Robbins, M.M. (2007). Gorillas: diversity in ecology and behavior. In *Primates in Perspective*, eds. C.J. Campbell, A. Fuentes, K.C. MacKinnon, M. Panger and S.K. Bearder. Oxford: Oxford University Press, pp. 305–21.

Roberts, M.L., Buchanan, K.L. and Evans, M.R. (2004). Testing the immunocompetence handicap hypothesis: a review of the evidence. *Animal Behaviour*, 68, 227–39.

Roberts, T.E., Davenport, T.R.B., Hildebrandt, K.B.P., *et al.* (2009). The biogeography of introgression in the critically endangered African Monkey *Rungwecebus kipunji. Biology Letters*, DOI: 10.1098/rsbl.2009.0741.

Rogers, M.E., Abernethy, K.A., Fontaine, B., *et al.* (1996). Ten days in the life of a mandrill horde in the Lopé Reserve, Gabon. *American Journal of Primatology*, 40, 297–313.

Romanes, G.T. (1893). *Darwin and After Darwin: An Exposition of the Darwinian Theory and a Discussion of Post-Darwinian Questions, Vol.1, The Darwinian Legacy*. London: Longmans Green and Co.

Rose, R.M., Bernstein, I.S. and Gordon, T.P. (1975). Consequences of social conflict on plasma testosterone levels in rhesus monkeys. *Psychosomatic Medicine*, 37, 50–61.

Rose, R.M., Gordon, T.P. and Bernstein, I.S. (1972). Plasma testosterone levels in male rhesus: effects of sexual and social stimuli. *Science*, 178, 643–5.

Roseweir, A.K. and Millar, R.P. (2009).The role of kisspeptin in the control of gonadotrophin secretion. *Human Reproduction Update*, 15, 206–12.

Rowe, N. (1996). *The Pictorial Guide to the Living Primates*. New York, NY: Pogonias Press.

Rowell, T.E. (1967a). A quantitative comparison of the behaviour of a wild and caged baboon group. *Animal Behaviour*, 15, 499–509.

Rowell, T.E. (1967b). Female reproductive cycles and the behavior of baboons and rhesus macaques. In *Social Communication among Primates*, ed. S.A. Altmann. Chicago, IL: University of Chicago Press, pp. 15–32.

Rowell, T.E. (1970). Baboon menstrual cycles affected by social environment. *Journal of Reproduction and Fertility*, 21, 133–41.

Rowell, T.E. and Chalmers, N.R. (1970). Reproductive cycles of the mangabey *Cercocebus albigena*. *Folia Primatologica*, 12, 264–72.

Rowell, T.E. and Dixson, A.F. (1975). Changes in social organization during the breeding season of wild talapoin monkeys. *Journal of Reproduction and Fertility*, 43, 419–34.

Ruckstuhl, K.E. and Neuhaus, P. (2002). Sexual segregation in ungulates: a comparative test of three hypotheses. *Biological Reviews*, 77, 77–96.

Ruckstuhl, K.E. and Neuhaus, P. (2005). Activity, synchrony and social segregation. In *Sexual Segregation in Vertebrates: Ecology of the Two Sexes*, eds. K.E. Ruckstuhl and P. Neuhaus. Cambridge: Cambridge University Press, pp. 165–79.

Ryan, M.J. (1985). *The Tungara Frog: A study in Sexual Selection and Communication*. Chicago, IL: The University of Chicago Press.

Saayman, G.S. (1970). The menstrual cycle and sexual behaviour in a troop of free-ranging chacma baboons (*Papio ursinus*). *Folia Primatologica*, 12, 81–110.

Sade, D. (1964). Seasonal cycle in size of testes in free-ranging *Macaca mulatta*. *Folia primatologica*, 2, 171–80.

Sadlier, R.M.F.S. (1969). *The Ecology of Reproduction in Wild and Domestic Mammals*. London: Methuen.

Sanderson, I.T. (1940). The mammals of the North Cameroons forest area. *Transactions of the Zoological Society of London*, 24, 623–725.

Sapolsky, R.M. (1982). The endocrine stress response and social status in the wild baboon. *Hormones and Behavior*, 16, 279–92.

Sapolsky, R.M. (1993). The physiology of dominance in stable and unstable social hierarchies. In *Primate Social Conflict*, eds. W.A. Mason and S.P. Mendoza. New York, NY: State University of New York, pp. 171–204.

Scaramuzzi, R.J., Campbell, B.K., Downing, J.A., *et al.* (2006). A review of the effects of supplementary nutrition in the ewe on the concentrations of reproductive and metabolic hormones, and the mechanisms that regulate folliculogenesis and ovulation rate. *Reproduction Nutrition and Development*, 46, 339–54.

Schaller, G.B. (1963). *The Mountain Gorilla: Ecology and Behavior*. Chicago, IL: University of Chicago Press.

Schank, J. (2000). Menstrual cycle variability and measurement: further cause for doubt. *Psychoneuroendocrinology*, 25, 837–47.

Schank, J. (2006). Do human menstrual cycle pheromones exist? *Human Nature*, 17, 448–70.

Schapiro, S.J. (1985). *Reproductive Seasonality: Birth Synchrony, Female–Female Reproductive Competition and Cooperation in Captive Cercopithecus aethiops and C. mitis*. PhD Dissertation. Davis, CA: University of California Davis.

Schneider, J.E. (2004). Energy balance and reproduction. *Physiology and Behavior*, 81, 289–317.

Schultz. A.H. (1969). *The Life of Primates*. London: Weidenfeld and Nicolson.

Schurmann, C. (1982). Mating behavior in wild orang-utans. In *The Orang-Utan: Its Ecology and Conservation*, ed. L.E.M. de Boer. The Hague: Dr. W. Junk, pp. 269–84.

Scruton, D.M. and Herbert, J. (1970). The menstrual cycle and its effect on behaviour in the talapoin monkey (*Miopithecus talapoin*). *Journal of Zoology, London*, 162, 419–36.

Setchell, J.M. (2005). Do female mandrills prefer brightly colored males? *International Journal of Primatology*, 26(4), 715–35.

Setchell, J.M. and Dixson, A.F. (2001a). Arrested development of secondary sexual adornments in subordinate adult male mandrills (*Mandrillus sphinx*). *American Journal of Physical Anthropology*, 115, 245–52.

Setchell, J.M. and Dixson, A.F. (2001b). Changes in the secondary sexual adornments of male mandrills (*Mandrillus sphinx*) are associated with gain and loss of alpha status. *Hormones and Behavior*, 39, 177–84.

Setchell, J.M. and Dixson, A.F. (2001c). Circannual changes in the secondary sexual adornments of semi free-ranging male and female mandrills (*Mandrillus sphinx*). *American Journal of Primatology*, 53, 109–21.

Setchell, J.M. and Dixson, A.F. (2002). Developmental variables and dominance rank in adolescent male mandrills (*Mandrillus sphinx*). *American Journal of Primatology*, 56, 9–25.

Setchell, J.M. and Wickings, E.J. (2004a). Social and seasonal influences on the reproductive cycle in female mandrills (*Mandrillus sphinx*). *American Journal of Physical Anthropology*, 125, 73–84.

Setchell, J.M. and Wickings, E.J. (2004b). Sexual swelling in mandrills: a test of the reliable indicator hypothesis. *Behavioral Ecology*, 15, 438–45.

Setchell, J.M., Abbott, K.M., Gonzalez, J.-P. and Knapp, L.A. (2013). Testing for post-copulatory selection for major histocompatibility complex genotype in a semi-free-ranging primate population. *American Journal of Primatology*, 75(10), 1021–33.

Setchell, J.M., Bedjabaga, I-B., Goossens, B., *et al.* (2007). Parasite prevalence, abundance and diversity in a semi-free-ranging colony of *Mandrillus sphinx*. *International Journal of Primatology*, 28(6), 1345–62.

Setchell, J.M., Charpentier, M. and Wickings, E.J. (2005a). Sexual selection and reproductive careers in mandrills (*Mandrillus sphinx*). *Behavioral Ecology and Sociobiology*, 58, 474–85.

Setchell, J.M., Charpentier, M. and Wickings, E.J. (2005b). Mate guarding and paternity in mandrills: factors influencing alpha male monopoly. *Animal Behaviour*, 70, 1105–20.

Setchell, J.M., Charpentier, M.J.E., Abbott, K.M., Wickings, E.J. and Knapp, L.A. (2009). Is brightest best? Testing the Hamilton-Zuk hypothesis in mandrills. *International Journal of Primatology*, 30, 825–44.

Setchell, J.M., Charpentier, M.J.E., Abbott, K.M., Wickings, E.J. and Knapp, L.A. (2010a). Opposites attract; MHC-associated mate choice in a polygynous primate. *Journal of Evolutionary Biology*, 23, 136–48.

Setchell, J.M., Kendall, J. and Tyniec, P. (2011a). Do non-human primates synchronise their menstrual cycles? A test in mandrills. *Psychoneuroendocrinology*, 36, 51–9.

Setchell, J.M., Knapp, L.A. and Wickings, E.J. (2006a). Violent coalitionary attack by female mandrills against an injured alpha male. *American Journal of Primatology*, 68(4), 411–18.

Setchell, J.M., Lee, P.C., Wickings, E.J. and Dixson, A.F. (2001). Growth and ontogeny of sexual size dimorphism in the mandrill (*Mandrillus sphinx*). *American Journal of Physical Anthropology*, 115, 349–60.

Setchell, J.M., Lee, P.C., Wickings, E.J. and Dixson, A.F. (2002). Reproductive parameters and maternal investment in mandrills (*Mandrillus sphinx*). *International Journal of Primatology*, 23, 51–68.

Setchell, J.M., Smith, T., Wickings, E.J. and Knapp, L.A. (2008). Social correlates of testosterone and ornamentation in male mandrills. *Hormones and Behavior*, 54, 365–72.

Setchell, J.M., Smith, T., Wickings, E.J. and Knapp, L.A. (2010b). Stress, social behavior and secondary sexual traits in a male primate. *Hormones and Behavior*, 58(5), 720–8.

Setchell, J.M., Vaglio, S., Moggi-Cecchi, J., *et al.* (2010c). Chemical composition of scent-gland secretions in an Old World monkey (*Mandrillus sphinx*): influence of sex, male status and individual identity. *Chemical Senses*, 35, 205–20.

Setchell, J.M., Vaglio, S., Abbott, K.M., *et al.* (2011b). Odour signals major histocompatibility complex genotype in an Old World monkey. *Proceedings of the Royal Society of London, Series B*, 278, 274–80.

Setchell, J.M., Wickings, E.J. and Knapp, L.A. (2006b). Signal content of red facial coloration in female mandrills (*Mandrillus sphinx*). *Proceedings of the Royal Society of London, Series B*, 273, 2395–400.

Setchell, J.M., Wickings, E.J. and Knapp, L.A. (2006c). Life history in male mandrills (*Mandrillus sphinx*): physical development, dominance rank, and group association. *American Journal of Physical Anthropology*, 131(4), 498–510.

Short, R.V. (1979). Sexual selection and its component parts, somatic and genital selection, as illustrated by man and the great apes. *Advances in the Study of Behavior*, 9, 131–58.

Short, R.V. (1985). Species differences in reproductive mechanisms. In *Reproduction in Mammals, Vol 4, Reproductive Fitness*, eds. C.R. Austin and R.V. Short. Cambridge: Cambridge University Press, pp. 24–61.

Silk, J.B. (1983). Local resource competition and facultative adjustment of sex ratios in relation to competitive activities. *American Naturalist*, 121, 56–66.

Silk, J.B. (1988). Maternal investment in captive bonnet macaques, *Macaca radiata*. *American Naturalist*, 132(1), 1–19.

Silk, J.B. and Brown, G.R. (2004). Sex ratios in primate groups. In *Sexual Selection in Primates: New and Comparative Perspectives*, eds. P. Kappeler and C. Van Schaik. Cambridge: Cambridge University Press, pp. 253–65.

Simmons, L.W. (2001). *Sperm Competition and its Evolutionary Consequences in the Insects*. Princeton, NJ: Princeton University Press.

Simpson, M.J.A. and Simpson, A.E. (1982). Birth sex ratios and social rank in rhesus monkey mothers. *Nature, London*, 300, 440–1.

Singh, D. (2006). The universal allure of the hourglass figure: an evolutionary theory of female physical attractiveness. *Clinics in Plastic Surgery*, 33, 359–70.

Singh, D., Dixson, B.J., Jessop, T.S., Morgan, B. and Dixson, A.F. (2010). Cross-cultural consensus for waist-hip ratio and women's attractiveness. *Evolution and Human Behavior*, 31, 176–81.

Singleton, M. (2012). Postnatal cranial development in papionin primates: an alternative model for hominin evolutionary development. *Evolutionary Biology*, DOI: 10.1007/s11692-011-9153-4.

Slob, A.K., Groeneveld, W.H. and Van der Werff Ten Bosch, J.J. (1986). Physiological changes during copulation in male and female stumptail macaques (*Macaca arctoides*). *Physiology and Behavior*, 38, 891–5.

Smith, J.T., Shahab, M., Pereira, A., Pau, K.Y. and Clarke, I.J. (2010). Hypothalamic expression of *KISS1* and gonadotropin inhibitory genes during the menstrual cycle of a non-human primate. *Biology of Reproduction*, 83, 568–77.

Smith, R.J. and Jungers, W.L. (1997). Body mass in comparative primatology. *Journal of Human Evolution*, 32, 523–59.

Sommer, V. (1993). Infanticide among the langurs of Jodhpur: testing the sexual selection hypothesis with a long-term record. In *Infanticide and Parental Care*, eds. S. Parmigiani and F. vom Saal. London: Harwood Academic Publishers, pp. 155–98.

Souquière, S., Bibollet-Ruche, F., Robertson, D.L., *et al.* (2001). Wild *Mandrillus sphinx* are carriers of two types of lentivirus. *Journal of Virology*, 75(15), 7086–96.

Stammbach, E. (1987). Desert, forest and montane baboons: multilevel societies. In *Primate Societies*, eds. B. Smuts, D. Cheney, R. Seyfarth, R. Wrangham and T. Struhsaker. Chicago, IL: University of Chicago Press, pp. 112–20.

Stander, P.E. (1997). Field age determination of leopards by tooth wear. *African Journal of Ecology*, 35, 156–61.

Staniland, I.J. (2005). Sexual segregation in seals. In *Sexual Segregation in Vertebrates: Ecology of the Two Sexes*, eds. K.E. Ruckstuhl and P. Neuhaus. Cambridge: Cambridge University Press, pp.53–73.

Stern, K. and McClintock, M.K. (1998). Regulation of ovulation by human pheromones. *Nature, London*, 392, 177–9.

Stone, A.I. (2014). Is fatter sexier? Reproductive strategies of male squirrel monkeys (*Saimiri sciureus*). *International Journal of Primatology*, 35, 628–42.

Strassmann, B.L. (1997). The biology of menstruation in *Homo sapiens*: total lifetime menses, fecundity, and non-synchrony in a natural fertility population. *Current Anthropology*, 38, 123–9.

Strassmann, B.L. (1999). Menstrual synchrony pheromones: cause for doubt. *Human Reproduction*, 14, 579–80.

Struhsaker, T.T. (1969). Correlates of ecology and social organization among African cercopithecines. *Folia Primatologica*, 11, 80–118.

Struhsaker, T.T. (1975). *The Red Colobus Monkey*. Chicago, IL: University of Chicago Press.

Swedell, L. (2006). *Strategies of Sex and Survival in Hamadryas Baboons: Through a Female Lens*. Upper Saddle River, NJ: Pearson Prentice Hall.

Szalay, F. and Delson, E. (1979). *Evolutionary History of the Primates*. New York, NY: Academic Press.

Telfer, P.T. (2006). *Molecular Phylogeny and Phylogeography of the Genus Mandrillus (Primates: Papionini)*. PhD Dissertation. New York, NY: Department of Anthropology, New York University.

Telfer, P.T., Souquière, S., Clifford, S.L., *et al.* (2003). Molecular evidence for deep phylogenetic divergence in *Mandrillus sphinx*. *Molecular Ecology*, 12, 2019–24.

Thibault, M. and Blaney, S. (2003). The oil industry as an underlying factor in the bushmeat crisis in Central Africa. *Conservation Biology*, 17(6), 1807–13.

Ting, N., Astaras, C., Hearn, G., *et al.* (2012). Ecologic signatures of a demographic collapse in a large-bodied forest dwelling primate (*Mandrillus leucophaeus*). *Ecology and Evolution*, 2(3), 550–61.

Tobler, R.E. (2008). *Female Reproductive Strategies and the Ovarian Cycle in Hamadryas Baboons*. MSc Dissertation. Wellington, NZ: Victoria University of Wellington, New Zealand.

Tobler, R., Pledger, S. and Linklater, W. (2010). No evidence for ovarian synchrony or asynchrony in hamadryas baboons. *Animal Behaviour*, 80, 829–37.

Trainor, B.C., Bird, I.M. and Marler, C.A. (2004). Opposing hormonal mechanisms of aggression revealed through short-lived testosterone manipulations and multiple winning experiences. *Hormones and Behavior*, 45(2), 115–21.

Treves, A. (2000). Prevention of infanticide: the perspective of infant primates. In *Infanticide by Males and its Implications*, eds. C.P. Van Schaik and C.H. Janson. Cambridge: Cambridge University Press, pp. 223–38.

Trivers, R.L. and Willard, D.E. (1973). Natural selection of parental ability to vary the sex ratio of offspring. *Science*, 179, 90–2.

Tsutsui, K. (2009). A key new neurohormone controlling reproduction, gonadotropin-inhibitory hormone (GnIH): biosynthesis, mode of action and functional significance. *Progress in Neurobiology*, 88, 76–88.

Tsutsui, K., Saigoh, E., Ukena, K., *et al.* (2000). A novel avian hypothalamic peptide inhibiting gonadotropin release. *Biochemistry Biophysics Research Communications*, 275, 661–7.

Tutin, C.E.G. and Fernandez, M. (1984). Nationwide census of gorilla (*Gorilla g. gorilla*) and chimpanzee (*Pan t. troglodytes*) populations in Gabon. *American Journal of Primatology*, 6, 313–36.

Tutin, C.E.G. and Fernandez, M. (1987). Gabon: a fragile sanctuary. *Primate Conservation*, 8, 160–1.

Tutin, C.E.G. and McGinnis, P.R. (1981). Chimpanzee reproduction in the wild. In *Reproductive Biology of the Great Apes: Comparative and Biomedical Perspectives*, ed. C.E. Graham. New York, NY: Academic Press, pp. 239–64.

Utami Atmoko, S. and Van Hooff, J.A.R.A.M. (2004). Alternative male reproductive tactics: male bimaturism in orang-utans. In *Sexual Selection in Primates: New and Comparative Perspectives*, eds. P. Kappeler and C. Van Schaik. Cambridge, UK: Cambridge University Press, pp. 196–207.

Utami Atmoko, S.S., Singleton, I., Van Noordwijk, M.A., Van Schaik, C.P. and Mitra Setia, T. (2009). Male–male relationships in orangutans. In *Orangutans: Geograpic Variation in Behavioral Ecology and Conservation*, eds. S.A. Wich, S.S. Utami Atmoko, T. Mitra Setia and C.P. Van Schaik. New York, NY: Oxford University Press, pp. 225–33.

Van Hooff, J.A.R.A.M. (1962). Facial expressions in higher primates. *Symposia of the Zoological Society of London*, 8, 97–125.

Van Hooff, J.A.R.A.M. (1967). The facial displays of the catarrhine monkeys and apes. In *Primate Ethology*, ed. D. Morris. London: Weidenfeld and Nicolson, pp. 7–68.

Van Noordwijk, M.A. and Van Schaik, C.P. (1988). Male careers in Sumatran long-tailed macaques (*Macaca fascicularis*). *Behaviour*, 107, 24–43.

Van Noordwijk, M.A. and Van Schaik, C.P. (2000). Reproductive patterns in eutherian mammals: adaptations against infanticide? In *Infanticide by Males and its Implications*, eds. C.P. Van Schaik and C.H. Janson. Cambridge: Cambridge University Press, pp. 322–60.

Van Schaik, C.P. (2000a). Infanticide by males: the sexual selection hypothesis revisited. In *Infanticide by Males and its Implications*, eds. C.P. Van Schaik and C.H. Janson. Cambridge: Cambridge University Press, pp. 27–60.

Van Schaik, C.P. (2000b). Vulnerability to infanticide by males: patterns among mammals. In *Infanticide by Males and its Implications*, eds. C.P. Van Schaik and C.H. Janson. Cambridge: Cambridge University Press, pp. 61–71.

Vandenbergh, J.G. (1965). Hormonal basis of the sex skin in male rhesus monkeys. *General and Comparative Endocrinology*, 5, 31–4.

Vandenbergh, J.G. (1969). Endocrine coordination in monkeys: male sexual responses to the female. *Physiology and Behavior*, 4, 261–4.

Vandenbergh, J.G. and Drickamer, L. (1974). Reproductive coordination among free-ranging rhesus monkeys. *Physiology and Behavior*, 13, 373–6.

Vasey, P.L. (2004). Sex differences in sexual partner acquisition, retention and harassment during female homosexual consortship in Japanese macaques. *American Journal of Primatology*, 64, 397–409.

Vasey, P.L. and Duckworth, N. (2006). Sexual reward via vulvar, perineal and anal stimulation: a proximate mechanism for female homosexual mounting in Japanese macaques. *Archives of Sexual Behavior*, 35, 523–32.

Vasey, P.L., Chapais, B. and Gauthier, C. (1998). Mounting interactions between female Japanese macaques: testing the influence of dominance and aggression. *Ethology*, 104, 387–98.

Verdonck, K., Gaethofs, M.C., Carels, C and de Zegher, F. (1999). Effect of low-dose testosterone treatment on craniofacial growth in boys with delayed puberty. *European Journal of Orthodontics*, 21, 137–43.

Viera, V.M., Nolan, P.M., Côté, S.D., Jouventin, P. and Groscolas, R. (2008). Is territory defence related to plumage ornaments in the king penguin, *Aptenodytes patagonicus*? *Ethology*, 114, 146–53.

Waitt, C., Gerald, M.S., Little, A.C. and Kraiselburd, E. (2006). Selective attention toward female secondary sexual color in male rhesus macaques. *American Journal of Primatology*, 68, 738–44.

Waitt, C., Little, A.C., Wolfensohn, S., *et al.* (2003). Evidence from rhesus macaques suggests that male coloration plays a role in female primate mate choice. *Proceedings of the Royal Society of London, Series B*, 270, S144–6.

Wallen, J., Winston, S., Gaventa, M., Davis-Dasilva, M. and Collins, D.C. (1984). Periovulatory changes in female sexual behavior and patterns of ovarian steroid secretion in group living rhesus monkeys. *Hormones and Behavior*, 18, 431–50.

Wallis, J. (1985). Synchrony of estrous swelling in captive group-living chimpanzees (*Pan troglodytes*). *International Journal of Primatology*, 6, 335–56.

Wallis, J., King, B.J. and Roth-Meyer, C. (1986). The effect of female proximity and social interaction on the menstrual cycle of crab-eating monkeys (*Macaca fascicularis*). *Primates*, 27, 83–94.

Wallis, S.J. (1981). The behavioural repertoire of the grey-cheeked mangabey *Cercocebus albigena johnstonii*. *Primates*, 22(4), 523–32.

Wallis, S.J. (1983). Sexual behaviour and reproduction of *Cercocebus albigena johnstonii* in Kibale Forest, Western Uganda. *International Journal of Primatology*, 4, 153–66.

Walusinski, O. (2010). Ed. *The Mystery of Yawning in Physiology and Disease. Vol.28, Frontiers of Neurology and Neuroscience*. Basel: Karger.

Watts, D.P. (2005). Sexual segregation in non-human primates. In *Sexual Segregation in Vertebrates: Ecology of the Two Sexes*, eds. K.E. Ruckstuhl and P. Neuhaus. Cambridge: Cambridge University Press, pp.327–47.

Weller, L. and Weller, A. (1993). Multiple influences of menstrual synchrony: kibbutz room-mates, their best friends, and their mothers. *American Journal of Human Biology*, 5, 173–9.

Weller, L., Weller, A. and Avenir, O. (1995). Menstrual synchrony: only in room-mates who are close friends? *Physiology and Behavior*, 58, 883–9.

Westneat, D.F., Hasselquist, D. and Wingfield, J.C. (2003). Tests of association between the humoral immune response of red-winged blackbirds (*Agelaius phoeniceus*) and male plumage, testosterone, or reproductive success. *Behavioral Ecology and Sociobiology*, 53, 315–23.

White, E.C., Dikangadissi, J.T., Dimoto, E., *et al.* (2010). Home-range use by a large horde of wild *Mandrillus sphinx*. *International Journal of Primatology*, 31, 627–45.

Whitehead, H. (2003). *Sperm Whales: Social Evolution in the Ocean*. Chicago, IL: University of Chicago Press.

Wickings, E.J. and Dixson, A.F. (1992a). Development from birth to sexual maturity in a semi free-ranging colony of mandrills (*Mandrillus sphinx*) in Gabon. *Journal of Reproduction and Fertility*, 95, 129–38.

Wickings, E.J. and Dixson, A.F. (1992b). Testicular function, secondary sexual development, and social status in male mandrills (*Mandrillus sphinx*). *Physiology and Behavior*, 52, 909–16.

Wickings, E.J., Bossi, T. and Dixson, A.F. (1993). Reproductive success in the male mandrill (*Mandrillus sphinx*): correlations of male dominance and mating success with paternity, as determined by DNA fingerprinting. *Journal of Zoology, London*, 231, 563–74.

Wickler, W. (1967). Socio-sexual signals and their intraspecific imitation among primates. In *Primate Ethology*, ed. D. Morris. London: Weidenfeld and Nicolson, pp. 69–147.

Wild, C., Morgan, B.J. and Dixson, A. (2005). Conservation of drill populations in Bakossiland, Cameroon: historical trends and current status. *International Journal of Primatology*, 26(4), 759–73.

Wildt, D.E., Doyle, U., Stone, S.C. and Harrison, R.M. (1977). Correlation of perineal swelling with serum ovarian hormone levels, vaginal cytology and ovarian follicular development during the baboon reproductive cycle. *Primates*, 18, 261–70.

Wingfield, J.C. (2005). The concept of allostasis: coping with a capricious environment. *Journal of Mammalogy*, 86(2), 248–54.

Wingfield, J.C., Hegner, R.E., Duffy Jr., A.M. and Ball, G.F. (1990). The "challenge hypothesis": theoretical implications for patterns of testosterone secretion, mating systems and breeding strategies. *American Naturalist*, 136, 829–46.

Wingfield, J.C., Lynn, S. and Soma, K.K. (2001). Avoiding the "costs" of testosterone: ecological bases of hormone-behavior interactions. *Brain, Behavior and Evolution*, 57(5), 239–51.

Wood, K.L. (2007). *Life history and behavioral characteristics of a semi-wild population of drills* (Mandrillus leucophaeus) *in Nigeria*. PhD Dissertation. Amherst, MA: Biology Program, University of Massachusetts.

Young, W.C. and Orbison, W.D. (1944). Changes in selected features of behavior in pairs of oppositely sexed chimpanzees during the menstrual cycle and after ovariectomy. *Journal of Comparative Psychology*, 27, 107–43.

Zahavi, A. (1975). Mate selection – a selection for handicap. *Journal of Theoretical Biology*, 53, 205–14.

Zahavi, A. and Zahavi, A. (1997). *The Handicap Principle: A Missing Piece of Darwin's Puzzle*. New York, NY: Oxford University Press.

Zinner, D., Alberts, S.C., Nunn, C.L. and Altmann, J. (2002). Significance of primate sexual swellings. *Nature, London*, 420, 142–3.

Zinner, D., Schwibbe, M.H. and Kaumanns, W. (1994). Cycle synchrony and probability of conception in female hamadryas baboons, *Papio hamadryas*. *Behavioral Ecology and Sociobiology*, 35, 175–83.

Zuckerman, S. (1932). *The Social Life of Monkeys and Apes*. New York, NY: Harcourt, Brace and Co.

Zuckerman, S. and Parkes, A.S. (1939). Observations on the secondary sexual characters in monkeys. *Journal of Endocrinology*, 1, 430–9.

Index

Note: all headings refer to the mandrill unless otherwise stated. Page numbers in *italics* refer to figures and tables.

Printed in the United States
By Bookmasters